SO-BWT-538

HEAT EXCHANGE FLUIDS AND TECHNIQUES

HEAT EXCHANGE FLUIDS
AND TECHNIQUES

M.W. Ranney

NOYES DATA CORPORATION

Park Ridge, New Jersey, U.S.A.

1979

Published in the United States of America by
Noyes Data Corporation
Noyes Building, Park Ridge, New Jersey 07656

TJ
260
.R 33

Library of Congress Cataloging in Publication Data

Ranney, Maurice William, 1934-
    Heat exchange fluids and techniques.

    (Chemical technology review ; no. 143)
    Includes index.
    1. Heat-transfer media--Patents. 2. Heat
exchangers--Patents. 3. Heat--Transmission--Patents.
I. Title. II. Series.
TJ260.R33   621.4'022   79-20336
ISBN 0-8155-0778-X

FOREWORD

The detailed, descriptive information in this book is based on U.S. patents, issued since January 1975, that deal with heat exchange fluids and techniques, and their potential for energy saving.

This book serves a double purpose in that it supplies detailed technical information and can be used as a guide to the U.S. patent literature in this field. By indicating all the information that is significant, and eliminating legal jargon and juristic phraseology, this book presents an advanced, technically oriented review of heat exchange fluids and techniques.

The U.S. patent literature is the largest and most comprehensive collection of technical information in the world. There is more practical, commercial, timely process information assembled here than is available from any other source. The technical information obtained from a patent is extremely reliable and comprehensive; sufficient information must be included to avoid rejection for "insufficient disclosure." These patents include practically all of those issued on the subject in the United States during the period under review; there has been no bias in the selection of patents for inclusion.

The patent literature covers a substantial amount of information not available in the journal literature. The patent literature is a prime source of basic commercially useful information. This information is overlooked by those who rely primarily on the periodical journal literature. It is realized that there is a lag between a patent application on a new process development and the granting of a patent, but it is felt that this may roughly parallel or even anticipate the lag in putting that development into commercial practice.

Many of these patents are being utilized commercially. Whether used or not, they offer opportunities for technological transfer. Also, a major purpose of this book is to describe the number of technical possibilities available, which may open up profitable areas of research and development. The information contained in this book will allow you to establish a sound background before launching into research in this field.

Advanced composition and production methods developed by Noyes Data are employed to bring these durably bound books to you in a minimum of time. Special techniques are used to close the gap between "manuscript" and "completed book." Industrial technology is progressing so rapidly that time-honored, conventional typesetting, binding and shipping methods are no longer suitable. We have by-passed the delays in the conventional book publishing cycle and provide the user with an effective and convenient means of reviewing up-to-date information in depth.

The table of contents is organized in such a way as to serve as a subject index. Other indexes by company, inventor and patent number help in providing easy access to the information contained in this book.

15 Reasons Why the U.S. Patent Office Literature Is Important to You —

1. The U.S. patent literature is the largest and most comprehensive collection of technical information in the world. There is more practical commercial process information assembled here than is available from any other source.

2. The technical information obtained from the patent literature is extremely comprehensive; sufficient information must be included to avoid rejection for "insufficient disclosure."

3. The patent literature is a prime source of basic commercially utilizable information. This information is overlooked by those who rely primarily on the periodical journal literature.

4. An important feature of the patent literature is that it can serve to avoid duplication of research and development.

5. Patents, unlike periodical literature, are bound by definition to contain new information, data and ideas.

6. It can serve as a source of new ideas in a different but related field, and may be outside the patent protection offered the original invention.

7. Since claims are narrowly defined, much valuable information is included that may be outside the legal protection afforded by the claims.

8. Patents discuss the difficulties associated with previous research, development or production techniques, and offer a specific method of overcoming problems. This gives clues to current process information that has not been published in periodicals or books.

9. Can aid in process design by providing a selection of alternate techniques. A powerful research and engineering tool.

10. Obtain licenses — many U.S. chemical patents have not been developed commercially.

11. Patents provide an excellent starting point for the next investigator.

12. Frequently, innovations derived from research are first disclosed in the patent literature, prior to coverage in the periodical literature.

13. Patents offer a most valuable method of keeping abreast of latest technologies, serving an individual's own "current awareness" program.

14. Copies of U.S. patents are easily obtained from the U.S. Patent Office at 50¢ a copy.

15. It is a creative source of ideas for those with imagination.

CONTENTS AND SUBJECT INDEX

INTRODUCTION

Considerable attention and much discussion is being given to our national and world energy resources. Problems of fuel and power supply are now the concern of the home owner, the industry executive, as well as many of us who in the past have taken our heating and power supplies for granted. Perhaps a dividend of more recent energy crises has been the resulting reexamination of our energy reserves and the manner in which we presently use and waste energy. It is becoming increasingly obvious that a more efficient use of fuels will help to offset increases in fuel costs and will help to guarantee unimpeded economic development for a long time to come.

The technical literature is crowded with conventional heating and heat exchange systems. Industrial heating furnaces, for example, are usually classified according to (1) the purpose for which the material is heated; (2) the nature of the transfer of heat to the material; (3) the method of firing the furnace; or (4) the method of handling material through the furnace. Home heating furnaces, on the other hand, are usually characterized by fuel consumed and the medium being heated—such as hot air, steam, hot water, etc. Most furnaces or boilers generate a relatively hot exhaust medium which is ultimately conveyed to the environment or, in the case of more sophisticated systems, to recycling apparatus. In the case of hot exhaust gases, for example, a great deal of energy is wasted or not recaptured by simple discharge of these gases at elevated temperatures into the atmosphere.

The uses of energy in the United States during the past century have been developed during a period of cheap energy. As a result many uses have been designed that are quite wasteful from an energy point of view, but economic in the overall point of view. With the recent sharply rising costs of energy, many designs for energy use that were economically attractive in the past are becoming economically unsound. As a result buildings that were not insulated in the past because it was cheaper to waste fuel than to pay for insulation are now being insulated, designs for processes that were basically wasteful of energy are undergoing redesign, and the like.

For example, it is well known in the art to generate steam by burning a source of

1

energy and then expanding the steam through a turbine to generate electricity. This process is basically wasteful of energy because in heating water it requires one Btu of heat to raise one pound one degree Fahrenheit; at the point water is converted from liquid to vapor, approximately 1,000 Btu of heat are required for each pound of water vaporized; upon becoming dry vapor the steam may be raised to higher temperatures with modest increases in heat added per pound of steam; the steam is expanded through a turbine which in turn generates electricity; the remaining steam is condensed to water by giving up approximately 1,000 Btu per pound to the cooling towers and cooling ponds. It is this latter heat loss that significantly diminishes the overall efficiency of the process where efficencies in the order of 33% are common.

As a result of continuously increasing oil prices being set by OPEC and the agreements concluded at the summit meeting in Tokyo of June 1979, all advanced, developed nations must reduce their energy requirements over the coming decade. The resources of the world's research, development and engineering community are, now increasingly, being focused on this critical energy conservation issue.

The manner in which the efficient utilization of thermal energy affects our industries, personal life and indeed, our ability to sustain a comfortable lifestyle is amply indicated by the broad subject matter of this book as reflected in the table of contents. The efficient transfer of energy, through heat exchange techniques covers subject matter from storage of blood to the maximization of the use of geothermal and solar energy as well as thousands of industrial applications.

This book, directed to a broad coverage and understanding of the many diverse uses of heat transfer techniques is based on the recent United States patent literature. The worldwide nature of this effort is reflected by the observation that about 20% of the processes described in the book were developed outside the United States.

GENERAL HEAT EXCHANGER
CONSTRUCTION AND DESIGN

INTRODUCTION

Heat exchangers are employed in various chemical engineering processes such as power plants, oil refineries, chemical reactors and energy retrieval systems. Generally, heat exchanger design has focused upon means to transfer the greatest amount of heat per unit surface area of the exchanger. The transfer of heat is governed by the equation: $Q = AU\Delta Tm$, where Q = the heat transfer rate; A = the heat exchanger surface area; U = the overall heat transfer coefficient; and ΔTm = the log-mean-temperature of the heat exchanger. Thus, the heat transfer rate is directly proportional to the surface area of the heat exchanger and the log-mean-temperature difference of the fluids in the exchanger.

In its simplest form, a first fluid enters the exchanger at a high temperature while a second fluid enters at a lower temperature. Heat is exchanged between the two fluids, either traveling in cocurrent or countercurrent paths. Unless thermodynamic transition occurs in one or more of the fluids, the high temperature fluid continues to cool while the lower temperature fluid continues to heat throughout the path of the exchanger.

When thermodynamic transition occurs, for example, within the fluid to be heated, that fluid does not change temperature during the heat transfer process. Such a phenomenon occurs when an ice-water mixture is at equilibrium in which heating of the mixture does not raise its temperature above 0°C until all of the ice is melted. Similarly, water will boil at 100°C at one atmosphere pressure without a temperature rise until the liquid water has dissipated. Refining high grade gasoline from crude oil utilizes the phenomenon of thermodynamic transition for fractional distillation.

The use of a fluid which undergoes thermodynamic transition in a heat exchanger environment can be most useful. As can be seen from the above equation, the heat transfer rate is proportional to the log-mean-temperature difference of the fluids within the exchanger. Thus, when one fluid undergoes thermodynamic transition, energy is transferred to that fluid without a temperature

3

increase so that the log-mean-temperature difference of the two fluids can be increased with respect to heat exchangers wherein no thermodynamic transition occurs. Looked at differently, a most economical heat exchanger is one where the temperature difference between the two fluids is kept at a maximum to minimize the surface area for a given heat transfer rate. When the temperatures of the two fluids begin to approach one another, the heat exchanger requires a maximum surface area and, thus, heat transfer per unit area is at an extremely low point. Upon approaching the theoretical limit wherein the fluids are at the same temperature, energy transfer would cease and the heat exchanger would no longer have practical utility.

TUBULAR CONSTRUCTION

Control of Steam Distribution in Air Cooled Condenser

F.V. Huber; U.S. Patent 3,882,925; May 13, 1975; assigned to Ecodyne Corporation has developed a method for attaining an even distribution of steam following through the heat exchange tubes of an air cooled steam condenser. The method resides in the return of the cooling air, which has passed across initial lengths of the heat exchange tubes, back across the tubes to give an opposite temperature differential in the succeeding tube lengths so as to mollify the temperature effects which cause a maldistribution of the steam in the condenser. A preferred example includes partitions and return hoods separating the heat exchange tubes along their lengths into air flow directing passages. The cooling air is directed sequentially through these passages, changing flow direction as it passes from one passage to the next. Referring to Figure 1.1, an air cooled steam condenser, constructed in accordance with this process, is indicated generally at **10**.

Figure 1.1: Air Cooled Steam Condenser

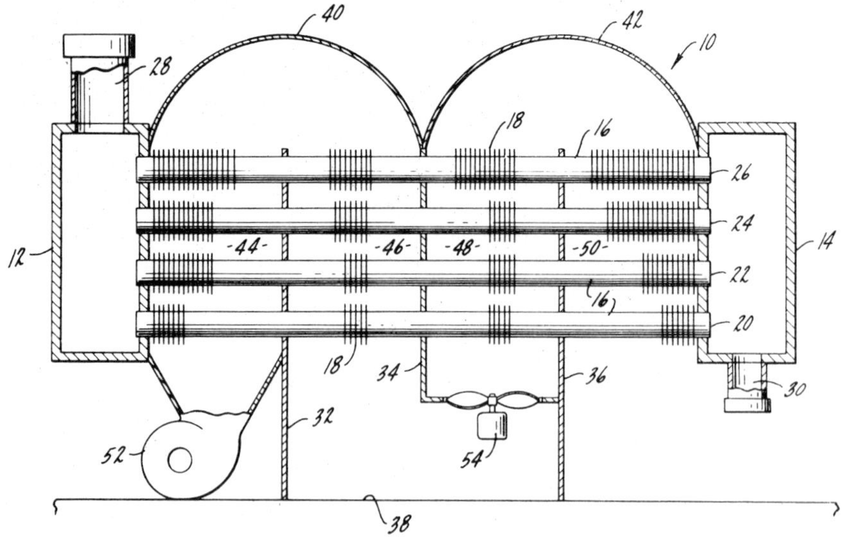

Source: U.S. Patent 3,882,925

Condenser 10 includes an inlet header 12 at one end, an outlet header 14 at the other end, and heat exchanger tubes 16 extending between and thus connected in parallel to one another. Helical fins 18 are preferably provided around tubes 16 to increase heat transfer. Tubes 16 are mounted parallel to each other in a plurality of rows 20, 22, 24 and 26 spaced one above the other. Tubes 16 are mounted to have a slight incline from inlet header 12 to outlet header 14 so that condensate flows or drains into header 14. A steam inlet pipe 28 directs steam, such as the exhaust from a steam turbine, to be condensed into header 12. A condensate outlet pipe 30 directs condensate from header 14 to service.

In accordance with this process, a plurality of transverse partition walls 32, 34 and 36 intersects the tubes 16 and divides same into four substantially equal lengths. The upper edges of partition walls 32, 34 and 36 extend a short distance above the tubes 16 in row 26. The lower edges of partition walls 32 and 36 extend downward and contact the floor 38. The lower edge of partition wall 34 extends downward a short distance below the tubes 16 in row 20 and is spaced from the floor 38. A pair of curved return hoods 40 and 42 respectively span the distance between the upper edge of partition wall 34 and the inlet header 12 and the outlet header 14. The upper edges of partition walls 32 and 36 are respectively spaced from the return hoods 40 and 42. Side panels extend from the floor 38 to a point at or above the return hoods 40 and 42 so as to define four closed cooling air flow passages 44, 46, 48 and 50 in series with one another.

A blower or fan assembly 52, of conventional construction, is provided immediately below and in closing relationship to flow passage 44. Blower assembly 52 draws ambient cooling air and directs it up through passage 44 and then serially through passages 46, 48 and 50 in cooling relationship to the heat exchange tubes 16. A booster blower or fan assembly 54 may be provided to increase the cooling air flow rate leaving passage 46 and entering passage 48.

In operation, steam to be condensed enters inlet header 12 through inlet pipe 28 and then passes through the heat exchange tubes 16 at substantially equal flow rates. The steam condenses as it passes through the tubes 16 and enters outlet header 14 as condensate for discharge through outlet pipe 30. The cooling air is initially drawn vertically upward into passage 44 by blower assembly 52 and successively passes around the lengths of tubes 16 within passage 44 in rows 20, 22, 24 and 26. As the cooling air passes around each successive row of tubes 16, the temperature of the cooling air rises and consequently the heat transfer drops, resulting in a reduced amount of steam being condensed in each successive row of tubes 16.

The cooling air leaving passage 44 contacts hood 40 and is directed in reverse flow down through passage 46 successively around the lengths of tubes within passage 46 in rows 26, 24, 22 and 20. This reverse flow of the cooling air through passage 46 results in an increase in the amount of steam condensed in the tube lengths within passage 46 in the rows which had reduced condensation in the initial passage of the cooling air through passage 44. The cooling air is then directed upward through passage 48 and back down passage 50 in a similar manner and causes a similar result to the lengths of tubes 16 within these passages as that caused by the passage of the cooling air respectively through passages 44 and 46.

Changing the direction of the cooling air flow passing across the succeeding lengths of the tubes **16** effectively mollifies the temperature differential effects which tend to cause maldistribution of the steam in the condenser. The number of cooling air passes may be varied as needed to alleviate the distribution problem. Under most conditions, two passes, as provided by this process, should suffice. One or more booster fans **54** may be provided at intermediate points in the system to attain additional cooling air flow rate capacity.

Pressure Staged Heat Exchanger

D.Y. Cheng; U.S. Patent 4,072,182; February 7, 1978; assigned to International Power Technology, Inc. describes a pressure staged heat exchanger in which at least two evaporators are separated by staged pumps. As the fluid to be heated enters the heat exchanger, it increases in temperature until it reaches its thermodynamic transition point at a pressure below the final desired exit pressure of the fluid. During the first thermodynamic transition, the fluid has only partially changed its thermodynamic phases. The two-phase fluid is then pressurized to a final designed pressure. At this point, the thermodynamic transition temperature has changed and the heated fluid begins to undergo a second thermodynamic transition.

This takes place while the fluid enters into thermodynamic transition in a second higher temperature section and continues thermodynamic transition until the fluid, for example, in a liquid state is vaporized. Once vaporized, the temperature of the vapor begins to increase and, in the case of water, superheated steam exits the heat exchanger.

The pressure staged heat exchanger described in the previous paragraphs in the case of liquid to vapor transition was described as having two evaporators separated by a single stage pump. However, a pressure staged heat exchanger can be designed with a multitude of evaporators separated by a multitude of staged pumps. The number of such stages depends upon design characteristics such as additional energy costs in operating multiple pumps, the surface area reduction of the exchanger, the nature of the fluids employed in the energy transfer and the payoff in weight, cost and energy recovery efficiency.

Alternate Heating and Cooling of Heat Exchanger Systems

O.U. Schafer; U.S. Patents 4,020,895; May 3, 1977 and 4,026,347; May 31, 1977 describes a heating and cooling apparatus with thermal recovery with two heat exchangers and a method of alternately heating and cooling same, in which, during the warm-up phase, the cool liquid returning from a heat exchanger, and, during the cool-down phase, the hot liquid returning from a heat exchanger, are delivered to reservoirs from which, during the warm-up phase, hot liquid is taken for the hot liquid source, and, during the cool-down phase, cold liquid· is taken for the cold liquid source, into which reservoirs both during the warm-up phase and during the cool-down phase amounts of liquid of different temperature are taken from the heat exchanger. The warm-up and cool-down phases of the second heat exchanger are out of step with the warm-up and cool-down phases of the first heat exchanger by a fraction of the length of these phases. The liquid from the first heat exchanger is used during the warm-up phase for the preheating of the second heat exchanger, and during the cool-down phase for the preliminary cooling thereof.

Figure 1.2: Alternate Heating and Cooling of Heat Exchanger System

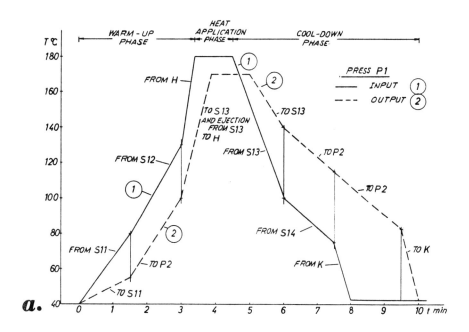

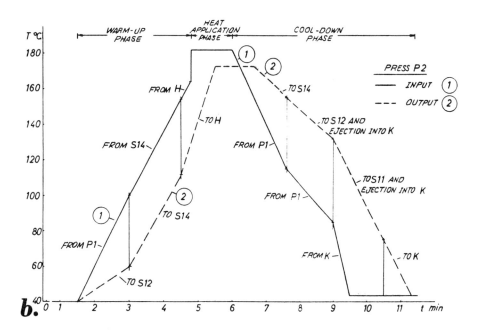

(continued)

Figure 1.2: (continued)

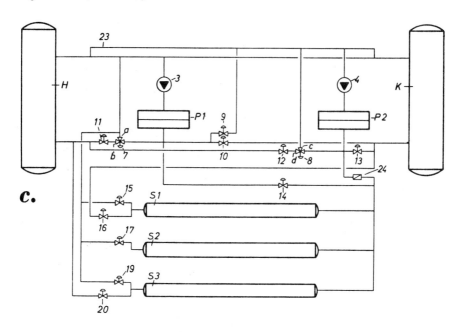

c.

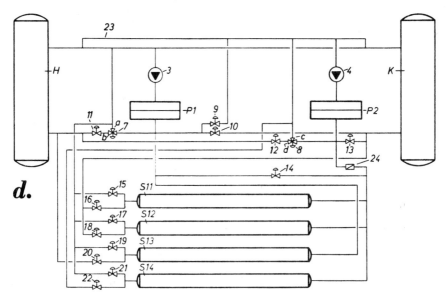

d.

(a)(b)   Temperature curves of water inlet and outlet temperature of two heat exchangers
         which can be connected together within a heating and cooling installation.
(c)   Heating and cooling installation which has two heat exchangers and three reservoirs.
(d)   Heating and cooling installation which has two heat exchangers and four reservoirs.

Source: U.S. Patent 4,020,895

In the temperature diagrams of Figures 1.2a and 1.2b for two heat exchangers, in this case two presses or two independently operated sections of a single press, of a heating and cooling apparatus, the solid curve **1** represents the water temperature at the point of admission to the press, i.e., the input temperature of the press, and the broken curve **2** represents the water temperature at the point of emergence from the press, i.e., the return temperature. The two temperature diagrams are to be considered in conjunction with one another, and they correspond to the conditions as they occur in the heating and cooling apparatus represented in Figure 1.2d.

Three phases are to be distinguished, namely a warm-up phase in which the press is heated from a low temperature of about 40°C to a high temperature of about 165°C, a heat application phase in which the press is maintained at this high temperature, and a cool-down phase in which the press is cooled from this high temperature back down to the low temperature. For ease in understanding, the individual steps of the process will be specified by stating whence the water being fed to the press comes and where it will go after leaving the press. The apparatus represented in Figure 1.2d will be taken as the basis. In this diagram, H = the hot water source; K = the cold water source; P1 = press 1; P2 = press 2; and S11 to S14 = four different reservoirs.

The size of the reservoirs corresponds to approximately one-third of the volume of the liquid that is circulated during the warm-up phase.

An important feature of this process consists in the fact that, in contrast to the state of the art, two heat exchangers, i.e., in this case two presses **P1** and **P2** (they may also be sections of presses), are provided, and that, as it can be seen from a comparison of Figures 1.2a and 1.2b, the warm-up and cool-down phases of the second press **P2** are out of step with the warm-up and cool-down phases of the first press **P1**, and the liquid from the first press **P1** is used to preheat the second press **P2** during the warm-up phase in the first press **P1** or during part of the heat application phase of the latter, and that the liquid from the first press **P1** is used for the preliminary cooling of the second press **P2** during the cool-down phase in the first press **P1**. In this manner the mixing losses, which otherwise occur in the operation of a press and are due to the entry into and departure from the reservoirs of quantities of liquid of different temperatures, are divided between two presses, so that the percentage of the mixing losses for the overall process is reduced.

In addition, this increases the economically useful temperature range of the recycled water in comparison to a press whose size corresponds to the sum of the sizes of presses **P1** and **P2**.

The sections marked off in the temperature diagrams of Figures 1.2a and 1.2b correspond to the process steps XI to XIX of the process that will be explained with reference to Figure 1.2d.

In Figure 1.2c, there is represented a heating and cooling apparatus in accordance with the process, with two presses **P1** and **P2** and three reservoirs S1, S2 and S3. The apparatus contains, in addition to the presses and reservoirs, a hot water source **H**, in this case the reservoir of a recirculating water boiler, a cooler **K**, in this case the cold water reservoir of a water cooler, pumps **3** and **4**, the former associated with press **P1** and the latter with press **P2**, and each disposed at the point of admission to its respective press, three-way regulating valves **7** and **8**, reversing valves **9** to **17** and **19** and **20**, and connecting lines generally designated as **23**. In addition, a nonreturn valve **24** is provided.

With the apparatus described, by means of the operation of the valves as specified in the table below, steps I through VIII of the process can be performed.

Circuit	7	8	9	10	11	12	13	14	15	16	17	19	20	S1	S2	S3
							Position of Valve								Reservoir Temp. Range*	
I	a- b+	c- d+	-	-	-	-	+	+	+	-	-	-	-	120/80	165/120	165/140
II	a- b+	c- d+	+	-	-	-	-	-	-	-	+	-	-	40/60	165/120	165/140
III	a- b+	c- d+	+	-	-	+	-	-	-	-	-	-	-	40/60	40/60	165/140
IV	a+ b+	c- d+	-	-	+	-	-	-	-	-	-	-	+	40/60	40/60	165/140
V	a- b+	c+ d+	-	-	-	+	-	-	-	-	-	+	-	40/60	40/60	150/120
VI	a- b+	c- d+	+	-	-	-	-	-	-	-	+	-	-	40/60	40/60	165/140
VII	a- b+	c- d+	+	-	-	-	-	-	+	-	-	-	-	40/60	165/120	165/140
VIII	a+ b-	c+ d-	-	+	-	-	+	-	-	-	-	-	-	120/80	165/120	165/140

Note: + signifies that valve is open; – signifies that valve is closed.

*°C at start of process step.

Upon the changeover from cooling to heating, the liquid is circulated in the following manner:

I with the circuit cold water source **K**, second heat exchanger **P2**, cold water source **K** running, the liquid is circulated in the circuit first reservoir **S1**, first heat exchanger **P1**, first reservoir **S1**; then

II in the circuit second reservoir **S2**, first heat exchanger **P1**, second heat exchanger **P2**, second reservoir **S2**; then

III in the circuit hot water source **H**, first heat exchanger **P1**, second heat exchanger **P2**, hot water source **H**; and finally

IV in the circuit hot water source **H**, first heat exchanger **P1**, hot water source **H**, on the one hand, and in the circuit hot water source **H**, second heat exchanger **P2**, reservoir **S3**, hot water source **H**, on the other hand.

Upon the changeover from heating to cooling, the liquid is circulated in the following manner:

V first, with the circuit hot water source **H**, second heat exchanger **P2**, hot water source **H** still running, liquid is circulated in the circuit third reservoir **S3**, first heat exchanger **P1**, third reservoir **S3**; then

VI in the circuit second reservoir **S2**, first heat exchanger **P1**, second heat exchanger **P2**, second reservoir **S2**; then

VII   in the circuit cold water source **K**, first heat exchanger **P1**, second heat exchanger **P2**, first reservoir **S1**, cold water source **K**; and lastly

VIII in the circuit cold water source **K**, first heat exchanger **P1**, cold water source **K**, on the one hand, and in the circuit water source **K**, second heat exchanger **P2**, cold water source **K**, on the other.

In this series of operations of the process, the warm-up and cool-down phases of the presses **P1** and **P2** are out of step. The return liquid from the first press **P1** is used for preheating the second press **P2** during the warm-up phase and for precooling it during the cool-down phase. In the table above, the circuits I through VIII correspond to the process steps I through VIII.

In the case of the apparatus represented in Figure 1.2c, as indicated in the above table, in process step I, i.e., at the beginning of the warm-up in press **P1**, the hot water is brought from reservoir **S1** into this press, which in some cases can result in unacceptable thermal stresses. On the other hand, in this step of the process the hot water is displaced by the cold water delivered from the press **P1**. The temperature difference is thus relatively high at the interface between the two amounts of water and thus the mixing losses are also relatively high.

If an additional reservoir is adopted it is possible to store amounts of liquid of different temperature ranges between the low temperature and the high temperature in separate reservoirs or separate sections of reservoirs and, upon the changeover from cooling to heating, to bring into press **P1** the amount of liquid contained in the reservoir of lowest temperature, followed by the amounts of liquid of the next higher temperatures, so as to raise the temperature of press **P1** more gradually, and, upon the changeover from heating to cooling, to bring the contents of the reservoir of highest temperature into press **P1**, followed by the amounts of liquid of successively lower temperature, so as to reduce the temperature of press **P1** more gradually.

In this manner, excessive thermal stresses in the press are avoided, and on the other hand, since in the charging or discharging of the reservoirs the temperature differences between the amounts of liquid being stored and released are not so great, the mixing losses are diminished, so that the thermal recovery can be still further improved. This principle is shown in the heating and cooling apparatus represented in Figure 1.2d.

The apparatus represented in Figure 1.2d, differs from the one in Figure 1.2c, in that it contains four reservoirs **S11** through **S14** instead of three reservoirs **S1** through **S3**, and additional reversing valves **18**, **21** and **22**. The rest of the parts of the apparatus are the same as those in the apparatus of Figure 1.2c, and are therefore identified in the same manner. For the performance of the process, steps XI through XIX of the process are executed by operating the valves as indicated in the table below.

. . . . Position of the Valve																. . . Reservoir Temp. Range*			
Circuit	7 8 9 10	11	12 13	14	15	16 17	18	19	20 21 22	S11	S12	S13	S14
XI	a- c- b+ d+	−	− −	−	+	+ +	−	−	− − −	120/70	140/120	165/140	165/140
XII	a- c- b+ d+	+	− −	−	−	− −	+	−	− − −	40/60	140/120	165/140	165/140

(continued)

	Position of the Valve																Reservoir Temp. Range*			
Circuit	7	8	9	10	11	12	13	14	15	16	17	18	19	20	21	22	S11	S12	S13	S14
XIII	a− b+	c− d+	−	−	−	−	−	−	−	−	−	−	−	+	~	+	40/60	40/60	165/140	165/140
XIV	a+ b+	c− d+	~	−	+	+	−	−	−	−	−	−	−	−	−	−	40/60	40/60	140/100	110/70
XV	a− b+	c+ d+	~	−	−	+	−	−	−	−	−	−	+	−	−	−	40/60	40/60	140/100	110/70
XVI	a− b+	c− d+	+	−	−	−	−	−	−	−	−	−	−	+	−		40/60	40/60	165/140	110/70
XVII	a− b+	c− d+	+	−	−	−	−	−	−	−	+	−	−	−	−	−	40/60	40/60	165/140	165/140
XVIII	a− b+	c− d+	−	+	−	−	−	−	−	+	−	−	−	−	−	−	40/60	140/120	165/140	165/140
XIX	a+ b−	c+ d−	−	+	−	−	+	−	−	−	−	−	−	−	−	−	120/70	140/120	165/140	165/140

Note: + signifies that valve is open; − signifies that valve is closed.

*°C at start of process step.

Upon the changeover from cooling to heating, the liquid is circulated in the following manner:

XI   with the circuit cold water source **K**, second heat exchanger **P2**, cold water source **K** running, the liquid is circulated in the circuit first reservoir **S11**, first heat exchanger **P1**, first reservoir **S11**; then

XII   in the circuit second reservoir **S12**, first heat exchanger **P1**, second heat exchanger **P2**, second reservoir **S12**; then

XIII   in the circuit hot water source **H**, first heat exchanger **P1**, third reservoir **S12**, hot water source **H**, on the one hand, and in the circuit fourth reservoir **S14**, second heat exchanger **P1**, fourth reservoir **S14**, on the other hand; then

XIV   in the circuit hot water source **H**, first heat exchanger **P1**, hot water source **H**, on the one hand, and in the circuit hot water source **H**, second heat exchanger **P2**, hot water source **H**, on the other hand.

Upon changeover from heating to cooling, the liquid is circulated as follows:

XV   with the circuit hot water source **H**, second heat exchanger **P2**, hot water source **H** running, the liquid is circulated simultaneously in the circuit third reservoir **S13**, first heat exchanger **P1**, third reservoir **S13**; then

XVI   in the circuit fourth reservoir **S14**, first heat exchanger **P1**, second heat exchanger **P2**, fourth reservoir **S14**; then

XVII   in the circuit cold water source **K**, first heat exchanger **P1**, second heat exchanger **P2**, second reservoir **S12**, cold water source **K**; then

XVIII   in the circuit cold water source **K**, first heat exchanger **P1**, cold water source **K**, on the one hand, and in the circuit cold water source **K**, second heat exchanger **P2**, first reservoir **S11**, cold water source **K** on the other; and finally

XIX   in the circuit cold water source **K**, first heat exchanger **P1**, cold water source **K**, on the one hand, and in the circuit cold water source **K**, second heat exchanger **P2**, cold water source **K**, on the other.

In the above table, the circuits XI through XIX correspond to process steps XI through XIX. As the table shows, quantities of liquid of low temperature are used for preheating press 1 or press 2, as the case may be, and high temperatures are used for the preliminary cooling of press 1 or press 2.

In the apparatus represented in Figures 1.2c and 1.2d, two separate presses P1 and P2 are the heat exchangers. Instead of two or even more presses, a single press can also be provided, having two or more sections which can be heated and cooled independently of one another.

Partial Extraction and Utilization of Steam Energy

A.C.M. Asfura; Patent Reissue 29,790; October 3, 1978 describes a process of handling fluids which increases heat transfer efficiency in all types of heat transmission equipment. This process is applicable both to single and multiple units; in both cases, heat transfer is enhanced and steam consumption reduced.

Basically, the process consists in extracting, along with the condensate, some of the steam from inside of a steam heated unit, separating this steam from the condensate and noncondensable gases, before and after feeding this extraction steam to another unit working at a lower pressure.

The process for handling fluids in the transfer equipment, consists in supplying steam to a unit or units by means of common pipes. Every unit has its own steam trap, modified with a steam extraction device. The thus modified steam traps discharge the condensate to a common pipe line which carries it into a "Flash Tank". Extraction steam is piped out from each modified steam trap into a common pipe header, which supplies this extraction steam, through a separator, to another group of units.

The second group of units has also individual modified steam traps, except that, in this case, both the extraction steam and the condensate are piped together to the "Flash Tank".

The "Flash Tank" has the purpose of supplying steam to the lowest pressure units. Since the condensate that is discharged to the "Flash Tank" is at a higher pressure, part of it flashes into steam when it enters the low pressure tank. This "Flash" steam, plus the extraction steam from the second group of units, is supplied to the lowest pressure units, thus resulting in the maximum utilization of the heat available in the fluids supplied to the machine.

It should be noted that all of the condensate is recovered and that no steam is wasted to the atmosphere, as it is in other conventional processes.

Pressure Control in Multistage Evaporation Unit

A process described by *A. Hoppe and W. Geistert; U.S. Patent 4,032,412; June 28, 1977; assigned to Deutsche Texaco AG, Germany* relates to a system for optimal pressure control in a multistage evaporation unit, wherein two or more evaporation stages, at least one of which is preceded by a heat source, are connected in series, and the fresh solution passes through heat exchangers in which it is heated by means of the vapors from the subsequent evaporation stages. The process is characterized in that a heat exchanger adapted to permit free flow of the vapor condensate through the heat exchanger inner portion or tube is arranged in the line through which the vapor leaves the higher-pressure evaporation stage.

The process solves the problem of pressure control by installing in the line through which the vapor effluent leaves the higher-pressure evaporation unit, a heat exchanger as a throttling means adapted to permit unimpeded passage of the vapor condensate and assume the pressure expanding function formerly performed by a pressure control valve. The effluent discharged at the outlet of the heat exchanger or condenser mainly consists of condensate which on account of the throttling means acts as a vapor or gas lock when the pressure in the pertinent evaporation stage is too low due to the vapor volume exceeding many times the condensate, and which only permits the passage of uncondensable gases present in small amounts.

In a preferred example of the system, the heat exchanger consists of a shell-and-tube heat exchanger, wherein the throttling is desirably effected in the tube portion, for example, by decreasing the cross section of the passage through the tubes in the flow direction.

Bent Heat Transfer Strips Attached to Tubes

G.A.A. Asselman and A.P.J. Castelijns; U.S. Patent 4,144,933; March 20, 1979; assigned to U.S. Philips Corporation describe an apparatus for heat exchange between a fluid and air which has at least one heat exchanger element comprising a plurality of fluid tubes to which are secured heat transfer strips which define between them air passages having length ≤25 mm, hydraulic diameter <2 mm, with each element being at an angle of at most 45° to the relative flow direction of the air before its entrance into the heat exchanger. At the rear of each element the strips are bent to extend in the discharge direction of the air issuing from the heat exchanger.

Thus, the process is characterized in that at least some of the strips or fins are bent at the rear of each of the elements, only so that the bent ends extend in the discharge direction of the air flow from the heat exchanger.

Without the need for separate baffle plates very good guidance of the air flow and consequently good distribution of the air flow over each of the elements are obtained.

Bending of the strips may readily be effected automatically during their manufacture.

Corrugated Ribs

I.M. Kalnin, V.N. Krotkov, T.M. Sutyrina, A.N. Sergeeva and O.A. Sergeev; U.S. Patent 4,141,411; February 27, 1979 describe a heat exchanger which includes ribs mounted transversely of the axis of the tube and uniformly spaced therealong. Each rib includes a plate with corrugations oriented along the flow of the fluid washing the tube. The ribs adjoin one another, their corrugations having lateral slits, the wall portions between these slits being offset relative to one another to provide passages for the flow of the fluid.

Owing to this structure the thickness of the boundary layer at each one of the portions is reduced, and the effectiveness of heat exchange is stepped up. Furthermore, a heat exchanger with these ribs is considerably more compact and needs less metal for its construction, as compared with known heat exchangers.

The tubular heat exchanger includes a tube **1** (Figure 1.3a) with ribs or fins **2** mounted perpendicularly to the axis of the tube **1** and uniformly spaced therealong. Each rib **2** is a plate whose width equals **a** and height equals **b** (Figure 1.3b), the plate supporting thereon corrugations **3** extending along the flow of the fluid medium washing in operation the tube **1** with the ribs **2**. Similar ribs **2** may alternatively have two or even more apertures for tubes, in which case they are simultaneously fitted over two or more tubes **1**. The ribs **2** are mounted on the tubes **1** and sealingly connected by any known technique employed with the known plain solid ribs.

Figure 1.3: Tubular Heat Exchanger

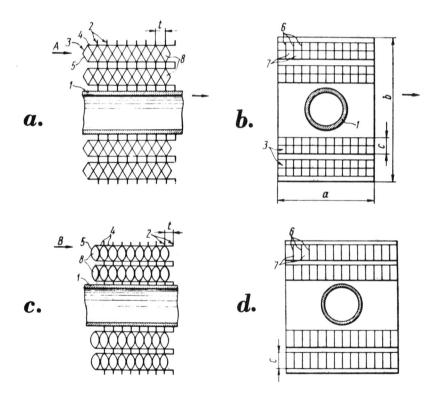

(a)   Longitudinal axial sectional view of a tubular
         heat exchanger with ribs.
(b)   View along arrow line **A** in Figure 1.3a.
(c)   Longitudinal axial sectional view of a portion
         of a tubular heat exchanger having ribs.
(d)   View along arrow line **B** in Figure 1.3c.

Source: U.S. Patent 4,141,411

The side walls **4** (Figure 1.3a) of the corrugations **3** are rectilinear and extend perpendicularly to the plane of the rib **2**. The corrugations **3** have apexes **5** through which transverse slits **6** are made. The portions **7** of the apexes **5** between the slits **6** are offset relative to one another in a direction transverse of the corrugation **3** and perpendicular to the plane of the rib **2**. The extent of this offsetting of the portions **7** should be no less than 1.5 to 2.0 mm. The greater the distance between the offset portions **7** of the apexes **5** of the corrugations **3**, the more compact the structure of the heat exchanger becomes.

Since the ribs **2**, in accordance with this process, adjoin one another, the maximum extent of the offsetting is limited in the case of the portions **7** of the apexes **5** by the spacing **t** of the ribs **2**, this spacing being selected to comply with the purity of the fluid medium washing the ribs, with the requirements as to the compactness of the heat exchanger and with the basic technological and economical calculations.

The portions **7** of the apexes **5**, offset relative to one another, are disposed to one side of the plane of the rib **2** and afford passages **8** for the flow of the medium, which are diamond-shaped at the slits **6**. However, depending on the actual technology of manufacture of the ribs **2**, these passages may also be either oval or circular to render the structure of the heat exchanger even more compact, which can be seen in Figures 1.3c and 1.3d.

As the gaseous medium flows along the surface of the walls of the corrugations **3** in operation of the heat exchanger, it leaps off these walls at the slits **6**, and, since the length of the portions **7** of the corrugations **3** between the slits **6** is small enough (as small as 3 to 4 mm), the thickness of the boundary layer of the flow, which builds up with uninterrrupted motion of the flow, is bound to be likewise small at the end of each portion **7**. This small thickness of the boundary layer which is the major opponent to heat transfer determines the high effectiveness of heat transfer. Furthermore, as the flow leaps off the short portions **7** of the walls of the corrugations **3**, it becomes turbulent, which promotes still further the effectiveness of heat transfer. As a result, the factor of heat transfer by the gaseous fluid at the corrugated portions of the ribs **2** is about two times greater than in the case of plain solid ribs.

Moreover, due to the action of the side walls **4** of the corrugations **3** and to the offset curving portions **7** of the apexes **5**, in the described structure the same spacing of the ribs **2** yields a greater heat transfer perimeter of the flow area afforded to the fluid, than in the hitherto known structures, which brings down the value of the hydraulic diameter and further enhances the effectiveness of heat transfer.

In the case of the ribs **2** shown in Figures 1.3b and 1.3d, with the width **c** of the corrugations **3** equalling the spacing **t** of the ribs **2** (Figures 1.3a and 1.3c), the surface of the corrugated portions of the ribs **2** is increased about three times over, due to the surface of the side walls **4** of the corrugations **3**. When the corrugated portions of the ribs **2** occupy about one-half of the area of the rib **2** (Figures 1.3b and 1.3d) in a plan view, the compactness of the heat transfer surface contacting the gaseous medium is increased approximately two times. The expression "compactness" refers to the total area of the heat exchange surface of the ribs **2** per unit of volume occupied by these ribs. With the portions **7** being of a semi-circular shape as a result of the offsetting, the surface of the apexes **5** of the corrugations **3** is increased by such offsetting 1.55 times. This corresponds to the total heat transfer surface of the rib being increased additionally 1.1 times. Thus, the total gain in compactness of the heat transfer surface, attained by the process in comparison with plain ribs arranged at the same spacing, is about 2.2 times.

Approximately the same gain is attained in the heat transfer perimeter of the flow area afforded to the gaseous medium, with corresponding reduction of the hydraulic diameter of the flow area. This means that the factor of heat transfer by the gaseous medium is increased by about 20%.

Coaxially Arranged Tubes

R.A. Pain; U.S. Patent 4,146,088; March 17, 1979 describes a heat exchanger for fluids comprising a plurality of coaxially arranged tubes of thermally conductive material, the tubes being spaced apart radially by end manifolds to form annular fluid flow passages. The design is characterized in that at least one of the manifolds includes sealing surfaces which bear against and form seals with the inside surfaces of the ends of the tubes.

It is preferred that the manifolds at either end of the heat exchanger are identical and each has a respective sealing surface for each tube.

It is further preferred that the manifolds are generally conical in shape and have progressively larger sealing surfaces spaced axially therealong for cooperating with the interior surfaces of progressively larger diameter tubes of the exchanger.

Referring to Figure 1.4, the exchanger comprises a pair of end manifolds **2** between which concentric heat conducting tubes **4** are disposed, a central bore **6** is provided through each of the end manifolds **2** and a long tension bolt extends through the bores **6** and through the innermost tube **4** and serves to hold the end manifolds and tubes in their correct position.

In the illustrated arrangement there are five tubes **4a**, **4b**, **4c**, **4d** and **4e** each of which is preferably formed from stainless steel and is provided with a continuous helical groove on its surface so as to improve its heat transfer properties. The spaces between adjacent tubes form annular fluid flow passages for the heat exchanger.

Both end manifolds **2** for the heat exchanger are identical and accordingly it is only necessary to describe the construction of one of the manifolds. It comprises a generally conical portion **8** integrally cast with tubular inlet/outlet spigots **10** and **12**. The spigots **10** and **12** permit connection of fluid conduits to the end manifold by conventional means. In the description which follows it will be assumed that the spigot **10** is used as an inlet for a first heat transfer fluid and that the spigot **12** is used as an outlet for the second heat transfer fluid, however, it is to be understood that the spigots **10** and **12** can be used interchangeably as inlets and outlets.

The conical portion **8** of the manifold includes an inlet chamber **14** and an outlet chamber **16** in communication respectively with the spigots **10** and **12**. The conical portion **8** further includes a central chamber **18** which is generally oval in shape and tapers in the same direction as the conical member **8**.

Figure 1.4: Heat Exchanger

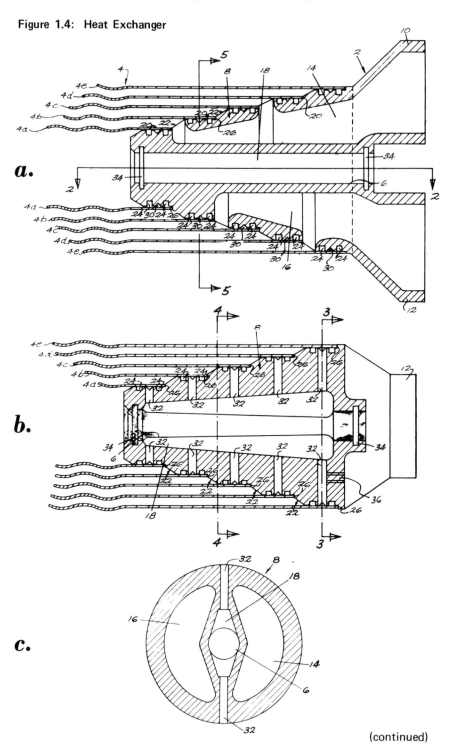

(continued)

Figure 1.4: (continued)

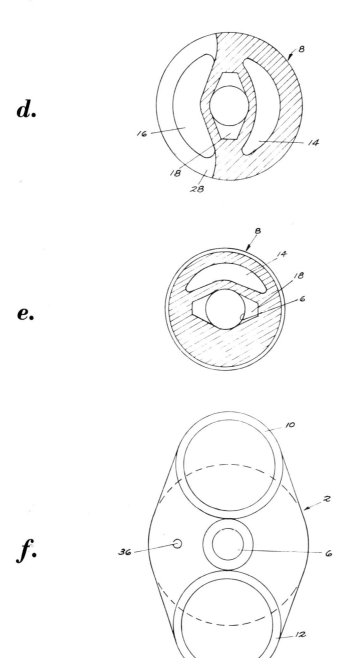

(continued)

Figure 1.4: (continued)

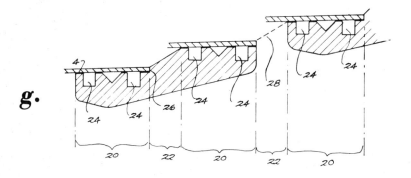

g.

(a)   Longitudinal cross section through the heat exchanger.
(b)   Longitudinal cross-sectional view taken along line 2–2 in Figure 1.4a.
(c)   Transverse cross section taken along line 3–3 in Figure 1.4b.
(d)   Transverse cross section taken along line 4–4 in Figure 1.4b.
(e)   Transverse cross-sectional view taken along line 5–5 in Figure 1.4a.
(f)   End view of the exchanger.
(g)   Detailed view of part of the sealing surfaces of the end manifold of
        the exchanger.

Source: U.S. Patent 4,146,088

The defining walls forming the central chamber **18** are machined out, where
necessary, to form the cylindrical bore **6** which extends through the manifold.

The external surface of the conical portion **8** of the manifold comprises a series
of generally cylindrical portions **20** spaced axially along the conical portion and
adapted to be inserted within respective ends of the tubes, the cylindrical portions **20** being interconnected by tapering transition portions **22**, as best seen in
Figure 1.4g. Each cylindrical portion has formed therein two spaced grooves
24 for receipt of O-rings for forming positive seals with the inner surfaces of
the tubes **4**. A shoulder **26** is formed at the leading edge of each of the transition portions **22** so as to form a seat against which the ends of the tubes **4** bear.

Access to the annular fluid passages defined between adjacent tubes **4** is by way
of recesses **28** formed into the transition portions **22**, as best seen from Figures 1.4a and 1.4d. It is usually desirable that the first and second heat transfer fluids are transmitted through alternate annular passages in the exchanger
and therefore it is desirable to arrange for alternate recesses **28** to open into
the inlet chamber **14** and for the intermediate recesses **28** to open into the outlet chamber **16**.

It has been found that the end manifold 2 of the abovementioned configuration
can be cast so as to weigh approximately one-third the weight of a comparable
manifold of the type described in Australian Patent 402788, thus very substantially reducing the cost of the manifolds and thus the heat exchangers.

An important further advantage can be obtained with the manifold construction described above. This is the ability to sense if any fluid leakage is occurring at the O-rings seated in the slots **24**. To this end, a V-groove **30** is machined into the cylindrical portion **20** between the slots **24**. Any fluid escaping past the O-rings in the first slot **24** will pass into the groove **30**. Radial ducts **32** are provided to communicate the grooves **30** with the central chamber **18** so that any fluid leaking past the O-rings will enter the central chamber **18** and thus cause the pressure within the central chamber **18** to rise sharply.

The end openings of the central chamber **18** are sealed against the central bolt of the exchanger by means of O-rings seated in slots **34**. A sensing hole **36** is provided so as to communicate with the central chamber **18** so as to permit sensing of any buildup of pressure within the central chamber, or, alternatively to simply observe fluid which has leaked so that the appropriate remedial action can be taken (Figure 1.4b).

By rotating the tubes relative to one another the relative positions of the helical grooves formed on their surfaces, shown in Figures 1.4a and 1.4b changes thereby substantially altering the effective fluid flow paths between the tubes. This permits an exchanger to be made having a variable NTU or θ valve.

Vapor-Type Heat Exchanger

J.R. Schieber; U.S. Patent 4,093,020; June 6, 1978; assigned to Betz Laboratories, Inc. describes a vapor-type heat exchanger in which cooling fluid conduit means are located within shell means and vapor inlet means and condensate outlet means are provided in the shell means to provide for the ingress and egress of a fluid to be cooled. By locating the vapor inlet means below the cooling fluid conduit means and the condensate outlet means above the same, it is ensured that the cooling fluid conduit means will be submersed in liquid.

Referring to Figures 1.5a and 1.5b, **1** indicates a heat exchanger according to the process. Element **2** refers to cooling fluid conduit means which are shown in Figure 1.5a as single tubes and as being connected for parallel flow of the cooling fluid, preferably water. Of course, as shown in Figure 1.5d, the tubes **2** can be easily connected for series flow. In use, the cooling fluid enters tubes **2** as shown schematically by arrow **3**, flows through the tubes, and leaves the same as schematically shown by arrow **4**. If, for example, the heat exchanger is to be used to evaluate corrosion and fouling of process heat exchanger surfaces, the tubes **2** are removably held within shell **5** by screw nuts **15** and **16**.

As is known to those skilled in the art, the tubes **2** could include a removable test specimen **17** used to evaluate corrosion and fouling. Shell **5** is provided in surrounding relationship to tubes **2**. Numeral **6** designates the inlet end of vapor inlet manifold **7** through which the heat exchange fluid to be cooled (hot fluid) is supplied to shell **5**. The inlet manifold **7** is preferably elongated and is preferably at least of the same length as shell **5**. Element **9** is a condensate outlet provided in shell **5**. By locating condensate outlet **9** above tubes **2**, and manifold **7** therebelow, it should readily be seen that shell **5** will fill up with condensate up to the level of outlet **9**. While outlet **9** is shown in the top of shell **5**, it could be located in an end wall thereof, as at **10** in Figure 1.5b, as long as it is located above tubes **2**. Accordingly, the tubes **2** will be submersed in liquid in use. A condensate drain **11** is provided in the bottom of shell **5** for draining condensate when the heat exchanger is not in use.

Figure 1.5: Vapor-Type Heat Exchanger

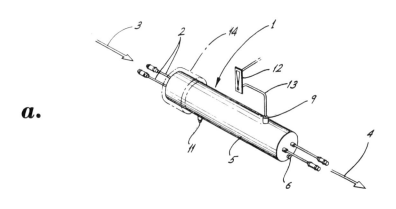

a.

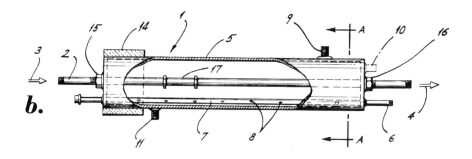

b.

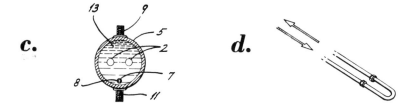

c.   **d.**

(a)   Schematic perspective view of heat exchanger.
(b)   Longitudinal side view of heat exchanger shown in Figure 1.5a
       with a portion of the shell broken away.
(c)   Cross-sectional view of heat exchanger taken along section **A—A**
       in Figure 1.5b.
(d)   Schematic view of cooling fluid conduit means modified for
       series flow.

Source: U.S. Patent 4,093,020

In order to control the heat input to the illustrated heat exchanger system, condensate outlet 9 is provided with a suitable valve 12 shown in Figure 1.5a. A rotameter, not necessary to the device, can be used to measure rate of condensate flow. As the incorporation of the illustrated valve arrangement into the condensate outlet conduit 13 is considered to be well within the knowledge and skill of the art, further details related to this matter are not included. By controlling the rate of flow of condensate from the shell, the rate of vapor input and thus heat input into the system is controlled.

The openings 8 in manifold 7 are arranged along the length thereof in such a manner as to provide jet mixing action of the condensate 13 (Figure 1.5c). The orientation of the openings with respect to the tubes 2 and the inside wall of shell 5 is important in this matter. The openings should be directed such that vapor being injected into condensate 13 will not impinge directly on tubes 2 or on the inside wall of shell 5 in such a manner as to dissipate vapor energy which would otherwise impart mixing motion to the condensate. If the openings are so directed for jet mixing action, the injected vapor will impart a swirling motion to condensate 13 to promote a uniform and constant condensate temperature and a uniform and constant heat transfer rate across the walls of tube 2.

The abovedescribed vapor-type heat exchanger would be utilized as a test heat exchanger using steam as the hot fluid and water as the cooling fluid to evaluate corrosion and fouling tendencies on the process heat transfer surfaces of a cooling water system as follows.

As already noted, the cooling fluid conduit means 2 may be connected in parallel (Figure 1.5a) or in series (Figure 1.5d). A parallel connection assures the same conditions in each tube if, for example, untreated and pretreated tubes are to be compared. A series connection requires one half as much cooling water for a given water velocity in the tubes, but the average water temperature in each tube will differ slightly. The test heat exchanger should be located at a readily accessible place for ease of installation, removal of test tubes and frequent observation. The unit is mounted horizontally. The water supply can be taken from the pump discharge of the process cold well where pressure is highest. The steam supply is taken from the top of a steam line, preferably one in active use. For quantitative data, the steam supply should be free of moisture and should be held at a substantially constant pressure.

The cooling fluid tubes should be the same metal as those being compared in the process heat exchanger system and the water velocity through the tubes in the test heat exchanger should be the same as that in the process heat exchanger system. The initial run normally duplicates the condition of the process system under typical conditions. Once initial results are obtained, additional runs are performed with modifications in operating procedures or in chemical treatment and controls. At the end of each run, the operator can remove the tubes, observe their condition and/or record pertinent data. During each run, typical data recorded includes flow rates, temperatures, needed adjustments, chemical controls and abnormalities in the process. In certain uses, it may be desirable to insulate the shell side 5 as partially indicated at 14.

OTHER HEAT EXCHANGER CONSTRUCTIONS

Vertical Counterflow Heat Exchanger

W.J. Darm; U.S. Patent 4,140,175; February 20, 1979 describes a counterflow heat exchanger with two sets of heat exchanger passages separated by heat exchanger plates disposed in substantial parallelism within the exchanger and supported so that the passages extend substantially vertically. The heat exchanger may be employed as an air-to-air or as an air-to-water heat exchanger and can be used to remove moisture or pollutants from hot exhaust air by condensation within the exchanger passages. When employed as an air-to-water heat exchanger, the water is sprayed onto the surfaces of the upper ends of one set of passages so that it flows down their length, while air is transmitted into the lower end of the other set of passages and caused to flow upward.

The ends of the heat exchanger plates are split into two end portions which are joined to different exchanger plates on opposite sides thereof to form the two sets of passages which allow the air and water to flow in opposite directions through such passages for counterflow heat exchange by direct lateral transfer through the thickness of the exchanger plate.

A preferred example of the heat exchanger apparatus is the air-to-water heat exchanger apparatus shown in Figure 1.6a and includes a counterflow type of heat exchanger 1. The heat exchanger includes a housing containing an assembly of heat exchanger plates shown in Figure 1.6b which form a first set and a second set of fluid passages separated by the exchanger plates.

Figure 1.6: Vertical Counterflow Heat Exchanger Apparatus

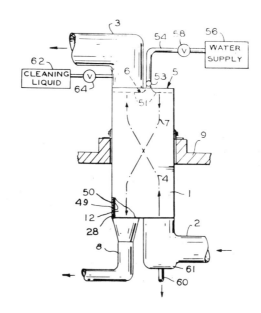

a.

(continued)

Figure 1.6: (continued)

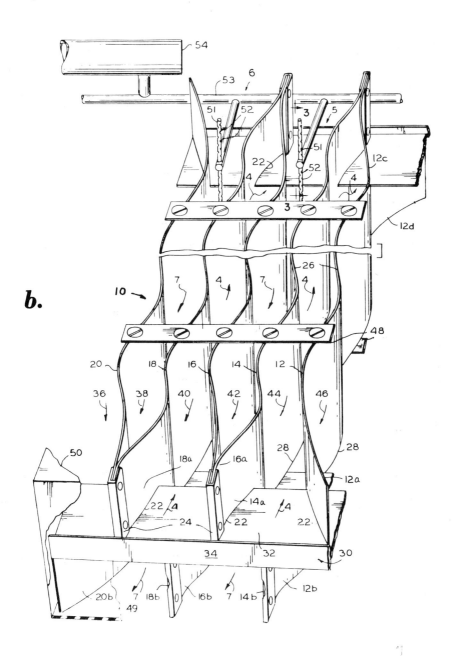

b.

(continued)

Figure 1.6: (continued)

c.

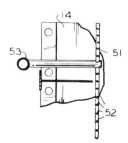

(a)   Diagrammatic elevation view illustrating an air-to-water heat
      exchanger apparatus with parts of the exchanger housing
      broken away for clarity.
(b)   Perspective view of a portion of the assembly of heat ex-
      changer plates and associated water sprayer provided in
      the heat exchanger apparatus of Figure 1.6a.
(c)   Horizontal section view taken along the line **3—3** of Fig-
      ure 1.6b.

Source: U.S. Patent 4,140,175

Heat exchanger **1** is supported so that the passages extend substantially vertically
between top and bottom ends. The bottom ends of one set of passages are connected to an air inlet conduit **2** which transmits air into such inlet from any
source, such as a source of hot humid exhaust air having a temperature of, for
example, 180°F dry bulb and 140°F wet bulb. The top ends of such one set of
passages are attached to an air outlet conduit **3** which discharges the air to the
atmosphere after it is cooled in the heat exchanger **1** while flowing upward
through the one set of passages in the direction of arrows **4**. Much of the moisture in the air condenses on the surfaces of the passages and flows downward
out of the heat exchanger apparatus through a drain. As a result, the discharged
air at outlet **3** is also of lower humidity and may have a temperature of 60°F dry
bulb and 60°F wet bulb.

The other set of passages in the heat exchanger have their top ends employed as
water inlets **5** by providing a water sprayer **6** either outside of the heat exchanger
and above the water inlets or inside the heat exchanger within such inlets. Water
is sprayed onto the surfaces of such other set of passages by sprayer **6** adjacent
the top ends thereof and is caused to flow downward in the direction of arrows
7 along such surfaces provided by one side of the heat exchanger plates. When
cooling hot exhaust air, cold water of, e.g., about 40°F is delivered to the heat
exchanger by sprayer **6** and hot water of, e.g., about 140°F is discharged into
a water outlet conduit **8** to the bottom ends of the other set of passages. As
shown in Figure 1.6a, the heat exchanger **1** is supported in a vertical position
and may extend through a hole in the roof **9** of building containing the source
of the hot exhaust air.

The heat exchanger plate assembly **10** within the exchanger is shown in Figure 1.6b and includes a plurality of exchanger plates **12, 14, 16, 18** and **20**. The plates are arranged in the assembly in substantial parallelism. Each of the plates has a length conforming substantially to the length of the exchanger with which the assembly is to be incorporated. The plates further may be provided with corrugations extending transversely of the plates, whereby any fluid, liquid or gas, moving through the exchanger is given a degree of turbulence.

Each of the plates at each of its ends is split with a cut **22** extending longitudinally of the plate. The cut parallels the longitudinal edges **26** and **28** of the plate, and normally is made about midway between the longitudinal edges, thus to divide the end of each plate into a pair of end portions or tongue segments, exemplified by tongue segments **12a** and **12b**, of equal width.

The tongue segments of each exchanger plate end are shown bent in reverse directions. Thus tongue segment **12a**, as shown in Figure 1.6b, is bent to curve outwardly (where it will meet with the housing of the exchanger which is assembled about the core element) and tongue segment **12b** below segment **12a** is bent to curve inwardly. Considering exchanger plate **14**, its upper tongue segment **14a** is bent inwardly, whereas its lower tongue segment **14b** is bent outwardly to meet tongue segment **12b**. Where adjacent ends of tongue segments meet, they may be fixed together using an overlying angle piece, such as angle piece **24**, secured in place as by crimping.

It will be noted that whereas the upper tongue segment **12a** of plate **12** is bent outwardly, and the lower one **12b** is bent inwardly, at the opposite end of the exchanger plate the upper tongue segment **12c** is bent inwardly whereas the lower one is bent outwardly. This same relationship holds true for the tongue segment at each set of ends of an exchanger plate. A divider member shown at **30**, including a horizontal wall expanse **32** and a vertical marginal flange **34**, may be inserted into cuts forming the tongue segments, at each set of ends of the exchanger plates. The divider member, when positioned as shown in Figure 1.6b, serves to separate end portions of channels defined on opposite sides of the various plates. Thus, and considering channel **40** in Figure 1.6b, the divider separates this channel where such is defined between tongue segments **18a, 16a**, from portions of channels **38** and **42** below the divider, defined between tongue segments **20b, 18b,** and **16b, 14b** respectively.

As shown in Figure 1.6b, one set of passages or channels **36, 40,** and **44**, at the end of the assembly pictured at the bottom of Figure 1.6b, open to the end of the assembly above the divider. At the opposite end of the assembly these channels open up to the end of the assembly below the divider. The reverse is true for the other set of passages or channels **38, 42,** and **46**, which at the end of the assembly pictured at the bottom of Figure 1.6b, open to the end of the assembly below the divider, and at the opposite end of the assembly, open to the end of the assembly above the divider.

With the arrangement, and assuming the presence of an encompassing casing, it should be obvious that one set of channels may be utilized for the passage of one body of fluid through the exchanger, and an alternating set for the passage of another body of fluid through the exchanger, with such bodies of fluid passing through multiple flow paths interspersed with each other.

In making an exchanger with a core element of the type described, and when it is remembered that typically a far greater number of exchanger plates are utilized than actually pictured in Figure 1.6b, it should be obvious that a problem arises with respect to positioning properly the adjacent exchanger plates where they extend in expanses between the ends of the plates. The plates usually are made of thin metal, and if corrugated transversely of their lengths, have considerable flexibility in a transverse direction. Further, they are easily twisted. Obviously, if the plates are not properly oriented in the completed exchanger with substantially uniform spacing existing between them where they extend throughout their length, the efficiency of the exchanger is affected.

Thus, according to this process, the various exchanger plates are arranged substantially as pictured in Figure 1.6b. During assembly, and to tie the various exchanger plates together, tying clips 48, which may be metal strips attachable to the edges of the plates, are assembled with the plates by fixedly attaching them to the edges of the plates at regions spaced along the length of the assembly. The clips are attached to each of the opposite sets of adjacent edges in the plate assembly. The tying clip 48 specifically illustrated has slotted depressions formed in it, the slots receiving edge portions of the plates which may be twisted slightly after being passed through the slots to hold the clip in place. The exchanger plates are also fastened together at their ends through joinder of the tongue segments with angle pieces 24.

As a next step in the manufacture of the exchanger, a set of adjacent longitudinal edges 26 or 28 in the exchanger plates assembly 10 is permanently fixed by embedding or encasing the edges in a resin slab 49. In preparing such a slab, reinforcing material, such as a fiber glass sheet, is laid down along the interior of a section of the metal housing 50 of the heat exchanger. Poured over this sheet is a mass of uncured, hardenable synthetic plastic resin material 49, such as a liquid epoxy resin, which is a thermosetting resin that cures to form a hard mass. The resin impregnates the reinforcing sheet, and the resin and sheet form a hardenable layer on the inner surface of the housing section. The housing section, which may be made of sheet metal, constitutes a form confining the resin layer during this stage of the manufacture.

The assembly of exchanger plates may then be fitted within the housing section, with a set of edges 26 or 28 of the plates, as well as any tying strips 48 connecting these edges, then pressed downwardly to be sunk into the layer 49 of resin and reinforcing material. After a period of time, on curing of the resin mass, a strong, rigid slab 49, of synthetic plastic results which bonds the longitudinal edges of the heat exchanger plates to the housing 50. Such a slab seals the edges of the plates, and provides insulation along one side of the exchanger, as well as forming a rigid structural element along one side of the plate assembly.

This operation may then be repeated by preparing in another housing section another layer of resin and reinforcing material similar to the one just described. The assembly of exchanger plates may then be inverted, and the edges in the opposing set of edges of the plates, together with tying strips, sunk into this layer of material. On hardening, another rigid slab is thus prepared encasing the edges of the opposing set, and forming a rigid unit of slab and various exchanger plates.

An article prepared as above includes the core element of assembled exchanger plates, having cured slabs of reinforced resin uniting and sealing the edges in the opposed set of edges of the plates. Encompassing the slabs a resinous material, and portions of the sides of the exchanger plate assembly, are housing sections which are joined together to form the completed housing.

As shown in Figures 1.6b and 1.6c, the sprayer means 6 may include a plurality of pairs of sprayer pipes 51 provided in each of the water passages 38, 42, and 46 extending across their width adjacent the upper ends. The sprayer pipes 51 spray water onto the surfaces of such passages adjacent their top ends through spray openings 52 spaced along such pipes. The water flows down the length of such passages in the direction of arrows 7 since they are supported in a vertical position. The sprayer pipes 51 may be of about ¼ inch diameter and are connected to larger header pipes 53 of one inch diameter which in turn are connected to a main line pipe 54 of 4 inch inner diameter.

The main line pipe 54 is connected to a water supply 56 through a valve 58. The water supply may be connected to the output of a water cooling tower whose input is connected to the water outlet conduit 8 of the heat exchanger to reuse the water after it has been cooled. The valve 58 may be automatically or manually adjusted to vary the flow of water through the heat exchanger and thereby control the temperature of the air discharged from such exchanger through conduit 3.

In addition, the heat exchanger apparatus of Figure 1.6a can be provided with a drain pipe 60 connected to a collecting basin 61 in the bottom of the air inlet conduit 2 to drain off any condensed water or pollutants, such as carbonaceous vapors which condense from the exhaust air as liquids onto the surfaces of the heat exchanger passages through which the exhaust air flows. Thus, some of the air pollutants may be chemical solvents which after condensing in the heat exchanger run down the heat exchanger plates and are recovered through the drain 60. In order to clean the surface of the heat exchanger plates and remove solids deposited on such plates, a source of cleaning liquid 62 is connected through a valve 64 to the air outlet conduit 3 at the top end of the air passages of the heat exchanger. Thus cleaning liquid is periodically injected into the conduit 3 and flows down the surfaces of the air passages for cleaning.

Alternatively, the cleaning liquid can be evaporated into the hot air and transmitted into the inlet conduit 2 so that such cleaning liquid condenses on the surface of the first set of heat exchanger passages as the air flows upward in the direction of arrows 4. The condensed cleaning liquid flows downward along the surfaces of the first passages and removes any solids deposited on such surfaces. The condensed cleaning liquid is collected in the basin 61 and is transmitted through the drain 60 to a suitable evaporator for reevaporating it into the hot air stream to provide a continuous cleaning operation.

Plate Type Heat Exchanger

A process described by *Y. Nakayama and N. Komano; U.S. Patent 4,146,090; March 27, 1979; assigned to Hisaka Works Ltd., Japan* relates to a plate type heat exchanger having a plurality of plates put together so as to define clearances through which two fluids pass for heat exchange.

According to the process, the distributing surface of the plate on the no-packing mount side is provided with distributing grooves having the same depth as the packing groove and equispaced in a direction perpendicular to the packing groove and the boundary between the distributing grooves and the packing groove is provided with a continuous projecting lateral wall defining the outer peripheral edge of the packing groove, the bottom of such distributing groove abutting against the upper surface of the lateral wall of the packing groove in the adjacent plate. On the other hand, in the packing-mounted plate, the continuous projecting lateral wall defining the outer peripheral edge of the packing groove is provided with recesses having the same depth as the packing groove and equispaced in the direction of the packing groove, the bottoms of such recesses abutting against the upper surface of the lateral wall of the packing groove in the adjacent plate.

According to another example of the process, in the double seal region, the packing groove in the plate having no packing mounted thereon has a smaller width than that of the packing groove in the packing-mounted plate and functions as a liquid passage groove.

Rotary Heat Exchanger

W.A. Doerner; U.S. Patent 3,866,668; February 18, 1975; assigned to E.I. Du Pont de Nemours and Company describes a rotary heat exchanger comprising an array of closely spaced parallel annular thermally conductive fins mounted coaxially for rotation as a unit. A plurality of thermally conductive heat exchange tubes extends longitudinally through the array of fins circumferentially about the rotational axis and the fins are dimensioned and spaced and rotationally driven at a speed operable to convey a gaseous fluid radially between the fins essentially by viscosity shear forces and accelerate such fluid substantially to the velocity providing optimum total heat exchange between the fluid and another fluid flowing through the heat exchange tubes.

Referring to Figure 1.7a, the rotary condenser comprises a cylindrical body or casing **1** of selected diameter and relatively short axial length having a continuous circumferentially extending wall **2** and axially spaced side walls **3** and **4**, respectively. Fixedly secured to and extending coaxially outward from the casing side wall **3** is a tubular shaft member **5** that is in communication with the interior of the casing **1** through an axial opening in the side wall **3**. The shaft **5** is rotatably mounting in bearings **6** and **7** and the shaft **5** and casing **1** are rotationally driven at the desired speed by means of an electric motor **M** driving a gear **8** which in turn drives a gear **9** secured on the shaft **5**.

Mounted outwardly adjacent the opposite side wall **4** of the casing **1** for rotation therewith, is an array of annular fins **10** arranged coaxially of the casing **1** in predetermined equally spaced parallel relation and defining internally thereof a coaxial inlet chamber **C** for gaseous heat exchange fluid to be discharged outwardly between the fins **10**. The fins **10** consist of separate or independent annual disk elements supported and secured in the desired closely spaced parallel relationship with respect to one another and the casing **1** by means of a plurality of heat exchange tubes or pipes **11** that extends longitudinally through the array of fins **10** circumferentially about the rotational axis thereof. The fins **10** and tubes **11** are fabricated of metal having high thermal conductivity such as, for example, copper or aluminum, and the fins preferably are bonded to the heat

Figure 1.7: Rotary Heat Exchanger

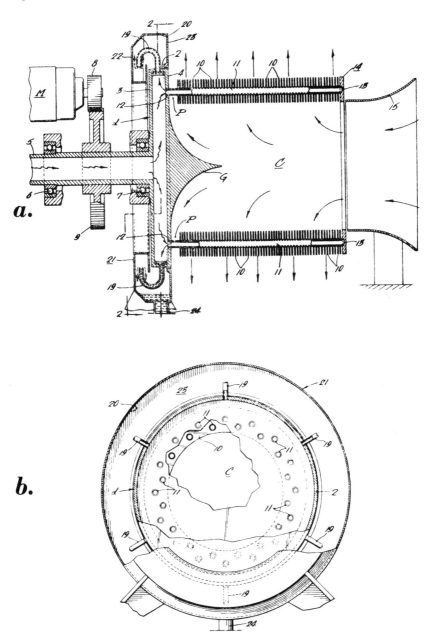

(a)   Sectional view diametrically of a rotary heat exchange condenser.
(b)   View, partially in section, taken on line 2–2 in Figure 1.7a.

Source: U.S. Patent 3,866,668

exchange tubes 11 by brazing, soldering or the like, to provide maximum thermal conductivity. The tubes or pipes 11 are arranged in equally spaced radially staggered relation circumferentially of the fins 10 and casing 1 as shown in Figure 1.7b. The inner ends of the tubes 11 are mounted and secured in corresponding openings 12 provided through the casing side wall 4 so that the interiors of the tubes 11 are in communication with the interior of the casing 1. The outer ends of the tubes 11 are mounted and secured in recesses 13 provided in an annular end ring 14 that is disposed coaxially of the apparatus adjacent the outermost of the fins 10. The end ring 14 effectively closes the outer ends of the tubes 11 and also supports them in the desired relationship.

As shown in Figure 1.7a, the outer radius of all of the fins 10 is the same and conforms substantially to the radius of the casing 1. The inner radius of all the fins is also the same. More particularly, the inner and outer radii of the fins 10 are predetermined and interrelated to provide an inner to outer radii ratio within a predetermined relatively narrow range of limits.

The array of annular fins 10 does not extend axially inward entirely to the casing wall 4 and the innermost of the fins 10 is spaced from the adjacent surface of the wall to provide between the wall 4 and fins 10 an annular passage P for the radial discharge of dirt, dust and other solid particles that become entrained and carried into the chamber C by the gaseous heat exchange fluid. Such particles have greater density and momentum than the gaseous fluid and tend to travel to the inner end of the chamber C. Thus, the particles are discharged outwardly through the discharge passage P and do not accumulate in the heat exchanger.

Extending coaxially into the chamber C from the casing wall 4 is a tapered guide member G having a curvilinear concave lateral circumferential surface that functions to guide the foreign solid particles outwardly through the discharge passage P and also to guide and distribute the gaseous heat exchange fluid outwardly between the fins. The guide member G may be formed as an integral part of the casing wall 4 as shown, or separately and secured thereto, as desired.

The outer diameter of the end ring 14 is substantially the same as the outer diameter of the fins 10 and the inner diameter of the ring is substantially the same as the inner diameter of the adjacent group of fins so as not to restrict the flow of fluid into the chamber C. An outwardly flared or bell shaped fluid intake member 15 is fixedly mounted on a stationary base or support 16 and disposed coaxially adjacent the outer face of the end ring 14. The smaller end of the intake member 15 adjacent the ring 14 has a diameter substantially the same as the inner diameter of the ring 14 to provide smooth uninterrupted flow of fluid inwardly through the member 15 and ring 14 to the chamber C.

In the condenser shown in Figure 1.7a, the vapor to be condensed enters the casing 1 through the tubular shaft 5 and then passes into the tubes 11 where the vapor is condensed by heat exchange with a gaseous cooling fluid, such as ambient air, discharged outwardly between the spaced fins 10 as herein described. The condensate thus formed in the tubes 11 flows back into the casing 1 from which it is discharged radially by centrifugal force generated by rotation of the condenser. In the arrangement shown, the condensate is discharged from the casing 1 through a plurality of U-shaped tubes 19 that form liquid traps which prevent discharge of the vapor directly from the casing and cause the vapor to be diverted into the heat exchange tubes 11. Upon leaving the U-shaped tubes 19 the vapor condensate is discharged radially outward against the inner surface of the cylindrical peripheral wall **20** of a stationary annular housing **21** that circumscribes the rotating casing **1** and has spaced apart side wall portions **22** and **23** which lie closely adjacent the peripheral portions of the casing side walls **3** and **4**, respectively. Condensate collecting in the housing **21** discharges therefrom through a drain **24**.

For optimum results the axial spacing of the fins, their speed of rotation and their inner and outer radii are correlated so that the gaseous fluid passing between the fins is accelerated to a velocity substantially less than the outer peripheral speed of the fins in order to retain the fluid between the fins **10** the longer time required to provide the optimum total heat flux or total heat exchange between the fluid passing between the fins and another fluid in the tubes or pipes **11**.

The axial spacing between the fins **10** or the relationship of the inner radius of the fins to their outer radius may vary within predetermined ranges or limits for any given range of speeds of rotation (rpm) of the condenser. The nature of the flow for rotational shear force devices is completely described by the Taylor number, N_{Ta}, where $N_{Ta} = d^2\, w/v$; d = distance between fins; w = angular velocity (radians per second); and v = kinematic viscosity.

It has been found that most efficient pumping occurs when $N_{Ta} = 3.25$. However, efficient fluid pumping does not lead to an efficient heat exchanger. Efficient pumping occurs when the energy transfer to the fluid is maximized. Efficient heat exchange depends upon both the fin area and the difference between the speed of the fins and the velocity of the fluid flowing between them. Thus, for heat transfer, the Taylor number is not adequate by itself to completely describe an optimum configuration. Thus, it has been found that for various combinations of inner radius (Ri) and outer radius (Ro) of the fins the Taylor number for an efficient heat exchanger will always be greater than 4.5. The precise values of Taylor number and the ratio of the inner to outer radii of the fins depend upon the thermodynamic and transport properties of the fluids exchanging heat and whether the heat transfer mechanism for the fluid in the heat exchange tubes is boiling, condensing or convective.

For heat transfer to or from air on the fin side to or from a boiling or condensing fluid within the tubes, it has been determined that the Taylor number for an efficient heat exchanger is within the range of from 5.0 to 10.0 and the inner to outer radii ratio of the fins is within the range of from 0.70 to 0.85. For optimum results the Taylor number will be in the neighborhood of 6.0 and the fin radii ratio in the neighborhood of 0.77, and these values constitute good starting points for the design of an efficient heat exchanger according to this process. The particular optimum design and operating conditions for any given heat exchanger installation can be determined by a person skilled in the art. It has been determined that the values of Taylor number and fin radii ratio for other gaseous fluids are essentially the same as the values stated for air.

A rotary condenser made according to the process is characterized by its comparatively small compact size and lightweight construction and the minimum power that is required to rotationally drive the heat exchanger at the desired speed. Also, the use of viscosity shear forces to convey the fluid between the spaced fins **10** with the inherent absence of the flow separation, produces a very low operating noise level free of cavitation such as frequently occurs when conventional lift forces are employed to accelerate a fluid.

By reason of these characteristics and advantages, the rotary condenser is particularly suited for use in high-performance closed Rankine cycle power systems having a rotary boiler to which the condenser can be directly mounted or coupled coaxially for rotation with the boiler as a unit.

Helical Guide

According to a process developed by *A. Zelnik; U.S. Patent 4,009,751; March 1, 1977; assigned to Slovenska vysoka skola technika, Czechoslovakia* heat and/or mass transmission between two or more phases is accomplished by a helical guide for both phases inside a column flowing either in countercurrent or in cocurrent. The phase with a higher specific mass flows within the column along a continuous track formed by an outer helix; the phase with a lower specific mass flows along an internal helix within the column. These flows are brought about by suitable elements which are built-in in a cylindrical column.

The process apparatus as shown in Figures 1.8a, 1.8b and 1.8c comprises a column 1 provided with a stable helical insert 2 on the internal wall of the column 1 so that it forms a continuous channel for the flow of the phase 12 with the higher specific mass. The column 1 furthermore has an internal helical insert 3 for the promotion of the rotational movement of the phase 13 with the lower specific mass. The apparatus also has an inlet neck 4 and an outlet neck 5 for the phase 13 with the lower specific mass and an inlet neck 6 with an outlet neck 7 for the phase 12 with the higher specific mass. The apparatus can also have an inlet neck 8 and an outlet neck 9 for cooling or heating medium if the hollow helical insert 2 is used as a heat transmitting surface. The arrangement can also be provided with a duplicator with an inlet neck 10 and an outlet neck 11 for the heat transmitting medium.

The phase with the higher specific mass is supplied via an inlet neck in the upper part of the arrangement into the space formed by the stable helical insert in the internal surface of the column. The first two or three turns of the insert are enclosed by a cover to prevent a direct dissipation of the phase with the higher specific mass into the stream of the phase with the lower specific mass. The phase with the lower specific mass is supplied via an inlet neck in the lower part of the arrangement; it flows through the central space of the column formed between the internal surface of the column and the helical insert, and leaves the arrangement via the outlet neck in the upper part thereof.

Due to centrifugal acceleration, the phase with the larger specific mass is forced into the space between the turns of the external stable helical insert and is accelerated in the course of its downward flow with respect to the phase with the lower specific mass, proceeding in countercurrent to it. The phase with the higher specific mass leaves the arrangement via the outlet neck in the lower part of the arrangement. The phase with the lower specific mass is caused to rotate by the internal helical insert.

The arrangement using a central tube for forming an annular space operates substantially similarly to the abovedescribed one, with the difference that the phase with the lower specific mass flows in a spiral course in the annular space formed

Figure 1.8: High Intensity Heat Transfer Column

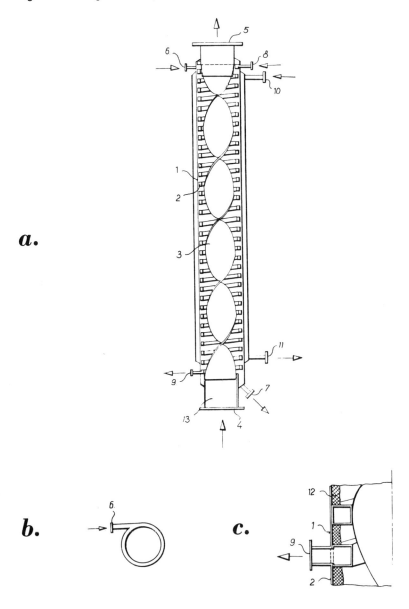

(a)   Cross-sectional elevation of column.
(b)   Sectional view of an inlet neck for the phase with the higher
         specific mass.
(c)   Sectional view of an outlet neck for the heat transfer medium
         if a hollow insert is used.

Source: U.S. Patent 4,009,751

by the central tube and the column, the central tube being provided with blade rosettes or with a helical surface respectively, which promote a rotational movement of this phase.

The arrangement can be applied for different operations and also for solutions of important problems of retaining emissions. In case of working with substantially dusty media, the arrangement has a self-cleaning effect, which is also one of its more significant advantages.

Helical Design

According to a process described by *P.C. Gaines, Jr.; U.S. Patent 3,893,504; July 8, 1975* a medium passed from a high temperature system is cooled by the steps of retarding the normal rapid expansion of the medium while maintaining a relatively high velocity while cooling same; passing the medium cooled by the first step through a coil section for further cooling same, and performing the first two steps in the presence of a cooling medium. A heat exchanger for carrying out the method has a shell providing a flow path for a cooling medium, and a coil arranged within the shell and including a first coil section which controls expansion of the medium to be cooled and maintains the velocity high enough to maintain any condensate formed in the coil to be rapidly moved along the coil in front of the expanding medium to be cooled thus preventing hammering by preventing spaced areas or slugs of liquid condensate from moving rapidly toward each other as gaseous pockets are condensed which normally cause reduced pressure areas.

A second coil section is connected to the first coil section to receive the medium to be cooled for further cooling. The first coil section has a pair of headers and a plurality of helical loops connected between the headers. The second coil section is formed by a plurality of helical loops connected.

Referring to Figure 1.9, a heat exchanger 10 has a shell 12 providing a flow path for a cooling medium, and a coil 14 arranged within shell 12. Coil 14 has means 16 controllably expanding a medium to be cooled for cooling same, and a means 18 connected to means 16 for receiving the medium to be cooled for further cooling the medium to be cooled.

Means 16 has a first coil section 20 with a plurality of helical loops 22. Coil section 20 has also a pair of headers 24, 25. Header 24 is adapted to be connected to a source of a medium to be cooled and header 25 is connected to means 18. Helical loops 22 are connected in parallel between headers 24, 25.

Means 18 has a second coil section 26 which is formed by a plurality of helical loops 28 connected together in series. It is to be understood that coil section 26 may be constructed from a length of tubing which is coiled in a known manner.

Shell 12 has a cylinder 30 with spaced end portions 32. A means 34 is provided for enclosing end portions 32, and for connecting shell 12 to a cooling medium flow system. Means 34 has a flange ring 36 connected to end portions 32 in a suitable, known manner, such as by welding or brazing, and a cover plate 38 attached to flange ring 36 as by, for example, bolt and nut assemblies 39. Each cover plate 38 is provided with a flange 40 arranged about an opening 41 also provided in each cover plate 38.

Figure 1.9: Helical Loop Heat Exchanger

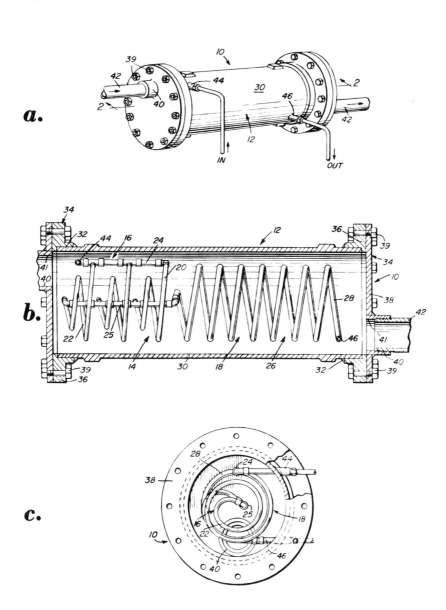

(a)   Perspective view showing a heat exchanger.
(b)   Sectional view taken generally along line 2—2 in Figure 1.9a.
(c)   End view, partly cut away and in section, of the heat exchanger
       according to Figures 1.9a and 1.9b.

Source: U.S. Patent 3,893,504

The orientation of inlet pipe in relation to the coil sections and outlet pipe may be varied by selective angulation of the end plates **38** and the heat exchanges may be vertically oriented if desired. Pipes **42** may be attached to flanges **40** as by screw threads or welding for connecting heat exchanger **10** to a cooling medium flow system. Such a system is conventional.

Coil section **20** has a pipe **44** arranged transversely with respect to cylinder **30** and passing therethrough. Pipe **44** is connected to header **24** for connecting header **24** to a source of a medium to be cooled. A pipe **46** is also arranged transversely with respect to cylinder **30** and passing therethrough, and is connected to the end of coil section **26** opposite the end connected to header **25** for passing cooled medium from the heat exchanger.

Heat exchanger **10** can be mounted directly into pipes **42** of the cooling or low temperature medium. It is then connected to a source of a medium passed from a high temperature system which may be a condensate from a high pressure steam system. Thus, header **24** will receive a condensate at high pressure and high temperature from the steam trap of a high temperature steam system. By means of a widely-spaced first coil section **20** and the larger second coil section **26**, the temperature of the condensate will be reduced without noise or hammer in the coils which is caused by the condensate flashing to steam as it expands and forming spaced areas of liquid and steam which upon further cooling are condensed and form reduced pressure areas that then cause the areas of liquid to move rapidly towards each other and resulting in a hammering noise. By retarding the expansion and maintaining a high velocity through the coil, the liquid condensate will move at a velocity sufficient to maintain the liquid in the form of a wall or continuous column as it is cooled thereby eliminating the hammering noise.

The header-type system used for coil section **20** has its plurality of small diameter coils of small size pipe dimensioned to retard expansion of the high temperature condensate and cool same. The spacing and sizing of loops **22** is such that the cooling medium surrounding coil section **20** will not rise in temperature so rapidly as to cause the cooling medium itself to be flashed to steam. After the temperature is reduced sufficiently in coil section **20**, the condensate is then run through the larger diameter, series-type coil section **26**, which is constructed of larger diameter pipe size to further cool the condensate and completely reduce the condensate temperature to that of the cooling medium. This condensate can then be either discharged into the cooling medium or connected to any external discharge point.

By mounting the coil sections **20, 26,** on pipes **44, 46,** respectively, complete freedom of movement required for rapid expansion and contraction of coil sections **20, 26,** in response to rapid temperature changes is realized.

The various elements of heat exchanger **10** may be constructed in a known manner from suitable, known materials. Material selection is determined by the specific requirements dictated by varying operating conditions involving the parameters of temperature, pressure, corrosiveness, and the like. The particular dimensions of a heat exchanger **10** are determined by the flow rates and temperatures of the cooling medium and the medium being cooled.

Rotating Rotor Design

M. Eskeli; U.S. Patent 3,981,702; September 21, 1976 describes a method and apparatus for transferring heat from a first fluid which is normally gaseous to a second fluid which is normally a liquid, by using a rotating rotor with passages for the first fluid extending from rotor center outward and with passages for the second fluid also extending outward within the rotor with the two fluids being in heat exchange relationship within the rotor with heat transferred from the first fluid to the second fluid, wherein the first fluid temperature is increased by compressing the first fluid within the rotor.

The two fluids are then returned in separate passages to the center of the rotor and discharged. In one form of the process, the rotor is mounted within a sealed casing with entry and exit for first fluid to the casing, and a heat exchanger for adding heat is provided.

Referring to Figure 1.10a, there is shown a cross section of one form of the heat exchanger. In the figure, **10** is casing, **11** is stationary heat exchanger for adding heat to the first fluid, **12** is rotor, **14** is rotor heat exchanger for the second fluid, **15** is first fluid entry to the rotor, **16** is shaft bearing and seal, **17** is rotor shaft, **18** and **19** are second fluid entry and exit to the rotor, **20** is second fluid distribution conduit, **21** is vane, **22** is first fluid nozzles, **23** and **24** are heating fluid entry and exit, **25** is first fluid space, **26** is thermal insulation, **31** is thermal insulation layer, **27** are bearings, **28** is rotor first fluid exit, **29** is vane on rotor exit side.

In Figure 1.10b, an end view with portions removed, is shown of the unit illustrated in Figure 1.10a. In the figure, **10** is casing, **11** is heat exchanger, **12** is rotor, **29** are vanes on exit side of rotor, **17** is rotor shaft, **23** is heating fluid conduit, **22** are rotor first fluid internal nozzles, **14** is heat exchanger, **20** is second fluid conduit, and **30** indicates direction of rotation for rotor.

In Figure 1.10c, another form of the heat exchanger is shown in cross section. In the figure, **40** is support base, **41** is shaft bearing, **42** is rotor shaft, **43** is friction reducer bearing, **57** is friction reducer opening to allow discharge of first fluid, **44** is first fluid discharge nozzle from rotor, **45** is friction reducer disc, **46** is rotor, **47** is rotor vane on rotor discharge side, **48** indicates space between rotor and friction reducer disc, **49** is first fluid passage at rotor periphery, **50** is rotor heat exchanger, **51** is second fluid conduit, **52** is first fluid inlet to rotor, **53** is second fluid inlet to rotor shaft passage, **54** is second fluid discharge, **55** is shaft bearing, **56** is vane, **59** is thermal insulation layer.

In Figure 1.10d, an end view of the unit shown in Figure 1.10c is shown, with portions removed to illustrate interior details. In the figure, **45** is friction reducer disc, **44** are first fluid exit nozzles, **57** are openings in friction reducer disc to allow first fluid discharge from rotor, **42** is rotor shaft, **40** is base, **49** is fluid passage with vanes **47**, **50** is rotor heat exchanger for transferring heat from first fluid to second fluid, **51** are second fluid conduits, and **58** indicates direction of rotation for rotor.

In Figure 1.10e, a pressure-enthalpy diagram for a typical first fluid is shown. In the figure, **60** is pressure line and **61** is enthalpy line, **62** are constant pressure lines, **63** are constant entropy lines and **64** are constant temperature lines.

For the unit of Figure 1.10a, the cycle within rotor with external heat addition is **65** to **66** compression with cooling, **66** to **67** expansion which is isentropic, in nozzles **22** at rotor periphery, and **67** to **69** expansion in rotor vanes, with heat addition from **69** to **65**. For the unit shown in Figure 1.10c, the compression is from **65** to **66**, expansion in rotor exit side vanes from **66** to **68**, and expansion in rotor exit nozzles from **68** to **69**. If the rotor of Figure 1.10c were within a closed casing, then heat addition would be from **69** to **65**, the same as for the unit of Figure 1.10a.

In Figure 1.10f, a typical schematic diagram is shown to indicate typical application of the heat exchanger. In the figure, **70** is heat exchanger connected to drive means **71** via a power transmission shaft, **72** is a heat exchanger to remove heat from the second fluid which is circulated via conduits from the heat exchanger, **73** is circulation pump for the second fluid, **74** is a supply of cool fluid to heat exchanger **72** and **75** is discharge of hot fluid from heat exchanger **72**, **76** and **77** are heating fluid entry and exit to the heat exchanger **70** stationary heat exchanger.

Figure 1.10: Rotating Rotor Heat Exchanger

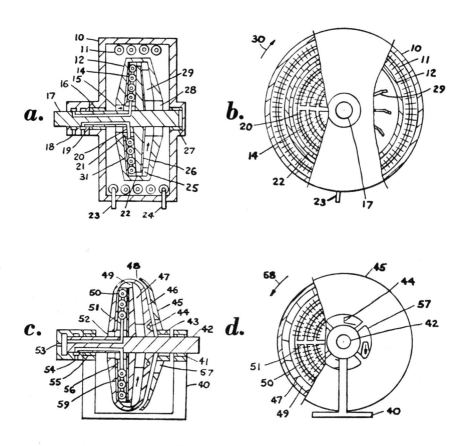

(continued)

Figure 1.10: (continued)

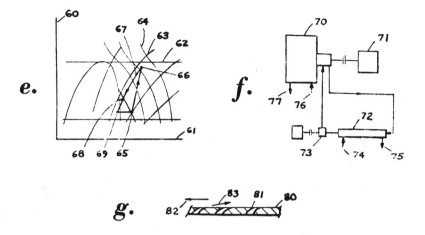

(a)   Cross section of one form of the heat exchanger.
(b)   End view of the unit shown in Figure 1.10a.
(c)   Cross section of another form of the heat exchanger.
(d)   End view of the unit shown in Figure 1.10c.
(e)   Typical pressure-enthalpy diagram for the first fluid
          used for the heat exchanger.
(f)   Typical schematic diagram for the application of the
          heat exchanger.
(g)   Detail of nozzles.

Source: U.S. Patent 3,981,702

In operation, power is supplied to the heat exchanger rotor shaft from an external power source, causing it to rotate. First fluid enters the rotor via entry at rotor center and is compressed by centrifugal action on the first fluid by rotor with accompanying temperature increase. The second fluid is circulated within the rotor in separate fluid passages in heat exchange relationship with the first fluid and heat is transferred from the hot first fluid to colder second fluid. The first fluid is then passed to inward extending exit passageways near the periphery of the rotor, and discharged from the rotor near the rotor center. The second fluid is also passed along its own passageways to exit passages located within the rotor shaft and from there to exit.

The compression of the first fluid is with cooling for best efficiency, with heat being transferred during compression from first fluid to second fluid, and heat may also be transferred after compression from the first fluid to the second fluid. The expansion in the exit side vanes is normally isentropic except for heat gains from rotor walls. The first fluid passageways for the rotor area for compression, is usually thermally insulated to prevent heat loss to rotor walls and to surrounding space, and also to prevent overheating the rotor structure which could lead to rotor failure.

There are two methods shown to reduce work input to rotor, while still assuring the propelling of the first fluid through the rotor. In Figure 1.10a, there are shown nozzles at rotor periphery, item **22**, and these nozzles are arranged to pass the first fluid backward in direction away from direction of rotation so that the first fluid absolute velocity will be less than the tangential velocity of the rotor in the area of nozzles; this velocity reduction will provide for a suitable pressure differential to assure that the first fluid will be transported through the rotor. The fluid will be discharged from nozzles **22** to space **25** for velocity adjustment before entry of the first fluid to exit side inward extending passageways formed by vanes **29**. It should be noted that the actual exit velocity for the first fluid relative to the nozzles **22**, is small, often in the area of 50 to 100 feet per second for most gases, and the nozzles are usually of the converging type. However, the nozzles **22** are sized and shaped to provide for maximum exit velocity for the pressure differential existing between entry and exit of the nozzles.

The second method for assuring that the first fluid will be transported through the rotor is shown in Figure 1.10c, where exit nozzles are provided for the first fluid leaving the rotor. Usually the exit nozzles **44** are located at a distance from rotor center that is greater than the radius of entry opening to the rotor for the fluid; this provides for needed pressure differential to assure that the first fluid passes through the rotor, while simultaneously producing thrust on the rotor wheel to reduce the work input to the rotor wheel from external sources. The nozzles **44** are oriented to discharge the first fluid tangentially backward away from the direction of rotation, thus producing thrust. The velocities for the first fluid leaving the exit nozzles **44** are usually higher than those for nozzles **22**, and nozzles **44** are sized and shaped to obtain highest attainable exit velocity for the first fluid leaving the rotor nozzles, for the pressure difference available.

It should be noted that the cooling of the first fluid during compression, as shown in Figure 1.10e, increases the first fluid pressure gain for the entry side of the rotor, and thus reduces the requirement for the need for other means to assure passage of the first fluid through the rotor. By cooling during compression the first fluid, and by using the other described means to reduce work input to the rotor, very low work input values to the rotor can be realized, thus providing for a very economical heat and cooling generating source.

The friction reducing disc shown in Figure 1.10c, item **45**, is intended to reduce friction losses on rotor external surfaces especially in the areas where rotor surface speeds are high. This is done by allowing the disc **45** to rotate at its own speed; this speed can be regulated approximately to a correct value by setting the space between the rotor wall and the disc so that by gas friction the disc will rotate at approximately one-half the speed of the rotor. This speed for the disc will reduce, theoretically, the rotor friction loss by approximately 40%, from the value without the disc. This is due to the lower velocity differential between rotor surface and surrounding fluid, and similarly between the disc and surrounding fluid.

In Figure 1.10c, one such friction reducer is shown, mounted on one side of the rotor; the discs may be mounted on both sides of the rotor, and similarly also for the rotor of Figure 1.10a. By this means, the friction losses can be reduced for the rotors by nearly half, thus improving the overall performance of the heat exchanger. Alternatively, the friction reducer discs may be omitted, and it should be noted that for lower rotor speeds, the discs are not of a great value.

In Figure 1.10g, a detail of nozzles **22** is shown. **80** is rotor center wall, **81** is a nozzle, **82** indicates direction of rotation of rotor and **83** indicates fluid leaving the nozzles **81**. It should be noted that the fins of heat exchanger coils **14** could be slanted backward to also add to the pressure differential to transport the first fluid through rotor; this is also illustrated in Figure 1.10b, with the slanting of the fins which act as vanes, the nozzles **22** could be deleted, or the two methods used in combination as shown in Figure 1.10b.

Noise Dampening in Liquid Heaters

A process described by *W. Brandl, Z. Koula and P. Nirk; U.S. Patent 3,877,511; April 15, 1975; assigned to Stotz & Co., Switzerland* relates to an improved method of, and apparatus for, beneficially suppressing the formation of noise at heating equipment operating according to vacuum vaporization techniques, through the use of relatively simple and inexpensive means.

As far as the apparatus aspects of this development are concerned, the special apparatus is manifested by the features that obstruction or dam-up means which oppose the formation of vapor bubbles are arranged in the vacuum vessel or container.

It has been found to be advantageous to employ as the obstruction or dam-up means, for instance granular material. This can be of coarse grain size. Better results can be realized, however, if there is employed as the obstruction or dam-up means a granular material having a grain size of less than 5 mm, advantageously in the range of 1.5 to 5 mm and preferably in the range of 1.5 to 2 mm.

Granular materials suitable for the purposes of the process are for instance the following: sand, glass grains, metal spheres or balls and chips, ceramic balls or spheres, furthermore, balls or spheres formed of plastic, which are specifically heavier than the liquid and resistant thereto throughout the prevailing operating temperatures, for instance those formed of polytetrafluoroethylene. Particularly suitable is quartz sand of the specified grain size.

By virtue of the obstruction or dam-up means proposed by this development, there is realized a surprising sound dampening effect which can be readily demonstrated by the following tabulated sound measurements which have been carried out for different grain sizes under otherwise similar conditions.

Dam-Up or Obstruction Means	Noise Level (dB)
None	70–80
Sand, grain size (mm)	
>10	~50
5–10	~46
1.5–2	<42

As a general rule, the sound dampening effect increases with decreasing grain size. However, there is a lower limit of the grain size which is governed by the requirement that with too small grain size the condensed heating liquid flows poorly to the heating surfaces.

Advantageously, a packing in the vacuum vessel formed of sand of the specified grain sizes is covered at the top with sand of a coarser granulation or grain size or a sieve. In this way relative movement of the grains with respect to one another is dampened, and thus, wear of both the equipment walls as well as the grains is reduced.

The fill height of the granular obstruction or damming-up or dam-up means advantageously extends up to at least approximately the height of the liquid level at the vacuum vessel, preferably above the liquid level.

According to another preferred example of the process, there is employed as the obstruction or damming-up means, a compact, yet porous body which fills the vacuum vessel at least approximately to the height of the liquid level, preferably beyond the liquid level.

Suitable as the compact porous body there can be in particular advantageously employed: ceramic sintered bodies; metallic sintered bodies; molded bodies formed of porous natural stone, such as tuff; sintered bodies formed of plastic which throughout the prevailing operating temperatures are resistant to the liquid, for instance polytetrafluoroethylene; plastic foams resistant to the liquid over the prevailing operating temperatures; molded bodies formed of glass fibers adhesively bonded together by a suitable binder or binding agent resistant to the liquid at the prevailing operating temperatures.

During fabrication of the equipment of this development the compact, porous body can be either introduced into the vacuum compartment as a prefabricated molded body or, otherwise, it can be formed in situ in the vacuum compartment by sintering, bonding or foaming the appropriate raw materials. The last mentioned fabrication technique has the advantage that there is automatically formed a body of the desired dimensions, so that no further processing or other work is necessary.

In comparison to other small component-obstruction or damming-up means, such type compact bodies possess the beneficial result that they remain in place in the vacuum vessel even during transport. Consequently, the equipment of this development can be finish-produced at the factory and subsequently transported to the desired site without undertaking any special measures.

Electrically Heated Unit for Transfer Fluids

A process described by *L.J. Palm and R.B. Palm; U.S. Patent 3,885,125; May 20, 1975; assigned to Fulton Boiler Works, Inc.* involves a method of electrically heating a heat transfer fluid prior to circulation through a heat exchange system wherein the heat transfer liquid is circulated helically in a continuously swirling bodily flow through an elongated substantially unobstructed annular chamber. The helical flow is caused by the angle and position of entry of the fluid into the vessel. The swirling fluid is exposed to axially elongated heating elements that are disposed in the annular chamber and that are sufficiently small relative to the width of the annular chamber to transfer heat with the fluid while accommodating the continuously swirling bodily flow without turbulence.

In the system shown generally in Figure 1.11a, thermal fluids, particularly liquids other than water, such as, mineral oils; diphenyl-diphenyl oxide mixtures; chlorinated biphenyls; silicones and silanes; polyglycols; and polyphenyl ethers and esters are pumped by a circulating pump **10** into a fluid vessel **11**.

This vessel is made of heat conductive material and consists of an inner annular shell 12 concentric with and surrounded by an outer annular shell 13. The outer shell 13 is of larger diameter than the inner shell 12, therefore, a space or path is defined through which the fluid is to flow. The cold fluid enters the fluid vessel through an inlet 14 placed at the bottom of the vessel 11. The pressure created by the circulating pump 10 forces the fluid to flow upward through the annular vessel until it reaches a fluid outlet 15 at the top of the vessel. The thermal fluid is heated to the desired temperature as it passes through the annular vessel.

The annular fluid vessel 11 is contained within the heating unit 16. The unit includes an outer steel jacket 17 and an inner steel jacket 18 with an insulation layer 19 between them. There is a space left between the inner jacket 18 and the outer annular shell 13 of the heat transfer vessel 11. The fluid vessel is attached to the base of the heating unit by angle supports 20 (only one shown). Heat is transferred to the moving fluid by hot gases ignited at the top of the heating unit 16.

Air is taken in through the inlets 21 in a blower assembly 22 and mixed with a gaseous ignition fuel in the burner assembly 23. The mixture is deflected downward through the air blast tube 24 and the funnel shaped air deflector 25. After passing the deflector 25, the gas-air mixture is ignited and combusts in the inner annular shell 12 of the heating vessel 11.

The burner assembly 23 shown in Figure 1.11a is located at the top of the heating unit 16; alternative construction would be to locate the burner at the bottom or in the middle of the unit.

As the hot blend goes downward through the interior of the inner annular shell 12, it gives up some of its heat into the shell. The hot gases are forced downward under the vessel 11.

As shown in Figure 1.11a, and in greater detail in Figure 1.11d, the hot gas after passing under the fluid vessel travels upward through a secondary flue pass. The secondary flue pass consists of the annular opening between the external shell 13 of the fluid vessel 11 and the inner jacket 18 of the insulation layer 19. Equally-spaced vertical ribs or fins 26 are joined to the circumference of the outer shell 13 of the fluid vessel 11. These ribs or fins 26 are effective in absorbing the heat from the gases rising through the secondary pass. The hot gases pass up through the secondary flue and give their remaining heat into the conductive outer annular shell 13 of the fluid vessel 11. Therefore, both sides of the fluid vessel are heated. The upward rising hot gases or products of combustion leave the system through a flue outlet 27. When the heated fluid reaches the top of the fluid vessel, it is forced through an outlet 15 into the external heating system.

The heated thermal fluid is circulated to the external system and then is recirculated from the external system through the pump 10 and inlet 14 after it performs its heating function. A pressure indicator and pressure fluctuation reliever 28 is provided on the return path of the fluid to the heating unit.

In accordance with the process, as shown in Figures 1.11b and 1.11c, improved efficiency and evenness of heat exchange are produced by the flow relationships occurring within the heater. The thermal fluid to be heated is pumped into the annular fluid heating vessel **11** through an inlet **14** which is tangential to the fluid flow path defined by the annular vessel **11** and at a 90° angle to the vertical axis of the annular vessel.

This tangential entry path causes the thermal fluid to come into and flow through the fluid vessel with a spinning or swirling motion. The entire volume of fluid rotates and mixes around the vessel. The fluid is therefore induced to spin around and between the annular shells **12, 13** of the fluid vessel **11** in a helical path.

To heat the fluid, the burner assembly gives a circular or whirling movement to the gaseous heat exchange medium as it passes downwardly of the interior of the inner annular shell **12** of the heating vessel **11**. The circular movement of the gas plus the natural tendency for heat to rise, slows the downward movement of the flame; thereby, efficiently heating the inner annular shell **12**.

When the hot gas reaches the bottom of the interior of the annular heating vessel, it turns upward to make a complete second pass around the exterior of the outer shell of the heating vessel, thereby, transmitting additional heat to the outer annular shell and consequently to the fluid.

The flow relationship shown in Figures 1.11b and 1.11c produces maximum heat transfer because of the smoothness of flow of the thermal fluid through the annular path and also because of the length of flow through the vessel caused by the rotational movement. This ideal fluid flow is exposed to double pass heat which takes maximum advantage of the heating ability of the gaseous medium.

The design of the heating vessel is ideal for heating thermal fluids due to the even distribution of two-pass heat and the minimal restriction of the moving fluid. The minimal restriction of the fluid in the heating vessel results in a low pressure drop.

The annular vessel can be constructed with the following dimensions depending on the heating system required:

(1) length of the vessel, 24 to 96 inches;

(2) outer diameter of the vessel, 12 to 48 inches;

(3) distance between the inner and outer walls of the vessel, 1 to 10 inches; and

(4) inlet diameter, 1.5 to 3 inches.

The flow rate of the fluid is controllable through the vessel, and as such is dependent upon the distance between the inner and outer walls of the vessel.

In addition, the flow rate must be kept above a minimum level in order to keep the thermal fluid from burning or scorching. This heater will operate at a minimum flow rate of one foot per second and can be adjusted to a maximum of 10 or 15 feet per second. This feature allows the thermal fluid heater to be used for a wide variety of applications.

Figure 1.11: Electrical Heating Unit for Heat Transfer Fluids

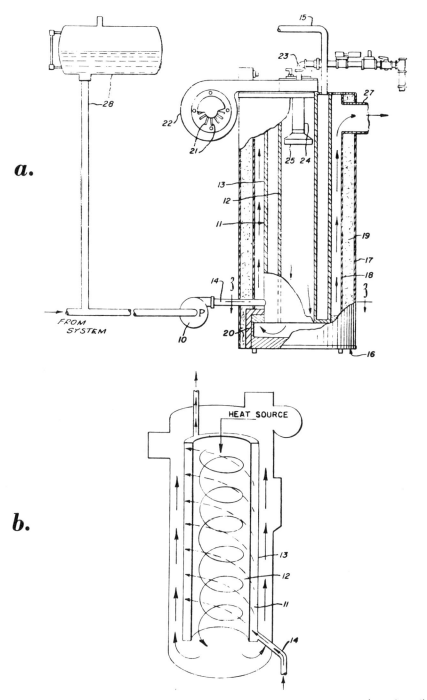

(continued)

Figure 1.11: (continued)

c.

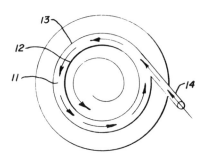

d.

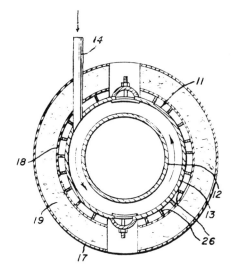

(a) System diagram showing the heater unit operatively connected
      to the various external components that complete a practi-
      cal operating example.
(b) Fragmentary perspective view showing the flow relationships
      occurring within the heater.
(c) Transverse sectional of the perspective of Figure 1.11b.
(d) Transverse sectional view through the heater taken along line
      3–3 in Figure 1.11a.

Source: U.S. Patent 3,885,125

PARTICULATE EXCHANGE MEDIUM

Fluidized Granulate Columns

D.G. Klaren; U.S. Patents 4,119,139; October 10, 1978 and 3,991,816; Nov. 16, 1976; both assigned to Gustav Adolf Pieper, Netherlands describes a method for operating a heat-exchanger of the type in which one of the heat-exchanging components flows as a liquid in a vertical upward direction through a system of granulate-containing tubes, thereby keeping this granulate fluidized.

In this process, it is of importance that the granulate has a composition and/or a shape and/or a crystallographic structure, which has a strong influence on increased scale formation. It is thus avoided that the walls of the tubes have to be cleaned periodically. Often such a granulate may be obtained by selecting its chemical composition in accordance with the substances with precipitate from the liquid. While flushing the granulate in the chamber the finest fractions will be drifted up most, whereas the coarsest fractions remain deposited or suspended near the bottom of the chamber. By opening the blow-off conduit near the bottom of the chamber it has been found to be possible to carry away the coarsest fractions without having a considerable quantity of finer material carried along as well. In this way it is possible now to very quickly reduce the grain size of the granulate inside the tubes without it being necessary to dismount and to empty the heat-exchanger.

It is noted that indeed the most obvious alternative consists in that the heat-exchanger is dismounted, the tubes are put upside down, which causes all granulate presently inside the tubes to fall out, whereupon the heat-exchanger, after having been put in the correct position again, is supplied with a new filling. Especially in connection with bigger heat-exchangers, which are part of, for instance, an expansion evaporator, this method would be prohibitive for the use of a fluidized granulate. It further has been found that supplementing with finer fractions and backfeeding of the granulate into the tubes may be performed very simply during the reduction of the speed of flow through the tubes. A certain mingling of the various grain fractions occurs.

In Figure 1.12, the heat-exchanger according to the process is demonstrated schematically, in which a granulate mass is used having only single composition and the granules are subject to grow as a consequence of a possible evaporation of the liquid.

The heat-exchanger consists of four compartments **1** in which heat is transferred upon a liquid which flows through tubes **2** in upward direction. Tubes **2** are fixed between two tube plates **3a** and **3b**.

The tubes **2** open into a chamber of outlet box **5**. At the bottom-side of the tubes **2** an inlet box **6** is provided from which liquid is carried through narrow openings into the tubes **2**. Inside the tubes **2** a granulate **4** is provided, which during operation is fluidized by the upward flowing liquid, and thereby expands into the outlet box **5**. The narrow inlet openings **7** prevent the escape of granulate into the inlet box **6**.

During normal operation the liquid to be heated is supplied to the heat-exchanger through conduit **8** and through valve **10** and pump **9**. The discharge of the heated liquid takes place through conduit **14** through the opened valve **15**. The circuit further includes a circulating conduit **13** with in it a pump **21** and valves **12** and **16**. During normal operation valves **12** and **16** are shut and pump **21** is not functioning. Also near the bottom of the outlet box **5** a blow-off conduit **17** connects with in it a valve **19**, whereas in the top of the outlet box **5** a supply-conduit **18**, also with a valve **20**, opens. During normal operation, valves **19** and **20** are shut too.

Figure 1.12: Fluidized Granulate Heat-Exchanger

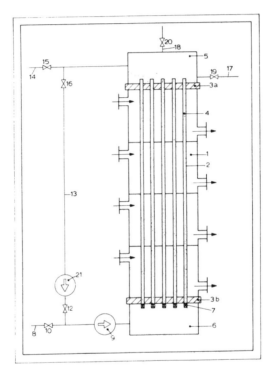

Source: U.S. Patent 4,119,139

If after some time the average grain size has increased to such an extent that the operation of the heat-exchanger deteriorates, which may be deduced from a decreasing temperature of the liquid discharged through conduit **14**, measures are to be taken in order to convey the most severely grown grains from the heat-exchanger and to supply the filling with smaller grains. Thereto first of all valves **10** and **15** are shut and valves **12** and **16** are opened. The flow of the heat transferring medium, which flow transfers to the tubes through the compartments **1**, may be shut off, however, whether or not this is necessary will have to be determined from case to case considering the circumstances. After pump **21** has been put into operation, whereas pump **9** remains operating, the liquid which is present inside the heat-exchanger is circulated through the circulating circuit with such an increased speed, that all granulate present in the tubes **2** is forced into the outlet box **5**. It thereby is advisable to increase the speed of flow sufficiently in order that the medium sized and the smaller grains in the top of outlet box **5** start suspending, whereas the bigger grains remain near the bottom of outlet box **5**. If thereupon valve **19** is opened the part of the flow of circulating liquid having a relatively reduced speed, will discharge mainly only the coarse grains out of the system through blow-off conduit. Then valve **19** is closed again.

Simultaneously with the blowing off, or shifted in phase therewith, valve **20** may be opened, allowing finer grains of the granulate to be fed into the system. If this happens, the speed flow of the circulating liquid may, either by reducing the speed of pump **21**, or by gradually closing one or more of the valves **12** or **16**, be reduced gradually, which causes the granulate to fall back into tubes **2** again. The heat-exchanger thereupon is suitable for normal operation again, whereupon pump **21** is shut off, valves **12** and **16** are closed, and valves **10** and **15** are opened.

Various variations of the described system are possible. For instance, instead of only one blow-off conduit, a number of blow-off conduits **17** may be connected around the outlet box **5**. Also, for instance, pump **21** is not necessary, if pump **9** has the required capacity to effect the circulation with the necessary speeds.

Inert Solids of Specified Particle Size

A process described by *L. Plass; U.S. Patent 3,886,997; June 3, 1975; assigned to Metallgesellschaft AG, Germany* involves a method of operating closed heat exchangers supplied with liquid heat exchange fluids and provides that inert solids having an average particle diameter in the range of 10 to 150 microns are incorporated in the heat exchange fluid in an amount of 0.5 to 20% by volume, and the average velocity of the suspension is adjusted to correspond to a Reynolds number in the range of 10,000 to 70,000 based on a pipe diameter of 1 cm.

It is known that the rate at which heat is transferred from a gas to a wall and vice versa can be increased by an addition of particulate solids to the gas (*Chem. Ing. Techn.* 39, page 282, 1967) because the addition of solids having a high specific heat to gas having a low specific heat results in a gas-solid mixture having a higher specific heat, which is believed to be responsible for the increase of the heat-transfer rate. Where liquid heat-transfer fluids are used which always have a higher specific heat than the added solids, a liquid-solid mixture will result which has a lower specific heat than that of the original liquid so that a decrease of the heat-transfer rate would be expected. Experiments have shown, however, that considerable improvement can be achieved, contrary to the expectations, if the requirements of this process are fulfilled.

Reynolds number is defined as $Re = \bar{u} D/v$ where $\bar{u}$ is the average velocity in the main direction of flow in meters per second, D is a length, in meters, which defines the cross sections of the passages of the heat exchanger which are traversed, e.g., the pipe diameter, and v is the kinematic viscosity in $m^2\ sec^{-1}$. For this reason the dimensions of the passages of the heat exchanger must be taken into account in selecting the Reynolds number range which meets the requirements of the process. As applied to tubular passages having a diameter which differs from the unit value of 1 cm, the Reynolds numbers must be linearly changed accordingly. Where tubes are used which are, for example, 2 or 3 cm in diameter, the Reynolds number must lie between 20,000 and 140,000 or between 30,000 and 210,000, respectively.

The increase of the heat-transfer rate will be particularly large if, according to a preferred feature of the process, the inert solids incorporated in the heat-exchange fluid have an average particle diameter of 10 to 50 microns and are used in an amount of 1 to 6% by volume, particularly 2.5 to 6% by volume.

The solids to be added to the liquid heat-exchange fluid should suitably be of medium hardness. Excessively high hardness could result in damage to the wall surface of the heat exchanger. An insufficiently high hardness could result in a progressively increasing reduction of the particle size of the added solids. Particularly suitable solids are those having a density of more than 1 g/cm^3, such as sand, ore dust, and ground slag.

The method according to the process can be used in virtually all closed heat exchangers. These include, double-tube heat exchangers, annular-gap heat exchangers, plate heat exchangers and spiral heat exchangers. Particularly suitable are the heat exchangers which have heating or cooling passages that are free from baffles. The passages may have any desired cross section.

Rotating Heat Exchanger

A process described by *V. Duhem; U.S. Patent 4,146,975; April 3, 1979; assigned to Fives-Cail Babcock, France* relates to a rotating heat exchanger comprising an axle of rotation and first and second groups of tubes arranged about the axle of rotation, the tubes of the first and second groups having axes parallel to the axis of the axle of rotation.

Referring to Figure 1.13, there is shown a heat exchanger comprising central tube 18, a first group of tubes 10 arranged about the central tube and a second group of tubes 12 arranged about the first group of tubes. The axes of the tubes are parallel to each other, the two groups of tubes being arranged concentrically about the central tube and the axes of the first and second groups of tubes defining respective cylinders which are coaxial with the axis of the central tube.

The central tube is an axle of rotation arranged to carry and drive the groups of tubes 10 and 12, a support plate 14 and 16 being affixed to axle of rotation 18 at each end of tubes 10 of the first group and the ends of the tubes of the first group being affixed to support plates 14 and 16. The axle of rotation has stub shafts 11 and 13 extending from respective ends of tube 18 and these stub shafts are journaled in bearing boxes 21 and 23 to enable the heat exchanger to rotate. The axle of rotation is driven by motor 25 which is coupled to stub shaft 11 by gear box 27.

One of the ends of tubes 12 of the second group is also affixed to support plate 14 while the other ends are supported by intermediate support plate 19 affixed to tubes 10 of the first group. Support plate 19 has openings wherein tubes 12 are glidably mounted, the components of the assembly 18, 10, 12, 14, 16 and 19 forming a rotary unit whose axle of rotation is horizontal or slightly inclined to facilitate movement of solid material from one end towards the other.

As may best be seen in Figure 1.13c in connection with support plate 19, at least one of the support plates is constituted by sheet metal disc 15 having a radius smaller than that of a cylindrical envelope generated by the rotation of tubes 12 of the second group and sheet metal elements 17 affixed to the periphery of the discs. In the illustrated case, disc **15** is hexagonal and the radius of the circle circumscribed about the hexagon is approximately equal to the distance between the axes of tubes **12** and of axle of rotation **18**. If desired, the disc may actually be circular. Sheet metal elements **17** are affixed to the sides of the hexagon by assembly on site. The structure of support plate **14** is identical to that of plate **19**. The construction will reduce the weight and make the apparatus less cumbersome in transport.

Figure 1.13: Rotating Heat Exchanger

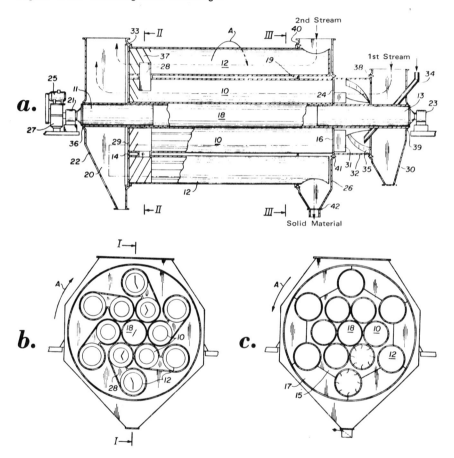

(a)   Longitudinal section of the heat exchanger, along line I-I of
          Figure 1.13b
(b)   Transverse section along line II-II of Figure 1.13a
(c)   Another transverse section along line III-III of Figure 1.13a

Source: U.S. Patent 4,146,975

As can be seen in Figure 1.13a, the ports of tubes **10** and **12** of the first and second groups communicate through ports **29** in support plate **14** with discharge chamber 20 defined between support plate 14 and hood 22 adjacent thereto. Fluid-tight friction joints 33 and 36 are mounted on stationary hood 22 and bear, respectively, on the periphery of support plate 14 and stub shaft 11.

At the same end of the heat exchanger, channel or conduit means 28 connect each tube 10 to a respective tube 12 for passing solid granular material from the tubes of the first group into the tubes of the second group. These channels or conduits are spaced from the end of the tubes to prevent the granular material from passing out of the tubes and into gas discharge chamber 20 as the material is transferred between the tubes. Helical channels 37 in the interior of the tubes between support plate 14 and conduits 28 guide the granular material from this zone towards the conduits. If desired, a single tube 10 could be connected to several tubes 12 or several tubes 10 could be connected to a single tube 12.

At their other ends, the ports of tubes 10 communicate with the interior of cylindrical sleeve 32 through ports 24 in support plate 16, the sleeve being arranged coaxial with axle of rotation 18 and for rotation with tubes 10 of the one group by affixing sleeve 32 to support plate 16. Means is mounted at this end of the heat exchanger for supplying a granular solid material into tubes 10, the illustrated supply means comprising delivery conduit 34 leading into sleeve 32, helical channels 35 in the interior wall of the sleeve for guiding the supplied material from conduit 34 towards the tubes and baffles 31 fixed to support plate 16 for guiding the material from helical channels 35 through ports 24 into the interior of tubes 10 (see chain-dotted lines in Figure 1.13a, indicating the path of the solid material).

The outer end of sleeve 32 is closed by hood 30 defining with support plate 16 a gas inlet chamber. Fluid-tight friction joints 38 and 39 are mounted on stationary hood 30 and bear, respectively, on the periphery of sleeve 32 and stub shaft 13.

The ends of tubes 12 of the second group opposite to the tube ends affixed to support plate 14 pass freely through openings in intermediate support plate 19 and open into a material discharge chamber defined by support plate 19 and adjacent hood 26. Fluid-tight friction joints 40 and 41 are mounted on stationary hood 26 and bear, respectively, on peripheries of support plates 19 and 16.

Helical entrainment elements may be mounted in the interior of tubes 10 and 12 for advancing the solid granular material from its feed end towards conduits 28 and from the conduits to the material discharge chamber when the treated material may be removed through screen 42 (see chain-dotted lines in Figure 1.13a).

To reduce heat exchange through the walls of the tubes between tubes 10 of the first group and tubes 12 of the second group, the two groups are spaced from each other in operations wherein the temperature of the gas circulating through one group of tubes differs substantially from that of the gas circulating through the other group of tubes. For instance, a hot gas may be circulated through tubes 10 to dry the granular material supplied to the heat exchanger while a cold gas is circulated through tubes 12 to cool the dried material. The radial gap between the two groups of tubes prevents the hot gas from being unduly cooled by the cold gas and the cold gas from being unduly warmed by the hot gas.

The operation of the apparatus will be obvious from the above description of its structure and will be further elucidated below.

Solid granular material to be treated is introduced through supply conduit **34** and passed by helical channels **35** in the interior wall of sleeve **32** and baffles **31** into revolving tubes **10** wherein they are displaced to the other end by helical channels in the interior walls of the tubes until they reach conduits **28** which transfer the material into revolving tubes **12** wherein they are displaced back by helical channels in the interior walls of the tubes until they reach the material discharge chamber defined by hood **26**. The treated material is discharged by gravity through screen **42**.

In the first group of tubes **10**, the material is treated during its passage through the tubes by a first stream of gas which is introduced into the gas inlet chamber defined by hood **30** and circulates through tubes **10** concurrently with the material, the spent gas being collected in gas discharge chamber **20** at the other end when it is evacuated by a flue (see broken line indicating first stream in Figure 1.13a). In the second group of tubes **12**, the material coming through conduits **28** from tubes **10** is subjected to treatment by a second and distinct stream of gas introduced through hood **26** and flowing through tubes **12** countercurrently to the flow of the material, the gas from the second stream also being collected in and evacuated from, discharge chamber **20** (see broken line indicating second stream in Figure 1.13a).

If desired, gas discharge chamber **20** may be divided into two compartments, the gases from the first stream coming from tubes **10** being collected in, and evacuated from, one of the compartments while the gases from the second stream flow from tubes **12** into the other compartment of gas discharge chamber **20** and are separately evacuated. In this case, it is also possible to reverse the direction of the gas flow in tubes **10** and/or in tubes **12**.

In all cases, steps must be taken to assure that the gas pressures in tubes **10** and **12** at connecting conduits **28** are substantially the same to avoid passage of gas between the tubes, i.e., to make certain that the two streams of gas are kept distinct and separate.

While the rotary heat exchanger has been illustrated as being supported at both ends by bearings, it could be supported by a bearing at one end while the other end rests on rollers, an annular race being provided on sleeve **32** or on axle of rotation **18** for engagement by the rollers. Obviously, both ends of the heat exchanger could be so supported.

The heat exchanger may be used in all instances where a product is to be subjected to two successive treatments, such as drying followed by cooling, i.e., consecutive treatments by hot and cold gases. Thus, it will find use in the manufacture of fertilizer, foundry sand, and agricultural and food products, such as sugar, for example.

Panel Bed

A process described by *A.M. Squires; U.S. Patent 3,981,355; September 21, 1976* relates to an improved method of countercurrently contacting gas and granular material with each other to effect heat exchange between them. Granular material at a given temperature is arranged in a bed having a plurality of transversely disposed, upwardly spaced, gas entry portions separated by interposed supporting members having outer and inner edges. The gas entry portions have gas entry faces having outer edges that are substantially contiguous with the outer edges of the supporting members.

A typical supporting member is gently curved and inclines downwardly and inwardly from its outer edge and then upwardly and inwardly toward its inner edge. The inner edge of a typical supporting member is either above the inner edge of the superjacent free surface of granular material, borne by the aforementioned typical member, or when below, a line drawn through these edges is inclined at an angle of less than about 45° to the horizontal.

The bed has gas exit portions spaced horizontally apart from the inner edges of the supporting members. Gas at a different temperature is caused to flow forwardly in a substantially continuing flow during the contacting through the gas entry portions of the granular material bed and outwardly from the gas exit portions to effect an exchange of heat between the gas and granular material. Thereafter, a transient flow of gas is caused to move in the direction in reverse of the aforementioned flow of gas.

The transient reverse flow produces first a rise (at a given rate of rise) and subsequently a fall in the pressure difference between the gas exit portions and the gas entry portions. This difference should remain greater than a first critical minimum difference for a time interval less than about 150 milliseconds, this first critical minimum difference being that difference at which a steady flow of gas in the reverse direction just produces a localized spill of granular material from the gas entry faces.

The pressure difference produced by the transient reverse flow should peak to a top value beyond a second critical minimum difference, which is the pressure difference at which a transient flow of gas in the reverse direction, producing the second critical minimum difference at the aforementioned given rate of rise, just initiates a body movement of the granular material toward the gas entry faces to remove a portion of the granular material from the bed. The second critical minimum difference depends upon the rate of rise in the pressure difference, being larger the more rapid the rise. The aforementioned time interval is advantageously held within about 50 milliseconds if the process is also used to filter dust from a gas.

The method is useful for heating air to a high temperature by heat exchange of the air with a hot granular material. This material after the heat exchange would advantageously be reheated by heat exchange according to the method with hot combustion products of combustion of a fuel with air.

The heated air could advantageously be supplied to a combustion in a furnace for operation at extraordinarily high temperature, such as a furnace to supply hot gas to a magnetohydrodynamic device for generating electricity, or a furnace melting or sintering or heat treating a refractory metallic or nonmetallic material, or a blast furnace reducing a metal ore with coke. In this application of the method, air could be supplied at high temperature at far lower pressure loss and in equipment of far smaller size than can be supplied by the conventional air-heating stoves of the iron blast furnace, for example.

The method is useful for heating air to be used in combustion of a low-grade fuel, such as municipal solid waste or the solid wastes of many important industries, such as the wood pulp and papermaking industry, or waste liquors containing fuel values, such as the waste liquors from this same industry as well as many other waste liquors, such as liquors from the treatment of petroleum fractions with sulfuric acid, or waste gas streams containing fuel species. Such fuels often have higher heating values below 5,000 Btu/lb, and are difficult to burn unless the air is provided at a high temperature.

Granular solid employed for heating the air may advantageously be heated by the method by heat exchange with off-gases from the combustion of the waste fuel material. This combustion is often advantageously conducted in a fluidized bed, with the advantage that the method may be used to remove particulate matter from the effluent from the fluidized bed. Often it is advantageous to use the same granular solid material in the combustion fluidized bed as that used in the panel bed gas-solid heat exchange device of the process.

The process may be used for heating a solid by heat exchange with a gas and for thereafter storing the solid to store the heat for later use. It sometimes occurs that a hot gas stream is being supplied at a time when it is not convenient to use the heat in the gas. In such cases, the heat may be recovered by heat exchange to a cold granular solid, the hot solid may be stored, and the heat may be recovered from the solid for use at a later more convenient time.

An important example of this practice arises in the case of the proposal to store compressed air in the ground against a requirement for peakload electricity generation. The air would be compressed and stored in a deep cavern during the night, for example, with use of offpeak shaft power having a relatively low economic value. The air would be withdrawn from the cavern during the day and would be heated by combustion with a fuel and expanded through an expansion turbine to provide electricity when the demand for electricity is large.

Hot compressed air from an air compressor operating in the night could be cooled by the new heat-exchange method against cold granular solid, the heated solid could be stored, and the air could be reheated in the daytime by the new method against the hot solid withdrawn from storage, which would then be stored cold for reuse during the night. The net effect of the procedure would be a saving of fuel.

SURFACE TREATMENTS

Erosion Corrosion Resistant Aluminum Composite

W.H. Anthony and J.M. Popplewell; U.S. Patent 3,960,208; June 1, 1976; assigned to Swiss Aluminum Ltd., Switzerland describe a composite aluminum article having increased resistance to erosion corrosion in aqueous environments. The composition comprises an aluminum base alloy cladding consisting essentially of 0.8 to 1.3% zinc, 0.7% maximum silicon plus iron, 0.10% maximum copper, 0.10% maximum manganese, 0.10% maximum magnesium, balance essentially aluminum, bonded to at least one side of an aluminum base alloy core consisting essentially of manganese from 1.0 to 1.5%, chromium from 0.1 to 0.4%, copper from 0.05 to 0.4%, balance essentially aluminum.

The excellent erosion corrosion resistance of the composite is highly desirable commercially. This property admirably lends the tubing of the process to use in heat exchange assembly such as in an aluminum radiator and the tubing of the process would result in a substantially longer useful life.

Example 1: Three alloys, A, B and C, were Durville cast and then homogenized at 1125°F for about 8 hours and air cooled. The composition of the resulting alloys is shown in the the following table.

Ingot	Si	Fe	Cu	Mn	Mg	Cr	Zn	Ti
A	0.21	0.41	0.11	1.18	–	–	0.11	0.005
B	0.19	0.35	0.20	1.18	–	0.21	0.11	0.007
C	0.03	0.24	–	–	–	–	1.03	0.005

...................Composition, % ....

Example 2: Ingots A and B of Example 1 were cut to 1.5 inches and then wire brushed and vapor degreased. Ingot C was hot rolled at 800°F to 0.25 inch gage using a 0.1 inch pass with reheating to 800°F with each second pass. The hot rolled material was then cold rolled to 0.050 inch gage. The 0.050 inch gage material of ingot C was then welded to each of the A and B ingot slabs on four sides to form A and B composites respectively, leaving one inch long openings in the weld across one of the shorter edges so that air could be expelled during further rolling of the composites. The composites were then heated to 800°F for five minutes and given skin passes of about a 3% reduction each with the partially opened edge facing in a direction opposite to the travel of the composites. The composites were then reheated to 800°F, hot rolled to 0.25 inch gage, and then cold rolled to 0.050 inch gage.

The cladding thickness of the A and B composites were then measured on mounted and polished sections and found to be 1.5 and 1.6 mils thick respectively.

The composites of Example 1 were then heated up and cooled down using a pit furnace in such a way to simulate the effect of a brazing step in a continuous aluminum radiator manufacturing line. This was done in order to allow for any possible interdiffusion effects which could result in reducing the electrode potential difference between the components of each composite during the aluminum radiator manufacturing. The heat up and cool down cycle is as follows. The composites were heated to 1150°F and cooled to 800°F within two minutes at a constant cooling rate and then quenched in water at 160°F.

Example 3: The composites of Examples 1 and 2 were cut into appropriate size specimens and subjected to impingement by a plurality of jets of an aqueous antifreeze material simulating the effects of long-term erosion corrosion in automobile radiators. Uncomposited Alloy A, further rolled to 0.050 inch after processing to 1.5 inches thickness in Example 1, and the composite A were employed as controls. The antifreeze material was a commercial, inhibited aqueous ethylene glycol containing a 45% nominal by volume ethylene glycol which was directed onto the samples at a temperature of about 200°F with the velocity of the jets at about 98 ft/sec. The test was carried out for six days.

At the end of the test the specimens were removed and rinsed in distilled water followed by solvent rinses in methanol and benzene. The samples were then chemically cleaned by immersing them in an aqueous bath of chromic plus phosphoric acids at 80°C. They were then rinsed in distilled water, dried and the depths of the resultant impingement craters measured. The depth of attack in the control composite comprising the A plus C material or composite A and the uncomposited alloy A material was found to be about three mils whereas the depth of attack in the composite comprising the B and C material or composite B was found to be about 1.8 mils maximum. The exposed core of the B composite or the B alloy was found to be substantially free of attack attesting to the galvanic protection afforded to the B alloy by the C alloy cladding of the composite whereas the exposed core material of the A composite or the A alloy had numerous small pits indicating that the galvanic protection afforded to the alloy by the C alloy cladding is practically nonexistent.

The cladding adjacent to the exposed core of the B composite was found to be substantially consumed thereby indicating cathodic protection was provided to the B alloy core whereas there was substantially less consumption of the cladding in the crater rim of the control A composite.

Example 4: This example illustrates the potential difference between the alloys of the composite of the process.

Durville ingots of the following composition were cast and homogenized and processed to 0.050 inch gage as in Example 1 and then subjected to a simulated brazed condition as in Example 2.

Ingot	Si	Fe	Cu	Mn	Mg	Cr	Zn	Ti
A	0.20	0.41	0.20	1.28	–	–	0.094	0.006
B	0.20	0.41	0.20	1.28	–	0.20	0.095	0.006

Specimens were cut from the A and B alloys and from 0.050 inch gage C cladding material of Example 1 for impingement testing as in Example 3. A portion of each specimen was passed through a special composite gasket of silicon rubber in the jet chamber of the jet tester without making electrical contact with the flange or leaking any antifreeze when the gasket was tightened. Special rubber inserts were employed so that the specimens were mounted without incurring any electrical leakage to the stainless steel jet tester chamber. In this manner it was possible to mount dissimilar specimens in jet test chambers and measure the current flow between them while they were subjected to antifreeze jet impingement at any temperature desired.

The current flow was measured by monitoring the potential drop across a two ohm resistor which shunted the electrodes externally. The value of the resistor was less than 0.5% of the total electrolytic resistance path in the antifreeze between the two test specimens. In this manner the current flow between Alloy C of Example 1 and Alloy A of this example and Alloy C of Example 1 and Alloy B of this example was monitored while the antifreeze impinged on the samples at 98 ft/sec. The temperature was cycled up and down from 40° to 105°C for three successive cycles. The direction of current flow throughout the cycling was such that the alloy C of Example 1 component remained anodic for both couples.

It was apparent that throughout the several cycles the current output of the Alloy B-Alloy C was about 5 times as great as the Alloy A-Alloy C. Thus an unexpectedly large difference to the protective cathodic current is provided by the Alloy C anode material coupled to Alloy B and this is especially true within the temperature range of 90° to 105°C where automobiles normally operate. In particular the Alloy A-Alloy C couple provided 16 microamps current in the descending leg of the third cycle at a temperature of 93.3°C (or 200°F) while the Alloy B-Alloy C couple provided 100 microamps at the same point.

Chrome Plating

K.H. Haller; U.S. Patent 4,054,174; October 18, 1977; assigned to The Babcock & Wilcox Company has found that chrome plating the internal surface of tubes, particularly in the burner or high heat absorbing zones of a furnace, markedly reduces the deposition of internal corrosion products. It is pertinent to clearly distinguish the common use of plated metals to protect the base metal from undergoing a chemical reaction in a corrosive environment and the use of internally plated tubes in a generally noncorrosive environment to inhibit the deposition of solids from the transporting liquid.

It is not the purpose to minimize chemical corrosion but to prevent or diminish the mechanical deposition of corrosion products formed elsewhere within the boiler cycle on selected areas within the boiler. Chrome plating has been used to provide corrosion resistance to such items as automobile bumpers and trim and noble metals such as gold and silver have been plated on base metals for oxidation resistance and decorative effects. However, in this process, a special effect has been determined when chrome plating is applied to the internal surface of tubes conducting a fluid in a high heat absorption zone of a furnace which is manifested by a great reduction in deposition of solid products from the fluid flowing within the tube as compared to an adjacent unplated portion of the same tube.

Internally Ridged Tube

A process described by *J.G. Withers, Jr.; U.S. Patent 4,007,774; February 15, 1977; assigned to UOP Inc.* relates to the improvement of heat transfer in heat exchangers and in particular to heat exchangers in which process fluids are circulating which have a tendency to coat or "foul" the inside tube surface. Such fouling coatings produce a thermal resistance which inhibits heat transfer and lowers the heat transfer coefficient of the tubing.

In the process, tubing having an internal ridge shape which resists deposits of a fouling layer on the upstream side of the ridging is utilized in combination with valve means which can be periodically actuated to reverse the direction of fluid flow. Since the rate of fouling deposition is sensitive to local turbulence levels, the flow reversal tends to remove at least a substantial portion of the fouling layer which had formed during the prior flow cycle on the downstream portion of the ridging. The plain end portions of the tubes which are usually provided for mounting the tubes in tube sheets will usually comprise an insignificant fraction of the tube length, albeit the inlet end will normally experience relatively turbulent flow due to the entry effect. However, in the ridged portion of the tube, the hydrodynamic pattern will change as each ridge convolution is encountered. The presence of a ridge is believed to produce a boundary layer separation which yields very active turbulence in the vicinity of the point where the boundary layer reattaches to the tube wall. In the less active zone under the separated boundary layer a fouling coating can form at about the same rate as in a plain tube while the relative turbulence in the region between the boundary layer reattachment point and the crest of the next ridge convolution will tend to keep the tube wall clean of any fouling layer. To be effective, the ridging has to be generally transverse to the tube axis, and preferably, the lead angle of the ridging, which can be single start or multiple start, should be less than 60°, as measured from a perpendicular to the tube axis.

Thus, the substitution of internally ridged tube for plain tube in a heat exchanger can be shown to provide an increase in overall heat transfer efficiency, not only for the expected reason that increased turbulence is provided by the ridges but because the ridges result in a lower fouling factor for the tube. By reversing the direction of flow periodically, heat transfer efficiency is enhanced since previously deposited fouling coatings are removed, at least to a substantial degree.

Shield Protection to Prevent Tube Erosion

H.E. Smith; U.S. Patent 4,142,578; March 6, 1979; assigned to Exxon Research & Engineering Co. describes an improved means for alleviating tube erosion due to fluid impingement in shell-and-tube heat exchangers. It has been found that such erosion can be effectively prevented by the installation on individual tubes adjacent the fluid inlet of tube shields which seat on the tubes and are secured at one end to a tube sheet or baffle and are slidably restrained on the other end by a bar or similar member fixed to an adjacent baffle or tube sheet. The tube shields employed may be constructed of angle iron, steel strips, channels, pipe segments, rods, or the like and need not be made of the same material as the tubes themselves.

Any difference between the thermal expansion of the shields and that of the tubes is compensated for by restraining one end of each shield so that it can slide with respect to the tubes. Experience has shown that the use of shields mounted in this manner provides ample protection for the first row of tubes adjacent the fluid inlet against erosion due to fluid impingement, eliminates the need for impingement plates and similar devices, allows the installation of more tubes per bundle than might otherwise be used and thus reduces heat exchanger costs, avoids the necessity for welding to the heat exchanger tubes, reduces inlet fluid pressure drop, and facilitates the rapid installation and replacement of shields as necessary.

Exchangers having triangular pitch or rotated square pitch tube arrangements, unlike those having the tubes arranged in an ordinary square pitch pattern, require protection for the tubes in both the first and second tube rows adjacent the fluid inlet. The tubes in the second row of such an exchanger are protected in accordance with the process by means of additional shields supported by the shields for the tubes in the first row or by associated supporting members. These second row tube shields may have cross-sectional configurations similar to those for the tubes in the first row and can be supported at each end by rods or other connecting members welded or otherwise connected to the first row shields. In a preferred case, the second row shields are formed from rods which are bent at each end to provide arms of sufficient length to permit welding to the first row shields.

Polyfluorinated Disulfides

A process described by *J.C. Deronzier, L. Foulletier, J. Huyghe and J.M. Niezborala; U.S. Patent 3,878,885; April 22, 1975; assigned to Commissariat a l'Energie Atomique and Societe Produits Chimiques Ugine-Kuhlmann, France* relates to a method for causing condensation in drops on heat-exchanger tubes as applicable in particular to installations for the distillation of seawater.

The precise aim of the process is to provide a method for initiating the formation of drops on the tubes of a heat exchanger by utilizing the action of a particularly advantageous agent for promoting condensation in drops. The method is characterized in that at least one chemical compound is incorporated in the feed vapor before this latter passes along the tube, the compound being constituted by a fluorinated derivative corresponding to the general formula $C_n F_{2n+1}(CH_2)_a X$ in which n represents a whole number comprised between 2 and 20, X represents a chemical function which is capable of causing the fluorinated product to adhere to the tube wall, and a is a whole number comprised between 2 and 20.

When the fluorinated derivatives are incorporated in the feed vapor, these compounds are fixed on the tube wall so as to form a coating which has excellent hydrophobic and deophobic properties. It is thus possible to cause condensation in drops not only of water vapor but also of vapors of organic liquids.

Among the fluorinated derivatives which are suitable for use can be mentioned the following products: $CF_3(CF_2)_n C_2H_4COOH$, $CF_3(CF_2)_n(C_2H_3Cl)OPO_3H_2$, $CF_3(CF_2)_n C_2H_4SO_2NH(CH_2)_6OH$, and $CF_3(CF_2)_n C_2H_4SO_2N(CH_3)C_2H_4OH$, with which excellent results are obtained on stainless steel walls, and $CF_3(CF_2)_n C_2H_4SH$, $CF_3(CF_2)_n(C_2H_3Cl)OPO_3H_2$, and $CF_3(CF_2)_n(C_2H_4)_5COOH$, with which excellent results are obtained on copper walls. The process can also be carried out by means of compositions having a base of fluorinated copolymers and polymers. In accordance with an advantageous feature of the process, there is employed as fluorinated derivative a polyfluorinated disulfide having the general formula $C_n F_{2n+1}(CH_2)_a SS(CH_2)_a C_n F_{2n+1}$ in which $C_n F_{2n+1}$ represents a straight or branched perfluorinated chain, and n and a are as defined above.

The incorporation of one or a number of polyfluorinated products in the vapor phase is carried out either by mixing with the liquid which is intended to form the feed vapor, in which case the fluorinated products are distilled under their own vapor tension; or by passing the vapor over a solution or emulsion of fluorinated products in water; or by mechanical injection of the fluorinated product or products into the feed vapor in the form of a mist, the product or products being dissolved in a solvent if necessary.

Polycrystalline Metal Whiskers

H.J. Schladitz; U.S. Patent 4,060,126; November 29, 1977 has developed a heat transfer element having a heat transfer surface to which are attached in heat conducting manner polycrystalline metal whiskers.

Polycrystalline metal whiskers [see for instance *Zeitschrift Fur Metallkunde*, Volume 59 (1968) No. 1, pages 16 to 22] can be produced with accurately determined diameters from about 0.1 μm and in similarly exact predetermined lengths of up to several centimeters. These very thin whiskers penetrate the laminar fluid boundary layer over the surface of the solid. Since the coefficient of heat conductivity of metals may be several hundred times that of liquids, even if only 20% of the surface of the solid is provided with whiskers, the heat conduction through the laminar boundary layer can be improved more than a hundredfold. A further improvement in thermal conductivity may be obtained if the whiskers extend through the laminar boundary layer into the flowing medium. Also, owing to the whiskers, turbulence is generated in the boundary layer itself, which in its turn contributes to improved heat transmission. Of course, the whiskers themselves increase the heat emitting or absorbing surface area of the solid.

Heat transfer elements according to this process preferably will be coated with polycrystalline metal whiskers which have a diameter from 0.1 μm to about 50 μm. The length to be adopted and the spacing between whiskers depends on the expected thickness of the boundary layer, i.e., on the viscosity of the flowing medium. If the spacing is sufficient for the medium readily to penetrate between the whiskers, turbulence is produced in addition. Owing to the great strength of polycrystalline metal whiskers, they can withstand the pressure of the flowing medium even when the free ends of the whiskers protrude beyond the boundary layer into the flowing medium.

The solid can take the form of a partition between two cavities for gaseous or liquid media, as for heat exchangers. In this case both sides of the partition are preferably provided with polycrystalline metal whiskers. The solid can also be an electric heat conductor, which is heated inductively, by direct passage of current, or by high frequency. The polycrystalline metal whiskers are attached to each surface exposed to the flowing medium to transfer heat thereto.

The metal whiskers are attached to the surface of the solid preferably by deposition of metal. Deposition of metal by current-free reduction of a metal compound and galvanic precipitation are suitable methods. Particularly favorable is the deposition of metal through thermal disintegration of a metal compound in the form of vapor, for instance, of a metal carbonyl. The separated metal is here not merely deposited in the zone of the points of contact of the whiskers with the surface of the solid, but also on the actual whiskers. The heat transmission can thus be improved if the metal used for separation and deposition is one having good heat conductivity, for example, silver. The whiskers could also be attached to the surface of the solid by electron beam welding.

If the largest possible quantity of whiskers per unit area are to be accommodated, it has been found desirable to use metal whiskers that consist at least partially of a ferromagnetic material. These are aligned parallel to one another during their attachment to the surface of the solid by means of a magnetic field. Conveniently these whiskers are of iron or nickel, but they may be metal whiskers with a ferromagnetic core and one or more casings of another metal, for example, copper.

Copper Bearing Base Metal Powders

A process described by *R.C. Borchert; U.S. Patent 4,101,691; July 18, 1978; assigned to Union Carbide Corporation* relates to a method for manufacturing an enhanced heat transfer device consisting of a metal substrate and randomly distributed metal bodies bonded to the substrate.

In this method base metal powder with particles of major dimensions less than 0.1 inch is provided and mixed with first liquid binder in proportion such that the weight ratio of base metal powder to first liquid binder is between 20:1 and 30:1, so as to form an adherent mass. Braze metal powder having a melting point lower than the base metal powder is also provided having particles of major dimensions such that the major dimension ratio of braze metal powder to base metal powder is between 1:60 and 1:3. The braze metal powder and the adherent mass are mixed in weight proportion such that the braze metal powder is between 10 and 30% of the braze metal powder plus the base metal powder, so as to form braze metal-coated mass.

A second liquid binder is applied on the metal substrate and the aforedescribed braze metal-coated mass is then applied on the second liquid binder coated metal substrate. The so-formed braze metal coated mass-metal substrate is heated sufficiently to remove the first and second binders, melt the braze metal and metal bond the base metal to the metal substrate thereby forming the metal bodies.

The base metal powder constitutes the bulk structure of the final metal bodies and the powder is preliminarily sized to provide the desired major or greatest dimension of the individual particles. The major dimension of the base metal powder particles should not exceed 0.10 inch and preferably should be within the range of 0.006 to 0.060 inch. Particle dimensions for these purposes correspond generally to the opening size of U.S. Standard series screen through which the particles will pass.

The adherent mass comprises base metal powder particles individually covered with a thin layer of first liquid binder. The first liquid binder may be a single component material, but preferably is a plural component material. A required component is a low volatility organic compound capable of wetting the base metal particles and capable of being removed from the base metal particles by vaporization and/or chemical decomposition without leaving an undesirable residue on the metal surfaces. A preferred low volatility organic compound is an isobutylene polymer having a molecular weight of at least about 90,000. A higher volatility compound which is a solvent for the low volatile organic compound may be included as another component of the first liquid binder. A preferred higher volatility compound is kerosene.

The braze metal-coated mass is produced by admixture of the braze metal powder with the adherent mass thereby obtaining a relatively dry flowable powder-like material. The braze metal powder is a metal or alloy material having a melting point lower than the base metal powder and capable of forming a metallic bond therewith and with the substrate. Its particle size is substantially smaller than that of the base metal powder, such particle size being defined as the major dimension of the particles. Preferably, the major dimension ratio of braze metal powder to base metal powder is between 1:30 and 1:3.

"Partially rigidizing" is a procedure for stiffening or setting the first liquid binder after formation of the braze metal-coated mass. The braze metal particles are thereby firmly attached to the base metal particles. A preferred method of partially rigidizing is by heating the braze metal-coated mass at temperatures of 150° to 200°F in order to vaporize higher volatility component from the first liquid binder.

Many combinations of base metal powder, braze metal powder and metal substrate may be employed usefully and effectively in the process. The following table lists several such combinations by way of illustration.

Example	Base Metal Powder	Braze Metal Powder	Metal Substrate
1	Cu Cu + 5-30% Ni	Cu + 8% P Ni + 11% P	Cu Cu + 1-2% Fe Cu + 5-30% Ni
2	Steel Cu	Ni + 11% P Ni + 10% P + 13% Cr	Carbon steel Low alloy steel 3.5-9% Ni
3	Steel Stainless steel 304 Stainless steel 316	Ni + 11% P Ni + 10% P + 13% Cr	Stainless steel 304 Stainless steel 316
4	Al	Al + 12% Si	Al

NOTE: Any of the base metal powders in an example may be used with
any of the braze metal powders or substrates in that example.

One combination of the foregoing table which is particularly useful is a copper bearing base metal powder, a phosphorus-nickel braze metal powder and a copper bearing metal substrate.

Example: A single layer of randomly distributed metal bodies was bonded to the inner wall of a tubular substrate. This single layer surface was prepared by first screening copper powder to obtain a graded cut, i.e., through 50 and retained on 60 U.S. Standard mesh screen from a copper powder. A portion of this graded cut weighing 2,070 g was placed into a large evaporating dish and subsequently slurried with 285 g of a 6 wt % isobutylene polymer, 6% kerosene and 88% benzene solution. After thorough mixing, sufficient of the benzene was evaporated on a hot plate to result in an adherent mass of copper particles and first liquid binder.

Phos-copper brazing alloy of 92% copper, 8% phosphorus, by weight. weighing 517 g which had been screened from phos-copper powder 1501 (New Jersey Zinc Company grade designation) to remove all particles larger than 325 mesh (U.S. Standard series sieves) was added to the adherent mass of copper base particles to formulate a ratio of 4 parts by weight copper to 1 part phos-copper.

Thus, the major dimension ratio of braze metal powder to base metal powder was about 1:7. This is based on a base metal powder dimension of 0.0117 inch corresponding to No. 50 U.S. Standard mesh screen opening, and a braze metal powder dimension of 0.0017 inch corresponding to No. 325 U.S. Standard screen opening. After thorough blending, the resultant dry mix of phos-copper coated copper base metal particles was allowed to stand at ambient temperature overnight. So treated, the particles of phos-copper brazing alloy were evenly disposed on and secured by the polyisobutylene coating to the surface of the copper particles. The powder was dry to the touch and free-flowing.

A CDA-192 (Copper Development Association designation) copper alloy tube with a 0.679 inch i.d. and a 0.735 inch o.d. was coated with second liquid binder composed of 30% polyisobutylene in kerosene by filling the tube with the binder followed by draining same from the tube to leave a thin adherent internal film of the binder on the internal tube wall. Next, the phos-copper coated copper base metal powder particles were poured through the tube thereby coating the internal tube surface substrate with a uniformly spaced single layer of braze metal-coated mass.

The external surface of the tube also was coated with a multiple layer of stacked copper particles integrally bonded together to form interconnected pores of capillary size in manner described in U.S. Patent 3,384,154 to R.R. Milton (porous boiling layer). The tube was then furnaced at 1600°F for 15 minutes in an atmosphere of disassociated ammonia, cooled and then tested for heat transfer characteristics as an enhanced heat transfer device.

The sensible heat transfer enhancement of the test device was determined by boiling Refrigerant 12 (dichlorodifluoromethane) at 48 psia on the exterior tube surface and by flowing water at higher temperature at 9 ft/sec through the internal surface covered with the single layer of bonded randomly distributed bodies. The boiling side heat transfer coefficient was already known, having been determined on a separate but similar boiling surface under the same conditions. The data of the test was reduced by extracting the known boiling side heat transfer resistance and the wall resistance in order to determine by difference the water side sensible heat transfer coefficient. The sensible heat transfer coefficient was found to be 2.55 times higher than obtained on a smooth surface metal substrate.

It should be noted that the use of an extremely volatile component of the binder such as benzene poses control problems which can usually be avoided by employing solvents of lesser volatility such as kerosene. In this example, the high volatility of the benzene makes it difficult to monitor and control the quantity of liquid binder remaining in the adherent mass during evaporation of a portion of the benzene. Moreover, continued vaporization of benzene occurs during addition and mixing of the braze metal powder and during application of the braze metal-coated mass to the substrate.

Unless special precautions are taken to limit such continued evaporation, then loss of some braze metal from the base metal may occur before brazing can be completed. Special precautions designed to limit continued evaporation and to maintain residual first liquid binder within the ratios 20:1 and 30:1 base metal powder to first liquid binder, may include chilling the adherent mass and the braze metal-coated mass, and handling the adherent mass and the braze metal-coated mass in an atmosphere with high content of benzene vapor.

A lower cost and less complex way to handle the above problem is, as stated previously, to substitute a solvent of lesser volatility, e.g., kerosene, for all or a portion of the highly volatile solvent, e.g., benzene. The preferred solvents should exhibit low vapor pressure at room temperature such that the content of residual binder in the adherent mass and in the braze metal-coated mass does not change appreciably during the specific procedures and periods of time employed in the practice of the method.

TUBE CLEANING

Scale Cleaning Liquid

A process developed by *R.L. Quintilliano; U.S. Patent 4,033,407; July 5, 1977; assigned to Hooker Chemicals & Plastics Corporation* relates to a method for more efficiently operating a heat exchanger wherein heat is exchanged between a scale-forming and a scale-cleaning liquid. The scale-cleaning liquid is passed periodically into the region of the heat exchanger wherein scale from the scale-forming liquid has accumulated, and the effluent is flushed from the heat exchanger.

In Figure 1.14, a heat exchanger **111**, which may be a titanium plate heat exchanger, in ordinary operation effects a heat exchange between fluid passing through the space schematized as the line **113** and the fluid passing through the space schematized by line **115**. With valves **105** and **125** in the closed position hot influent scale-cleaning liquid flows from line **109** through valve **129** and emerges after transferring heat in the heat exchanger **111** through open valve **131** and outline **127**.

Figure 1.14: Heat Exchanger Cleaning System

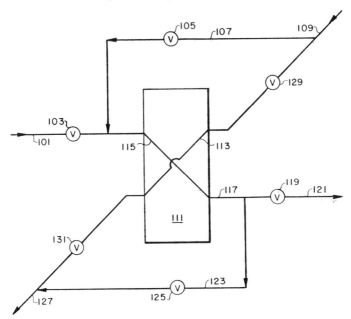

Source: U.S. Patent 4,033,407

Similarly, cold influent scale-forming liquid passes from line **101** through open valve **103** through the space schematized as line **115** inside heat exchanger **111**, wherein heat is transferred to the fluid inside line **115** and in which region scale forms. This heated scale-forming fluid then emerges out line **117** from heat exchanger 111, through open butterfly valve 119, and outline 121. During the normal operation of the heat exchanger 111, butterfly valves 105 and 125 are closed, and no fluid flow through bypass line 107 or bypass line 123 can occur.

When a cleaning operation is to begin, butterfly valves 129, 131, 103 and 119 are closed, preventing flow of fluids through the heat exchanger. Butterfly valves 105 and 125 are then opened, and hot influent scale-cleaning liquid from line 109 then flows through line 107, through butterfly valve 105, and into the heat exchanger 111 through line 115, which is to be cleaned of accumulated scale incrustations. Hot cleaning fluid effluent emerges from the heat exchanger 111 through line 117, passes through bypass line 123, through open butterfly valve 125, and out line 127 for a sufficient time to clean the scale particles lodged inside line 115. The time for effective cleaning can typically range from about ten minutes to several hours, with about 30 to 60 minutes generally sufficient.

After completing a cleaning operation, butterfly valve 105 is closed, thus isolating bypass line 107. Butterfly valve 103 is then opened, allowing cold influent liquid to pass through line 101, through butterfly valve 103, into line 115 inside heat exchanger 111, through lines 117 and 123, through butterfly valve 125, and out line 127 until line 115 has been flushed of hot cleaning solution. Butterfly valve 125 is then closed, isolating bypass line 123, and operation of the heat exchanger 111 can then be resumed by opening butterfly valves 119, allowing brine to flow through the heat exchanger 111, and opening butterfly valves 129 and 131, allowing hot influent fluid to pass through the heat exchanger 111.

In the preferred form of the process, the influent scale-forming liquid comprises cold or cool brine, which emerges as hot brine from the heat exchanger in normal operation. In a commercial operation for the manufacture of chlorine and sodium hydroxide by the electrolysis of brine, chlorine gas evolved is absorbed by an aqueous medium to which is added from about 0.1 to about 1 wt % of sulfuric acid. The purpose of the acidulation is to reduce the level of chlorine which may be absorbed, typically about 100 to about 1,000 ppm of chlorine. The resulting liquid, at a temperature of about 98° to about 100°C, comprises the influent scale-cleaning liquid of acidulated hot chlorine water, which emerges as cooled chlorine water solution of a temperature of about 30° to about 35°C.

It is necessary to employ a corrosion-resistant piping material such as, for example, titanium pipes, titanium lined steel pipes, or pipes made of an inert polymeric material, such as polytetrafluoroethylene or a corrosion resistant glass reinforced unsaturated polyester resin containing chlorendic acid and cured with styrene. The piping material must be resistant to chlorine water in the lines which ordinarily conduct chlorine water into and out of the heat exchanger, namely lines 109 and 127. Since the bypass lines 107 and 123 also conduct chlorine water, these must also be resistant to hot chlorine water. Piping designed for conducting exclusively hot brine and cold brine, namely lines 101 and 121, can be manufactured from mild steel, stainless steel, rubber-lined steel, concrete, or other pipes resistant to corrosion in concentrated brine solution. Since all valves employed in this process contact a chlorine-water solution on at least one side, these valves are preferably constructed of titanium. The butterfly valve type is particularly useful in this application.

Example: Brine at a concentration of about 25 wt % sodium chloride and at a temperature of 23°C was fed through a 6 inch diameter mild steel pipe at an approximate rate of 360 gpm into a Type 31 plate heat exchanger and emerged at a temperature of 60°C. A solution of chlorine in water, derived during chlorine processing and containing approximately 0.5 wt % of sulfuric acid and 200 ppm of chlorine, at a temperature of 99°C was fed into the heat exchanger at a rate of 215 gpm, and emerged at a temperature of 33.5°C.

A piping arrangement as set forth in Figure 1.14 was used, and all piping carrying a chlorine water solution, including bypass lines, was constructed of a corrosion resistant glass reinforced unsaturated polyester resin containing chlorendic acid, cured with styrene. Bypass pipes were 3 inches in diameter. A reducing coupling reduces the brine inlet and outlet to 4 inches diameter immediately adjacent to the heat exchanger. After 5 weeks of continuous operation, the transfer of heat through the heat exchanger was significantly reduced by scale incrustations formed inside the heat exchanger, and the pressure drop of brine increased from the normal differential of about 10 psi to 20 psi. The influent and effluent lines were valved off, and the bypass lines **107** and **123** were opened. After 45 minutes, scale incrustations were substantially removed, and the influent lines were reopened, giving a pressure differential of the brine line of 12 psi and a temperature differential of 36°C (compared with a previous temperature differential of 28°C).

Anhydrous Hydrochloric Acid

A process described by *G.L. Rounds; U.S. Patent 3,888,302; June 10, 1975; assigned to Kaiser Steel Corporation* relates to the removal of deposits from the interior passages of furnace regenerators.

The process involves the introduction of a deposit reactive material into the regenerator passages while maintaining the regenerator in service and substantially at operating temperatures where it reacts with the deposit to form reaction products which are readily removed at the operating temperatures of the regenerator. Preferably the reaction products are volatile at operating temperatures and pass off with the stack gases. Any remaining nonvolatile vestiges of the deposit are friable and readily flushed out of the regenerator passages by the flow of the gas passing therethrough.

In accordance with the process, the reactive material is added as a gas or finely divided liquid spray to the air, fuel or combination gases prior to the entrance into the regenerator. The total amount of reactive material introduced into the regenerator during the course of treatment is dependent upon the amount of deposited material to be reacted. The rate of introduction of reactive material is largely dependent upon the amount of time allowed for treatment, and the size of the equipment being treated. In the cleaning of coke oven regenerators good results are obtained when the oxide reactive material is introduced at a rate of about 2 cfm per regenerator unit for a total treatment time of about 12 hours.

Deposit reactive materials that meet the process requirements include materials such as hydrochloric acid, sulfuric acid, carbon monoxide and hydrogen sulfide. In addition hydrogen and chlorine in combination are useful deposit reactive materials.

Anhydrous hydrochloric acid is a highly preferred oxide reactive material for removing blast furnace gas deposits from regenerator passages. It is highly reactive with zinc oxide to form zinc chloride and water, both of which are vaporized at the flue temperatures of about 1100°C. Likewise, carbon monoxide is also a highly preferred oxide reactive material since it forms volatile metal carbonyls with the metal cations of the oxide deposit.

Gas Phase Reaction Mixtures

K. Jaschinski, W. Fuhr and K. Brandle; U.S. Patent 4,018,262; April 19, 1977; assigned to Bayer AG, Germany describe a process and apparatus for heat exchange with gas/solids mixtures which operates simply without clogging the heat exchanger due to solids buildup.

In the process, a solids-containing gas to be cooled or heated is introduced into a heat exchanger, and a flushing medium is introduced into the heat exchanger separately from the solids-containing gas, the flushing medium being directed angularly at the wall and sweeping about the wall whereby the wall is kept substantially free from solid deposits.

The process also provides an apparatus for indirect heat exchange between a gas-solids mixture and the walls of a cyclonelike, rotationally symmetrical heat exchanger, characterized by the fact that the apparatus includes a rotatably mounted arm which is adapted to the shape of the heat exchanger and which is formed with openings directed on to the heat-exchange surfaces.

By means of this process and apparatus, it is possible to cool hot gases containing solids with a tendency to adhere to walls, without any appreciable reduction in heat exchange compared with solid-free gases, in permanent operation and without any wear affecting the corners of the rotatable cleaning arm.

One example of the application of the process is the cooling of hot gas/solids mixtures of the kind formed during gas-phase reactions between metal or metalloid halides, for example, the chlorides of titanium, silicon, zirconium, iron, zinc, chromium or aluminum, and oxygen-containing gases at temperatures in the range from about 800° to 1500°C. The reaction gases consist essentially of halogens, for example, chlorine, and a certain proportion of inert gases, such as nitrogen and/or oxides of carbon and oxygen. By virtue of the process, it is possible to precipitate the hot gas/solids mixtures although the fine particles, which in some cases are used as pigments, frequently show a marked tendency to adhere and stick immediately after the reaction for reasons which have not yet been fully explained.

In connection with hot gas/solids mixtures of this kind, it has repeatedly been proposed to lower temperatures solely by the addition of cold gas. Unfortunately, the effect of adding relatively large quantities of cold gas is that the costs involved in the operation of following gas-cleaning apparatus and also general running costs are correspondingly increased, so that it is best to add cold gas in the smallest possible quantity. By virtue of the fact that the cooling surfaces are regularly flushed in accordance with the process, the walls remain substantially free from solids. Accordingly, it is possible to obtain high heat-transfer values in permanent operation as well. In addition, the edges of the cleaning arm, which are intended to form a substantially parallel line to the heat-exchange surface, are protected against abrasion. In a preferred case, it is possible by tangentially introducing the gas/solids mixture into the heat exchanger, and by virtue of the rotational flow generated in this way, to obtain high velocities at the exchange surfaces. It is advantageous for rotation of the cleaning arm, rotation of the gas/solids flow and addition of the flushing medium to take place in substantially the same direction. If the flushing gas is added in the direction of rotation, the walls to be cleaned are largely freed from solids before the outermost edge of the cleaning arm reaches this surface.

The flushing gas is preferably introduced in the form of individual jets, for example, through nozzles, bores or slots, in order to minimize the quantity of flushing gas. In this connection, it is advantageous to direct the flushing gas obliquely on to the cooling surfaces because, in this case, the cleaning effect of an individual jet is greater than it is in cases where the jet is directed perpendicularly on to the heat exchange surfaces.

Reverse Flow Heat Exchanger

J.W. Barr, Jr.; U.S. Patent 4,143,702; March 13, 1979; assigned to Sterling Drug, Inc. describes a heat exchanger which comprises an inner small tube and an outer surrounding shell in tube form. This is sometimes referred to as a double-pipe or tube-in-tube heat exchanger. Other types suitable to the process include those with side-by-side construction.

Liquids may flow in either direction in the tube in the double-pipe heat exchanger and a different liquid may flow in the opposite direction in the shell, the liquid in the shell surrounding the tube. Both the tube and shell are connected to a reactor and proper valving is provided so that influent will pass selectively through either tube or shell to the reactor and thermally conditioned liquid from the reactor back through the heat exchanger in the opposite part thereof, whatever the particular type of heat exchanger is utilized. This apparatus thus provides for reverse flow of sludge and heated liquid merely dependent on opening and closing certain valving.

After a predetermined time of normal operation, which may be a day, week, or month, or when a certain pressure drop is observed, the flow is reversed as to both sludge and heated liquid, and this causes the scale removal and increased efficiency while continuing the operation, without the use of chemicals.

The process is illustrated as applied to the double-pipe heat exchanger, but the principle is the same in other types also. In Figure 1.15a a double-pipe heat exchanger 10 is shown and this comprises the tube 12 and surrounding shell 14. The tube has openings at its ends at 16 and 18 and the shell has openings at 20 and 22. The reactor is indicated at 24 with the bottom entrance 26 and top exit 28. The influent enters through a pipe at 30 and can be directed selectively to tube entrance 16 or shell entrance 22 through valved pipes 32 or 34.

If the influent enters the tube 12 at 16, it exits at 18 and is directed through a valved pipe 36 into the reactor at 26, and thermally conditioned liquid or effluent exiting from the reactor at 28 and proceeding through valved pipe 38 to the shell entrance at 22, through the shell, and exits at 20 through valved pipe 40. This has been shown in solid lines.

Figure 1.15: Reverse Flow Heat Exchanger

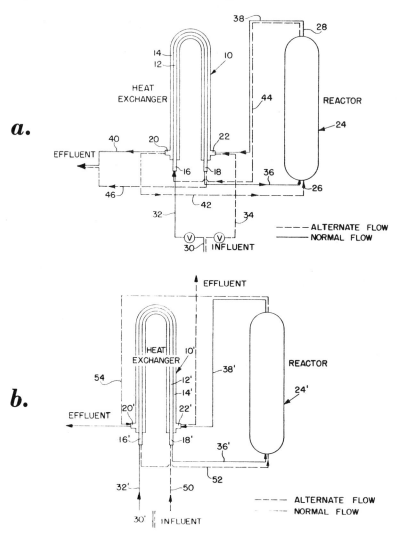

Source: U.S. Patent 4,143,702

By proper manipulation of the valved pipes, a reactor influent and effluent are interchanged in the tube and shell of the heat exchanger with flow direction in the tube and shell prior to interchange maintained. In this case the reactor influent enters the shell at **22** and exits at **20**. The influent exiting from the shell flows through pipe **42** and enters the bottom of the reactor at **26**. The exiting conditioned effluent leaves the reactor at **28** and enters the tube through pipe **44** at **16**, passing through the tube and exiting the system through pipe **46** from **18**. This has been shown in dashed lines.

This operation assumes all appropriate valving and piping to switch the flows as described.

Referring now to Figure 1.15b, a modification is shown. The numerals in Figure 1.15b are those of Figure 1.15a primed where the parts are the same. The heat exchanger 10' and reactor 24' are as before described, the heat exchanger having the tube 12' and surrounding shell 14', with respectively entrances and exits at 16', 18' and 20' and 22'. The influent at 30' reverses flow in the tube, i.e., the entrance (32') to the tube becomes the exit (52) to the reactor, and the exit (36') to the reactor becomes the entrance (50) to the tube.

A reversal occurs as to the flow of conditioned liquid in the shell 14', see pipes 38' and 54. It will be seen that the influent does not enter the shell at any time and the liquid from the reactor does not flow through the tube. The effluent exits only from the shell, at 20' or 22' selectively.

Unlike Figure 1.15a, the reactor influent remains in the tube, no interchange, but its direction of flow is reversed; the reactor effluent remains in the shell; its flow is also reversed and the heat exchanger is operated in the countercurrent mode. The process of descaling embodied in Figure 1.15a exploits an enhanced solubility due to both scale inverse solubility and the reduced concentration of the scale-forming ingredients of the influent, whereas the process of Figure 1.15b, which also reverses the large end-to-end temperature gradient, descales by inverse solubility only.

HEAT TRANSFER FLUIDS

INTRODUCTION

The use of heat transfer fluids is an essential part of many manufacturing operations. Typical operations include petroleum refining, chemical synthesis, asphaltic aggregate production, plywood lamination, plastic molding, etc. To permit operation at design capacity and to reduce heater hot spots, the heat transfer fluid should not cause serious fouling of the heat transfer surfaces.

Many organic heat transfer oils have been developed and have effected good heat transfer at elevated temperatures, typically 300° to 600°F. The most common and most economical of these organic oils is the petroleum hydrocarbon oil, although synthetic and chlorinated hydrocarbon oils are often used.

Many different types of materials are utilized as functional fluids and functional fluids are used in many different types of applications. Such fluids have been used as electronic coolants, diffusion pump fluids, synthetic lubricants, damping fluids, bases for greases, force transmission fluids (hydraulic fluids), heat transfer fluids, die casting release agents in metal extrusion processes and as filter media for air conditioning systems. Because of the wide variety of applications and the varied conditions under which functional fluids are utilized, the properties desired in a good functional fluid necessarily vary with the particular application requiring a functional fluid having a specific class of properties.

Of the foregoing, the use of functional fluids as heat transfer fluids has posed what is probably the most difficult problems in application. The requirements of a heat transfer fluid are as follows. The fluid should be liquid over a wide temperature range, and in general have a low vapor pressure so as to be utilized at atmospheric pressure. Such fluid should be operable as a heat transfer medium over an extended period of time at given temperatures, and should exhibit a high degree of thermal and hydrolytic stability. Thus, a heat transfer fluid is often required to operate at temperatures in the order of 700°F or higher over extended periods of time.

Such fluids, in addition, should be noncorrosive to metals with which they are in contact and in particular such fluids should be noncorrosive at the required operating temperature.

FLUID COMPOSITIONS

Phenoxybiphenyl and Phenoxyterphenyl Compounds

W.C. Hammann and R.M. Schisla; U.S. Patents 3,957,666; May 18, 1976; and 3,860,661; January 14, 1975; both assigned to Monsanto Company have found that functional fluids which have excellent physical properties and which are particularly suitable for use as heat transfer fluids are obtained through the use of phenoxybiphenyl and -terphenyl compounds having from 5 to 10 aromatic rings and from 2 to 8 oxyether linkages, characterized in that the number of aromatic rings is at least 2 greater than the number of oxyether linkages, and that at least 40% and preferably at least 50% of the total linkages are in the meta- position.

Particularly preferred compositions are those having 6 phenyl rings and 4 oxyether linkages, such as 3,4-bis(m-phenoxyphenoxy)biphenyl. Also preferred are mixtures of such di(phenoxyphenoxy)biphenyls wherein at least 40% of the total linkages in the mixture are in the meta- position, such as a mixture of 3,3'- and 3,4'-bis(m-phenoxyphenoxy)biphenyl.

The phenoxybiphenyl and -terphenyl compounds either singularly or as mixtures can be blended together with other compounds such as bis(phenoxy)biphenyl compounds, phenoxyphenoxybiphenyl compounds and analogs thereof as well as polyphenyl ethers and polyphenyl thioethers.

The phenoxybiphenyl and -terphenyl compounds as defined in this process have been found to be especially good functional fluid compositions having high thermal stability, high oxidative stability, high hydrolytic stability, low vapor pressure at elevated temperatures, wide liquid range, good viscosity characteristics, good lubricating properties and good metal compatibility when in contact with metal mechanical members of fluid handling systems. The functional fluids are particularly useful as inexpensive high temperature heat transfer fluids. These fluids are also generally useful as high vacuum diffusion pump oils, dielectric fluids, coolant moderators for nuclear reactors, and as high temperature lubricants and lubricant base stock.

Biphenyl Fluids

L.L. Jackson, W.F. Seifert and D.E. Collins; U.S. Patents 3,888,777; June 10, 1975 and 3,966,626; June 29, 1976; both assigned to The Dow Chemical Co., describe a three component heat transfer fluid by volume of 5 to 90% of diphenyl oxide, 5 to 50% of biphenyl and 5 to 90% of biphenylyl phenyl ether which is a superior heat transfer fluid. These fluids have a greater range of liquidity than known heat transfer agents while at the same time retaining desirable thermal stability characteristics.

Mixtures of diphenyl oxide, biphenyl and biphenylyl phenyl ether were prepared in the proportions shown in the following table.

Unless noted, the biphenylyl phenyl ether was a mixture of isomers which were about 70% ortho, 25% para and 5% meta. These fluids are compared to known fluids.

Heat Transfer Fluids of Diphenyl Oxide (DPO), Biphenyl (BP) and Biphenylyl Phenyl Ether (BPE)

| | . . . Composition (vol %) . . . | | | Liquidity Characteristics | |
| | | | | Freezing Point (°F) | Boiling Point (°F) |
Example	DPO	BP	BPE		
Compound A	100	—	—	82	495
Compound B	—	100	—	158	493
Compound C	—	—	100	99	662
Compound D*	73.6	26.4	—	54	495
1	10	15	75	0	590
2	5	10	85	35	615
3	5	15	80	34	600
4	5	20	75	35	590
5	20	5	75	35	590
	Dowtherm A	BPE			
6	20	80		19	600
7	25	75		15	590
8	30	70		−10	572
9	40	60		−7	—
10	50	50		−35	543
11	55	45		−41	—
12	65	35		−10	—
13	60	40		−27	—
14	70	30		1.5	516
15	75	25		29	509
16	80	20		35	—
17	50	50**		−40	544
18	50	50***		28	540

*Dowtherm A.
**m-BPE.
***o-BPE.

The proportions of the components of the heat transfer fluid may vary widely so long as each component is present in the amounts specified. In preferred compositions, about 30 to 70% by volume of the mixture is biphenylyl phenyl ether, with those fluids containing about 40 to 60% by volume of biphenylyl phenyl ether being especially preferred because of their desirable low temperature characteristics.

Also preferred in this process are those compositions where the diphenyl oxide and biphenyl are present in the proportion of the eutectic mixture, referred to as Dowtherm A. Of special interest are compositions of Dowtherm A and m-biphenylyl phenyl ether containing about 30 to 70 volume percent of Dowtherm A, with those fluids containing about 40 to 60% Dowtherm A being of particular importance because of their low freezing points.

It has been found by *L.L. Jackson, W.F. Seifert and D.E. Collins; U.S. Patent 3,931,028; Jan. 6, 1976; assigned to The Dow Chemical Co.* that the well-known heat transfer fluid consisting essentially of the eutectic mixture of diphenyl oxide and biphenyl is improved by the addition of monomethyl- or monoethylbiphenyl. The resulting fluids have very nearly the same boiling point and heat stability as the binary eutectic but have substantially lower freezing points.

In related work *L.L. Jackson, W.F. Seifert and D.E. Collins; U.S. Patent 3,907,696; September 23, 1975; assigned to The Dow Chemical Company* have found that heat transfer agents containing at least three components consisting of by volume about 5% to 90% of diphenyl oxide, 5 to about 50% of biphenyl and 5 to about 90% of polyphenyl ether having three or four aromatic nuclei, alkylated biphenyl or diphenyl oxide having 1 to 4 methyl or ethyl substituents, ethylbenzene oil and mixtures thereof are superior heat transfer fluids. These fluids have a wide range of liquidity while at the same time retaining desirable thermal stability characteristics.

Biphenyl Fluids Containing a Polyphenylphenol Mixture

W.D. Watson; U.S. Patent 4,054,533; October 18, 1977; assigned to The Dow Chemical Company describes heat transfer fluids consisting essentially of about by weight: 20 to about 40% diphenyl ether; 40 to about 60% of a 2-biphenyl-ylphenyl ether (2-bippe) and 4-biphenylylphenyl ether (4-bippe) mixture at a 2-bippe:4-bippe) weight ratio of at least about 2:1 and 12 to about 25% of a polyphenylphenol mixture.

These mixtures exhibit freeze points of $-18°C$ or less. These heat transfer fluids have excellent fluidity over a broad temperature range and display good thermal stability.

Tertiary Diamides

A process developed by *R.M. Thompson; U.S. Patents 3,910,847; October 7, 1975 and 3,915,876; October 28, 1975; both assigned to Sun Ventures, Inc.* provides tertiary diamides having the following structural formula:

$$R'-\underset{\underset{CH_3}{|}}{N}-\overset{\overset{O}{\|}}{C}-(CH_2)_n-\overset{\overset{O}{\|}}{C}-\underset{\underset{CH_3}{|}}{N}-R'$$

where n is 4 to 12
and where R' is

$$CH_3(CH_2)_m-\overset{\overset{H}{|}}{\underset{\underset{R}{|}}{C}}-(CH_2)_x-$$

where R is H, alkyl radical having C_{1-5}; m is 2 to 8; and x is 1 to 5. These diamides are water-white liquids at ambient temperatures and have useful physical properties for use as heat transfer fluids and plasticizers. The tertiary diamides can be prepared by reacting a normal paraffinic diacid with a secondary amine wherein one radical is a methyl. This general reaction is illustrated by the following equation:

$$HOOC(CH_2)_nCOOH + 2R'NHCH_3 \longrightarrow R'\underset{\underset{CH_3}{|}}{N}CO(CH_2)_nCO\underset{\underset{CH_3}{|}}{N}R' + 2H_2O$$

The normal paraffinic diacid of the equation can contain 6 to 14 carbon atoms; preferably 8 to 12. Accordingly n of the diacid of the equation equals 4 to 12, preferably 6 to 10. Examples of such acids are suberic, azelaic and sebacic. Also prepared are liquid tertiary diamides having the following structural formula:

$$
\begin{array}{c}
\quad\;\; CH_3 \quad\; CH_3 \qquad\quad CH_3 \quad\; CH_3 \\
\quad\;\; | \qquad\;\; | \qquad\qquad\quad | \qquad\;\; | \\
R-C-N-CH_2-C-(CH_2)_n-C-CH_2-N-C-R \\
\;\;\; || \qquad\qquad\; | \qquad\qquad\; | \qquad\qquad || \\
\;\;\; O \qquad\qquad CH_3 \qquad\; CH_3 \qquad\quad O
\end{array}
$$

where n is 0 to 10; R is $-(CH_2)_mCH_3$, and m is 4 to 12.

Preparations and the physical and chemical properties of the amines are given in Kirk-Othmer, *Encyclopedia of Chemical Technology.*

The tertiary diamides have many properties which make them excellent heat transfer media. Among these properties are that they are generally innocuous towards metals; have relatively good thermal stability, have high boiling points and are liquid at ambient temperature.

Silicone Fluids

K.O. Knollmueller; U.S. Patent 4,116,847; September 26, 1978; assigned to the Olin Corporation describes alkoxysilane cluster compounds having the formula:

$$
\begin{array}{c}
\qquad\quad R \quad\;\; R'' \quad\; R'' \qquad R \\
\qquad\quad | \qquad\; | \qquad\; | \qquad\quad | \\
[(R'O)_3SiO]_2-Si-O-Si-(O-Si-)_n-O-Si-[OSi(OR')_3]_2 \\
\qquad\qquad\quad | \qquad\; | \\
\qquad\qquad\quad R''' \quad\; R'''
\end{array}
$$

where n is an integer from 0 to 300; R is hydrogen, alkyl, alkenyl, aryl, aralkyl or $-OSi(OR')_3$; each R' is independently selected from alkyl, alkenyl, aryl or aralkyl with the proviso that at least a majority of R' radicals are sterically hindered alkyl groups having at least 3 carbon atoms; and R'' and R''' are independently selected from hydrogen, alkyl, alkenyl, aryl, aralkyl, hydroalkyl, and halo or cyano substituted alkyl, alkenyl, aryl, aralkyl and hydroalkyl. These silicone compounds have utility as functional fluids for use as heat transfer fluids, hydraulic fluids, brake fluids, and transmission fluids.

Perfluoroamino Ethers

S. Benninger and T. Martini; U.S. Patent 3,882,182; May 6, 1975; and S. Benninger and S. Rebsdat; U.S. Patent 3,891,625; June 24, 1975; both assigned to Hoechst AG, Germany describe the preparation of perfluorinated hydrogen-free amines having side chains which contain ether bonds. These compounds are formed by the electrolysis, carried out in anhydrous hydrofluoric acid, of the products of addition of hexafluoropropene to amino alcohols. The tertiary perfluoroamino ethers correspond to the following formula 1:

(1)
$$
\begin{array}{c}
(R_{F1})_xN-(CF_2CF-O-CF_2CF_3)_z \\
\qquad\quad | \qquad\quad | \\
(R_{F2})_y \quad R_{F3}
\end{array}
$$

where x, y and z each are integers; x and y each being 0, 1 or 2; z being 1, 2 or 3; the sum of $x + y + z$ always being 3; R_{F_1} and R_{F_2} each are linear and/or branched perfluoroalkyl radicals having from 1 to 10 carbon atoms; and R_{F_3} is a trifluoromethyl group or a fluorine atom.

The upper limit of the number of carbon atoms is not exactly a limit in principle, but when the number of carbon atoms is greater than 10, for example from 11 to 16, the properties characteristic for this class of compounds become gradually closer to those of perfluoroalkanes, especially in the case where z is 1.

It has been found that these tertiary perfluoroamino ethers which can be obtained by electrofluorination of tertiary hexafluoropropoxyalkylamines, are excellently suitable for the heat transfer media or dielectrics in electric systems. These applications require viscosities and dielectric constants as low as possible, and also chemical inertness to metals, oxides, oxygen and plastics.

The process for the preparation of tertiary perfluoroamino ethers of formula 1, comprises dissolving a compound of formula 2

$$(2) \qquad \begin{array}{c} (R_1)_x N - (CH_2CHOCF_2CHFCF_3)_z \\ | \qquad | \\ (R_2)_y \quad R_3 \end{array}$$

where R_1 and R_2 are linear and/or branched alkyl radicals having from 1 to 16, preferably from 1 to 10, carbon atoms, R_3 is H or CH_3 and x, y and z are as defined in formula 1, in anhydrous hydrofluoric acid, and electrolyzing the solution.

The starting substances of formula 2 are prepared by reaction of the corresponding amino alcohols or their amino alcoholates with hexafluoropropene.

The electrofluorination of the products (2) to form the tertiary perfluoroamino ethers of formula (1) is carried out in a usual Simons cell (see U.S. Patent 2,519,983), which consists of a vessel made from stainless steel and having a capacity of 1.5 liters, provided with double jacket, reflux condenser with device for alkaline scrubbing of waste gas connected to it, circulation pump and liquid level indicator. The electrodes are a package of parallelly mounted, electrically insulated nickel plates of alternating polarity having a total anode area of 30.8 dm^2.

The following examples illustrate the process of electrofluorination of partially fluorinated amino ethers of formula 2.

Example 1: A Simons cell was charged with 80 g of tris(hexafluoroproporyethyl)-amine and 1,400 g of anhydrous hydrofluoric acid. Within 44 hours, a further 300 g of starting material was added in small portions of about 20 to 30 g. At an electrolysis temperature of +5°C, an amperage of a constant 30 A and a voltage of from 4.9 to 6.0 volts and after a total of 51 hours, 237 g of perfluorinated crude product were obtained, corresponding to 43.1% of the theoretical yield, relative to the following reaction equation:

$$N(C_2H_4-OCF_2CHFCF_3)_3 + 15HF \xrightarrow{\text{30 faradays}} N(C_2F_4OC_3F_7)_3 + 15H_2$$

212 g of dry, stabilized products having a boiling range from 149° to 219°C were

isolated from the crude product. The stabilized product contained the following compound $N(CF_2CF_2OCF_2CF_2CF_3)_3$ as the main component, having the highest peak observed at m/e = 850 in the mass spectrum, corresponding to M − F.

The indicated structure was confirmed by the F − 19 NMR spectrum. About 71% of the fraction having the boiling range of 206° to 219°C/760 mm Hg (corrected) consisted of the cited product. About 17 area percent of the stabilized crude product consisted of a substance the highest mass peak of which is at m/e = 712, corresponding to M − 3F.

According to the NMR spectrum, the substance had the following structure: $(CF_3CF_2CF_2OCF_2CF_2)_2NCF_2CF_2OCF_3$.

Furthermore, substances of the following composition were isolated in the fluorination product: $C_2F_5N(CF_2CF_2OCF_2CF_2CF_3)_3$; 17 area percent, highest mass peak: m/e = 684, corresponding to M − F and 7.3 area percent, highest mass peak: m/e = 832 (M − F) of $(CF_3CF_2CF_2OCF_2CF_2)_2N(CF_2CF_2OCF_2CFHCF_3)$.

Example 2: In an electrolysis cell having a capacity of about 38 liters, the design is similar to that of the laboratory cell, 35.1 kg of anhydrous hydrofluoric acid and 2.2 kg of tris(hexafluoropropoxyethyl)amine were electrolyzed. In the course of 75 hours, 4.6 kg of amino ether were added in small portions. After a total of 92 hours of operation at an electrolysis temperature of 0°C, a current density of 0.3 A/dm² and a voltage of a maximum of 6.0 volts, a total of 4.92 kg of fluorination product was taken off the cell, which corresponds to about 55% of the theoretical yield, relative to the reaction equation of Example 1.

The qualitative composition was identical to that of the product described in Example 1, but the content of the main component was 57% higher than that of Example 1; this being a result of the decreased temperature.

Example 3: Within 81 hours, 582 g of n-butyl-bis(hexafluoropropoxyethyl)amine were electrolyzed in the electrolysis cell as described before at a temperature of 15°C, a current density of 2.1 A/dm² and a voltage of from 4.9 to 6.2 volts. A total of 546 g of fluorination product was obtained corresponding to 54% of the theoretical yield, relative to the following reaction equation:

$$C_4H_9N(C_2H_4OC_3F_6H)_2 \ + \ 19HF \ \xrightarrow{\text{38 faradays}} \ C_4F_9N(C_2F_4OC_3F_7)_2 \ + \ 19H_2.$$

The crude product treated with alkali had a boiling range of 130°C/760 mm Hg (corrected) and contained: $CF_3(CF_2)_3N(CF_2CF_2OCF_2CF_2CF_3)_2$; 60 area percent; highest mass peak at m/e = 765; corresponding to M − 2F and

$$\begin{array}{l} CF_2(CF_2)_3NCF_2CF_2OCF_2CF_2CF_3 \\ \qquad\qquad | \\ \qquad\quad CF_2CF_2OCF_3 \end{array}$$

10.2 area percent; highest mass peak at m/e = 634, corresponding to M − CF_3.

In a related process described by *S. Benninger, T. Martini and S. Rebsdat; U.S. Patent 3,962,348; June 8, 1976; assigned by Hoechst AG, Germany* mono- or polyhydric alcohols of the alkane, tetrahydrofuran or tetrahydropyran series or di- or trialkyleneglycols are dissolved in an aprotic polar solvent and reacted with

C_3F_6 in the presence of trialkylamines, or with tetrafluoroethylene to give the corresponding tetrafluoroethyl- or hexafluoropropyl ethers. The solution of the fluoroethers in anhydrous hydrofluoric acid is electrolyzed. Hydrogen free perfluoroethers are obtained.

Antifouling Additive for Hydrocarbon Fluids

R.L. Peeler and J.M. King; U.S. Patents 3,958,624; May 25, 1976 and 3,920,572; Nov. 18, 1975; both assigned to Chevron Research Company have found that heat exchanger fouling from an organic heat transfer medium can be substantially reduced by incorporating into the medium a combination of an oil-soluble barium overbased calcium sulfonate having a base ratio of at least 4 and an oil-soluble phenolic antioxidant.

The barium overbased calcium sulfonates having a base ratio below 4.0 and particularly below 3.5 must be used at too high a concentration in order to sufficiently suppress exchanger fouling.

Example 1: This example is presented to illustrate the preparation of a preferred dialkyl benzene sulfonate which is used to prepare the overbased metal sulfonates of this process.

Benzene is alkylated using a tetramer polypropylene fraction and HF alkylation catalyst, a reaction temperature of about 65°F, and efficient mixing. The hydrocarbon phase is separated, washed and fractionated. The lower alkyl fraction (boiling point range 318° to 478°F ASTM D 447 distillation) is collected as feed for the second stage alkylation with a mixture of straight chain 1-olefins. The average molecular weight of the above branched chain alkylbenzene is 164. This corresponds to an average of 6 carbon atoms per alkyl group in the mixture. The overall alkyl carbon atom content corresponding to the above boiling point range is the C_{4-9} range.

Using the above branched chain monoalkylbenzene and a substantially straight chain C_{17-21} 1-alkene fraction obtained from cracked wax, and hydrogen fluoride catalyst, the desired dialkylbenzene is produced in a stirred, continuous reactor. The 1-alkene feed has the following characteristics:

Average molecular weight	268
Average carbon atoms per alkyl group	19
Olefin distribution, wt %	
C_{17}	2
C_{18}	22
C_{19}	39
C_{20}	32
C_{21}	5
Reaction conditions	
LHSV	2
Temperature, °F	100
Monoalkylbenzene to α-olefin, mol ratio	2-1
Hydrocarbon to HF ratio, volume	2.3-1

After reaction the settled product is separated into an organic phase and a lower HF acid phase. The crude dialkylbenzene organic phase is washed and then fractionated by distillation.

A minor amount of forecut, mainly monoalkylbenzene, is collected up to an overhead temperature of about 450°F at 10 mm Hg. The balance of the distillate is the desired product, and has an average molecular weight of about 405. The difference between the average carbon atom content of the alkyl chain types is about 13.

The dialkylbenzene prepared as in Example 1 is charged to a stirred reaction vessel fitted for temperature control along with 130 neutral oil which is substantially free of sulfonatable material. The volume ratio of the two materials is 3½ to 4, respectively, and to this mixture is added, over a period of several hours, 2 volumes of 25% oleum. The reaction temperature is maintained at about 100°F. Two phases developed in the settled mixture, the lower being a spent mineral acid phase and the upper being the desired sulfonic acid phase.

The separated sulfonic acid-oil mixture is then neutralized with one volume of 50% aqueous caustic diluted with 15 volumes of aqueous 2-butanol. During the neutralization the temperature is maintained below 110°F and after completion the neutral solution is heated and maintained at 140°F during a second phase separation. Two phases developed, a lower brine-alcohol solution and an upper neutral alcohol-sodium sulfonate solution.

Example 2: The preparation of a neutral calcium sulfonate is illustrated in this example. A 3-liter glass flask is charged with 80 g of calcium chloride and 800 ml of water. Then 1,500 g of the sodium sulfonate as prepared by the method of Example 1 is charged to the flask. The contents are heated to 85°F under agitation and maintained at these conditions for 1 hour. After 1 hour, the contents are allowed to phase separate and the water layer drawn off. 800 ml of distilled water is admixed with the sulfonate and heated for one hour. The phases are allowed to separate and the aqueous phase drawn off. The sulfonate is washed three additional times with water and one time with an aqueous isobutyl alcohol solution. The mixture is heated to 112°C to remove any residual water and isobutyl alcohol. 500 ml of toluene is added to the sulfonate and the admixture filtered through a Celite 512 filter. The product is stripped at 185°C under 3 mm Hg vacuum to yield 740 g of neutral calcium sulfonate. Analysis of the product reveals 6.09 wt % sulfated ash, 1.93 wt % calcium and 0 base number.

Example 3: The preparation of a neutral barium sulfonate is illustrated by this example. A 3-liter, 3-neck round bottomed flask is charged with 170 g of barium chloride ($BaCl_2 \cdot 2H_2O$) in 800 ml of water and 1,500 g of the sodium sulfonate prepared by the method of Example 1. The contents are stirred and heated to 85°C for 1 hour. The product is then phase separated and washed four times in the same manner as described in Example 2. The yield is 665 g of barium sulfonate. An analysis of the product reveals 9.84 wt % sulfated ash; 4.7 wt % barium and 0 base number.

Example 4: This example is presented to illustrate an exemplary overbasing procedure in preparing an exemplary overbased calcium sulfonate. A 1-liter glass 3-necked flask is charged with 18.9 g of calcium hydroxide, 20 ml of methanol, 250 ml of a petroleum aliphatic thinner (6% aromatic hydrocarbons, 250°F initial BP and 310°F end BP), and 100 g of a calcium sulfonate prepared by the method of Example 1 except containing 1.64 wt % calcium. An additional

250 ml of the thinner is then added and the contents stirred. Carbon dioxide is bubbled through the mixture at room temperature and stopped when the uptake rate leveled off. A total of 14 g of CO_2 is taken up by the mixture. The product is heated to 130°C to remove methanol and water and thereafter the product is filtered through a Celite 512 filter. The thinner is stripped by heating to 180°C at two mm Hg vacuum. The yield of overbased metal sulfonate is 101 g. The basic calcium content of the product is 7.71 wt % with a base ratio of 5.8. The base number is measured to be 216 mg KOH/g.

Example 5: This example is presented to illustrate an exemplary overbasing procedure in preparing an overbased barium sulfonate. A two-liter glass flask is charged with 50 g of barium sulfonate prepared by the method of Example 3, 500 ml of xylene and 72 ml of methanol. Thereafter, 36.2 g of barium oxide is added to the mixture in three separate portions—each 10 minutes apart. Carbon dioxide is bubbled into the reaction mixture until a total of 12 g are taken up. Methanol and water are removed by distillation and the product filtered through a Celite 512 filter. The xylene is stripped from the system by heating to 180°C under a 2 mm Hg vacuum. The product has a barium base ratio of 4.7 and a base number of 160 mg KOH/g.

Example 6: This example is presented to illustrate the superior low fouling heat transfer fluids containing an overbased metal sulfonate. In this example, a series of heat transfer tests are performed which measure the fouling properties of oils containing varying amounts and varying kinds of metal sulfonates.

In the test, the sample oil is placed in the test apparatus and heated to a temperature of 600° to 650°F for a period of 240 to 504 hours. At the end of the test the oil is visually observed for deposit content. The heavier the deposit, the greater the fouling tendencies of the heat transfer oil. The rating is: heavy, representing an opaque deposit; medium, representing a brown translucent deposit; light, representing a lacquer or slight staining of the test apparatus walls; trace, representing trace amounts of deposits with only partial coverage of the surface; and none, representing no visual deposits.

The sample oils are prepared by incorporating varying amounts of a neutral or overbased metal sulfonate into a solvent refined Midcontinent 200 neutral oil. The neutral sulfonates are substantially prepared by the method of Examples 2 and 3. The overbased sulfonates are substantially prepared by the method of Examples 4 and 5.

The test apparatus comprises an elongated glass tube of 450 mm total length having an upper tubular section open at its top, 100 mm in length, with a 16 mm o.d. tubing; a middle section, 250 mm in length, with a 6 mm o.d. tubing and lower section 100 mm in length with a 16 mm o.d. tubing closed at its bottom end. The lower section is immersed in a salt bath which is maintained at a constant temperature of 600° or 650°F. The upper section is equipped with a cooling jacket so that water may be circulated through the jacket to cool the sample oil within the tube. The sample oil is placed within the tube so that the oil in the lower section is heated to elevated temperatures from the salt bath and travels upward through the small diameter middle section to the upper section where it is cooled. Since the upper section is open to the atmosphere, the oil is in contact with air in this section. The results from the various sample oils tested herein are displayed in the following table.

Heat Transfer Fouling Test

Test	Metal Sulfonate Additive Type	Base Ratio*	Concentration**	Test Temperature (°F)	Test Time (hr)	Fouling Deposit Amount
1	None	—	—	600	240	heavy
2	Neutral calcium sulfonate	0	7	600	240	heavy
3	Overbased calcium sulfonate	10.2	14	600	504	none
4	Overbased calcium sulfonate	5.7	14	600	504	none
5	Overbased calcium sulfonate	5.7	21	600	504	none
6	Overbased calcium sulfonate	5.7	7	600	504	none
7	Overbased calcium sulfonate	5.8	14	600	504	none
8	Overbased calcium sulfonate	5.8	29	600	504	none
9	Overbased calcium sulfonate	7.8	29	650	336	trace
10	Overbased calcium sulfonate	10.2	29	650	336	trace
11	Overbased calcium sulfonate	10.2	48	650	336	trace
12	Neutral barium sulfonate	0	14	600	240	heavy
13	Overbased barium sulfonate	0.6	7	600	240	medium
14	Overbased barium sulfonate	4.7	7	600	504	none
15	Overbased barium sulfonate	4.7	14	600	504	none
16	Overbased barium sulfonate	7.3	14	600	504	trace
17	Overbased barium sulfonate	5.0	7	600	504	none
18	Overbased barium sulfonate	5.0	14	600	504	none
19	Overbased barium sulfonate	4.5	7	600	504	trace
20	Overbased calcium sulfonate	3.85	21	650	336	heavy
21	Overbased calcium sulfonate	5.8	15	650	326	heavy
22	Overbased barium sulfonate	10.4	14	600	504	light medium

*The base ratio is the ratio of the chemical equivalents of excess metal in the product to the chemical equivalents of the metal required to neutralize the sulfonic acid.

**Concentration of metal sulfonate in oil as expressed in mmol/kg.

The above table illustrates the practice of the process in reducing fouling of a heat transfer oil. Test 1 illustrates the heavy deposits associated with the test oil without the presence of an antifouling additive. Tests 2 and 12 illustrate the relatively little effect of neutral calcium and barium sulfonates on the fouling properties both exhibiting heavy deposits. Tests 3 through 11 and 14 through 19 illustrate the substantial reduction of deposit formation by using an overbased calcium or barium sulfonate having a base ratio above 4. Tests 2, 12, 13 and 20 illustrate the problems of using neutral or overbased metal sulfonates having a base ratio below 4. Tests 21 and 22 are presented to illustrate that variations in the data occur with this test apparatus. Thus, the data as a whole must be viewed in order to observe the effect of the various additives.

Example 7: This example is presented to illustrate the effectiveness of the combination of various barium overbased calcium sulfonates and a representative phenolic antioxidant in improving the oxidation stability of a heat transfer oil.

In each of the tests, various amounts of an overbased metal sulfonate are added to a 200 neutral lubricating oil along with varying amounts of a hindered phenolic antioxidant 4,4'-methylenebis(2,6-di-tert-butylphenol) marketed by Ethyl Corporation under the brand name Ethyl 702. This solution is then subjected to an oxidation test. The test is conducted in a glass cell consisting of a 100 ml sample compartment, a stirring mechanism and a compartment holding solid sodium or potassium hydroxide pellets which serve as a drying agent. The pellets have only vapor phase contact with the liquid in the sample compartment. The cell is only open to a 1.5 liter bell jar filled with 99.85% oxygen kept at atmospheric pressure. In each test a 25 g sample of the compounded oil and 20 ppm of iron as iron naphthenate are charged to the sample compartment of the cell. The oil within the sample compartment is heated to a temperature of 340°F and maintained at that temperature under agitation. The time required for an equivalent 100 g of the test sample to remove 1,000 ml of oxygen is observed and reported in the following table.

Oxidation Tests

Test	Type*	Base Ratio	.....Overbased Metal Sulfonate ..... Concentration mmol/kg	Phenolic Antioxidant Concentration (wt %)	Oxidation Life (hr)
1	none	–	–	none	0.42
2	none	–	–	0.3	1.67
3	Ca o/b Ca	5.8	29	0.3	0.98
4	Ca o/b Ca	10.2	29	0.3	1.26
5	Ba o/b Ba	5.0	22	0.3	2.36
6	Ba o/b Ca	5.1	29	0.3	2.88
7	Ba o/b Ca	5.1	67	0.3	3.42
8	Ba o/b Ca	2.2	67	0.3	4.6**
9	Ba o/b Ca	5.3	27	0.3	3.0
10	Ba o/b Ca	5.3	67	0.3	3.63
11	Ba o/b Ca	5.5	29	0.3	3.45

*o/b is overbased.
**Commercial sulfonate mixture may contain additional antioxidants.

The above table illustrates a substantial improvement of the combination of a barium overbased calcium sulfonate (tests 6 through 11) over the calcium over-based calcium sulfonates (tests 3 to 4) and over the barium overbased barium sulfonate (test 5). The phenolic antioxidant alone exhibited an oxidation life of 1.67, but when combined with a barium overbased calcium sulfonate such as in test 10, the combination exhibited a more than two-fold increase in oxidation life.

Stabilization of Chlorobenzenes

W. Mahler; U.S. Patent 4,068,706; January 17, 1978; assigned to E.I. Du Pont de Nemours and Company has found that chlorobenzenes with 2 to 5 chlorine substituents particularly trichlorobenzenes, can be used in a heat transfer method when contained in steel or aluminum vessels and at temperatures in the range of 260° to 450°C by contact with a thermal stabilizing amount of a solid acid acceptor selected from a carbonate, phosphate, borate, or molybdate of an alkali metal or an alkaline earth metal or an oxide of zinc, cadmium or an alkaline earth metal.

Example 1: A small coupon of aluminum weighing approximately 0.20 g was sealed into each of two Pyrex glass tubes containing about 1 g of 1,2,4-trichlorobenzene. The metal coupon in each was completely immersed in the liquid. One of the tubes also contained 0.1 g CaO. Both tubes were kept in an oven held at 400°C. After four days the tube without CaO showed evidence of major attack of the liquid on the aluminum whereas the contents of the tube containing CaO were unchanged in appearance. The tube without CaO was then cooled to about 25°C and gave an indication of high gas pressure when opened. The aluminum coupon showed a weight loss of 89.8 mg after washing, lightly rubbing with a paper towel to remove the superficial coating and drying.

The tube containing CaO was kept at 400°C without any visible change of the Al until after 28 days when it was cooled to $-196°C$ and then opened in an evacuated system. Even after warming to 25°C the pressure rise was less than 1 mm Hg, indicating no significant amount of gaseous decomposition products. The aluminum showed little weight change (0.3 mg gain) when treated and measured in the same way as described above for the unstabilized fluid.

Example 2: In a manner similar to that of Example 1, coupons weighing about 0.2 g each were sealed into 2 cc Pyrex glass tubes containing 0.3 cc of 1,2,4-trichlorobenzene and 0.05 g of various inorganic solids to be tested as stabilizers. Sample tubes were kept at 400°C and were periodically examined for visual evidence of decomposition. When pure fluid without stabilizer was tested in this manner no change was apparent until after three days, when the completely black appearance of the contents of the tube indicated sudden and extensive reaction. With the following stabilizers present (0.05 g per gram of liquid) no visual change was noted after 28 days at 400°C when the tests were discontinued: $CaCO_3$, Na_2CO_3, Na_3PO_4, $Na_5P_3O_{10}$, $Na_2B_4O_7$, MgO, CaO and SrO. With 5% K_2CO_3 the liquid began to darken on the 20th day indicating the start of some decomposition. With 0.02 g ZnO the sudden evidence of reaction occurred after fourteen days.

Stabilization of Trichlorodifluorobenzenes

According to a process described by *W. Mahler; U.S. Patent 3,944,494; March 16, 1976; assigned to E.I. Du Pont de Nemours and Company* trichlorodifluorobenzene is thermally stabilized particularly in the presence of engineering metals by contacting it, especially when at temperatures greater than about 200°C, with solid alkaline earth carbonates or with alkali or alkaline earth borates in a sufficient amount. The method is particularly useful in Rankine cycle engines using trichlorodifluorobenzene as the working fluid, when the stabilizer is contained in the boiler.

Gas-Liquid Aerosol Fluids

A process described by *S. Wyden; U.S. Patent 3,886,760; June 3, 1975* involves indirect heat exchange using a fluid comprising a gas and a liquid combined or entrained together. This fluid can take the form of a foam or an aerosol and its effectiveness can be facilitated by means of electrostatic forces and other devices.

Thus the process consists of the use of a two-phase fluid, usually comprising a gas and a liquid together with one entrained within the other as an indirect coolant. Several advantages to this type of coolant are noteworthy. There is no problem of coolant liquids contaminating or damaging the material being cooled. The proportions of the liquid-gas phases can be varied to accomodate varying heat absorption requirements. The efficiency of this coolant will exceed that of the gas alone. Using liquid alone one cannot recover from the coolant the heat it has absorbed from the material because liquids are noncompressible. The dispersal of the liquid increases its surface area and thereby increases its heat absorption capacity per liquid unit while also serving as a heat sink for the gas vehicle. With this increased heat absorption capacity is retained the compressibility property of gases–permitting recovery of a portion of the heat absorbed by compression of the output coolant fluid, something not possible with a completely liquid coolant.

In one example, the gas is entrained in the liquid, usually by adding a surface tension lowering agent to the liquid. In terms of its discharge, except where bubbles would be of value, such a system would require heat recovery from the coolant and recycling of the coolant. Another approach is the entrainment of the liquid in the gas in the form of an aerosol. The use of an aerosol lends itself to several applications.

In one application, in contact with the surface that encases a material to be cooled, an aerosol is introduced which will absorb heat from the material. At the end of the pathway of contact between the coolant and material the cooling fluid is directed to a compressor to recover the absorbed heat. The formation and maintenance of the aerosol can be facilitated by the application of electrostatic forces to the aerosol generating systems and incorporated in the indirect contact pathway. Contact between the liquid particles and the contacting surface can be minimized by similar electrostatic forces on or near such surfaces and/or by a moving gas layer near the surface, generated by a compressed gas flow. Another means of minimizing contact of the liquid with the surface involves the texture of the contact surface, as, for example, the surface shapes involved in anisothermic heat transfer system or the nonwettability of the surface due to chemical treatment.

The heat-laden coolant discharge can be treated in several ways. First, much of the absorbed heat can be removed by compression, wherein the heat can be used for other applications. Permitting the compressed fluid to expand after heat exchange should further cool the fluid and cause precipitation and/or condensation of liquid particles, which if the aerosol is not to be reused, can be further precipitated by electrostatic precipitators or similar filter devices. The cooled gas, if it is air, might then be discharged into the atmosphere. There are some uses for the discharge. If an electrostatic system is employed, the discharge may consist of high charge density particles, especially if the coolant has traversed a long pathway. Such particles are known to have both military and nonmilitary applications including various acceleration devices. The heat recovery compressor can be used in conjunction with a focusing anisotherm heat exchange surface to generate an energy rich beam of molecules for diverse application including acceleration propulsion, and chemical reactions. Furthermore, the cooling effects of the expanding, spent cooling gasses can serve as a quenching medium in such chemical reactions.

ABSORPTION HEATING

2-(tert-Butoxymethyl)Tetrahydrofuran and Fluorocarbon Mixtures

H.R. Nychka, R.E. Eibeck and C.C. Li; U.S. Patent 4,042,524; August 16, 1977; assigned to Allied Chemical Corporation have developed an absorber pair for absorption heating and refrigeration which has a high coefficient of performance, has a relatively high flash point, operates at approximately atmospheric pressure and has low toxicity. In addition, both pair members can be readily and inexpensively manufactured. The high coefficient of performance is due to a strong affinity between the solute and solvent, good mutual solubility over the whole range of operating conditions, good absorbent volatility and a solute having a high latent heat of vaporization. The solvent in the absorber pair is a special compound, namely 2-(tert-butoxymethyl)tetrahydrofuran, of the formula

The absorption pair comprises from about 4 to 60% of a fluorocarbon solute selected from dichloromonofluoromethane, monochlorodifluoromethane, trifluoromethane and monochloromonofluoromethane which are dissolved in the 2-(tert-butoxymethyl)tetrahydrofuran compound by weight of the compound.

The 2-(tert-butoxymethyl)tetrahydrofuran solvent composition is readily and inexpensively prepared in high yield from tetrahydrofurfuryl alcohol, isobutylene and a catalytic amount of sulfuric acid.

The preferred fluorocarbon for use in the preferred absorption pair due to both high (COP)h and (COP)c is dichloromonofluoromethane. The quantity of fluorocarbon to be used in conjunction with 2-(tert-butoxymethyl)tetrahydrofuran solvent is from about 4 to 60% fluorocarbon by weight of solvent. The preferred quantity of fluorocarbon is from about 10 to 40% by weight of solvent.

The methods of absorption heating or refrigeration comprise absorbing gaseous lower alkyl fluorocarbon solute in 2-(tert-butoxymethyl)tetrahydrofuran solvent to release heat of solution in the vicinity of an area to be heated, or in the case of use in refrigeration, away from the area to be cooled; heating the resulting solution in the vicinity of an area to be heated, or in the case of use in refrigeration, away from the area to be cooled, to release gaseous fluorocarbon from the solvent; condensing the released fluorocarbon to form liquid fluorocarbon in the vicinity of an area to be heated, or in the case of refrigeration away from an area to be cooled; evaporating the liquid fluorocarbon at a location removed from the vicinity of the area to be heated, or in the case of refrigeration in the vicinity of an area to be cooled; and returning the evaporated fluorocarbon for reabsorption into the solvent.

Example 1: Preparation of 2-(tert-Butoxymethyl)Tetrahydrofuran —153 g of tetrahydrofurfuryl alcohol (THFA) and 9.8 g of sulfuric acid are added to a 250 ml flask provided with a dry ice packed condenser, a thermometer and agitator. 32 g of isobutylene are bubbled through the THFA over a period of 50 minutes at 60° to 65°C. The reaction mixture is then allowed to reflux for one hour. The resulting mixture is then analyzed by gas chromatography and is found to contain a 70% theoretical yield of 2-(tert-butoxymethyl)tetrahydrofuran, the balance of the mixture being unreacted tetrahydrofurfuryl alcohol and sulfuric acid with a small percentage of unreacted isobutylene.

Example 2: 40 wt % dichloromonofluoromethane solute in 2-(tert-butoxymethyl) tetrahydrofuran solvent by weight of solvent is introduced into an absorption heating apparatus consisting essentially of a generator, condenser, evaporator, and absorber. The condenser is cooled with water to maintain a temperature of 125°F in the condenser and absorber, and a gas flame is provided under the generator to obtain a generator temperature of 300°F. A throttling valve is provided between the condenser and evaporator which is adjusted to maintain an evaporator temperature of 45°F and a high pressure in the generator and condenser and a lower pressure in the evaporator and absorber. The heat or energy input provided by the gas flame is calculated by determining the volume of gas burned. The heat output is determined by measuring the temperature rise in a known volume of water which is recycled around the condenser and absorber. The (COP)h for the absorption system is calculated to be 1.540 and the (COP)c is calculated to be 0.527, both of which indicated high efficiency.

Dichloromonofluoromethane and Furan Compounds

C.C. Li; U.S. Patent 4,005,584; February 1, 1977; assigned to Allied Chemical Corporation describes a method and apparatus for absorption heating and an absorber pair for absorption heating which has a high coefficient of performance, has good stability, causes little corrosion, has a relatively high flash point, operates at approximately atmospheric pressure and has low toxicity. The high coefficient of performance is due to a strong affinity between the solute and solvent, good mutual solubility at absorber conditions and ease of separation at generator conditions, good absorbent volatility and a solute having a high latent heat of vaporization. The method of absorption heating comprises absorbing a lower alkyl fluorocarbon solute selected from the group consisting of dichloromonofluoromethane, monochlorodifluoromethane, trifluoromethane, and monochloromonofluoromethane in a furan-ring-containing solvent to release heat of solution in the vicinity of an area to be heated, heating the resulting solution to release the lower alkyl fluorocarbon from the solvent, condensing the released lower alkyl fluorocarbon to form liquid lower alkyl fluorocarbon, evaporating the liquid lower alkyl fluorocarbon at a location removed from the vicinity of the area to be heated, and returning the evaporated lower alkyl fluorocarbon to the vicinity of the area to be heated for reabsorption into the solvent. A special furan compound, namely, n-butyltetrahydrofurfuryl ether, finds special utility when used as a component of the absorption pairs of this process.

The preferred compositions for use in conjunction with the process comprise:

> from about 4 to 60 wt % of dichloromonofluoromethane dissolved in n-butyltetrahydrofurfuryl ether based on the weight of n-butyltetrahydrofurfuryl ether;

> from about 4 to 60 wt % of dichloromonofluoromethane dissolved in methyl-2,5-dihydro-2, 5-dimethoxy-2-furan carboxylate based on the weight of methyl-2,5-dihydro-2,5-dimethoxy-2-furan carboxylate; and,

> from about 4 to 60 wt % of a lower alkyl fluorocarbon selected from the group consisting of monochlorodifluoromethane, trifluoromethane and monochloromonofluoromethane dissolved in a furan-ring-containing solvent selected from methyltetrahydrofurfuryl ether, ethyltetrahydrofurfuryl ether, propyltetrahydrofurfuryl ether, n-butyltetrahydrofurfuryl ether and methyl-2,5-dihydro-2,5-dimethoxy-2-furan carboxylate based on the weight of such solvent.

The improved absorption apparatus of the process comprises known prior art absorption heating apparatus components in conjunction with one of the foregoing lower alkyl fluorocarbon solutes in one of the foregoing furan-ring-containing solvents as the absorption pair.

Example 1: An absorption pair, consisting of 40 wt % dichloromonofluoromethane solute in methyltetrahydrofurfuryl ether solvent by weight of solvent is introduced into an absorption heating apparatus consisting essentially of a generator, condenser, evaporator and absorber. The condenser is cooled with water to maintain a temperature of 125°F in the condenser and absorber, and a gas flame is provided under the generator to obtain a generator temperature of 300°F. A throttling valve is provided between the condenser and evaporator which is adjusted to maintain an evaporator temperature of 45°F and a high pressure in the generator and condenser and a low pressure in the evaporator and absorber. In Example 1, the generator and condenser is 60.2 psia and the pressure in the evaporator and absorber is 13.8 psia. The heat or energy input provided by the gas flame is calculated by determining the volume of gas burned and multiplying by the calories provided per unit volume of burned gas. The heat output is determined by measuring the temperature rise in a known volume of water which is recycled around the condenser and absorber. The COP for the absorption system is calculated to be 1.544 indicating a very high efficiency.

Example 2: Example 1 is repeated except ethyltetrahydrofurfuryl ether is substituted for methyltetrahydrofurfuryl ether. The COP is calculated to be 1.540.

Example 3: Example 1 is repeated except butyltetrahydrofurfuryl ether is substituted for methyltetrahydrofurfuryl ether. The COP is calculated to be 1.514.

Example 4: Example 1 is repeated except a saturated solution of ammonia in water at 125°F at absorber pressure is used as the absorption pair. The COP is calculated to be only 1.243 which is low compared to the systems of the process illustrated in Examples 1, 2 and 3.

M.E. Berenbaum, F.E. Evans, R.E. Eibeck and M.A. Robinson; U.S. Patent 4,072,027; February 7, 1978; assigned to Allied Chemical Corporation describe a stabilized heat exchange medium which includes a fluorocarbon, an absorbent and a tri-basic phosphite stabilizer. The preferred absorbents are asymmetrical furan derivatives containing at least 1 oxygen with a single bond to an adjacent carbon, and is most preferably an alkyltetrahydrofurfuryl ether. The fluorocarbon contains 1 or 2 carbons, 1 or 2 hydrogens and the remainder chlorine and fluorine. The tribasic phosphites are of the formula $(R_1'O)(R_2'O)(R_3'O)P$, wherein R_1', R_2' and R_3' are each independently alkyl, alkenyl, phenyl, alkylenephenyl, alkylenealkylphenyl or alkylphenyl.

The tribasic phosphite stabilizers provide stability for prolonged periods at high temperatures for the pairs consisting of fluorocarbons and furan derivatives of the process. It should be appreciated that stabilization under high temperatures for several months would correspond to many years of life in a heat pump system wherein any one aliquot of mixture is exposed to such high temperatures only when the heat pump is operating, and then only during a small portion of the cycle. For example, the heat exchange medium in the form of a rich liquor containing the high percentage of fluorocarbon would be subject to such high temperatures in the generator portion only, and would quickly be cooled by heat exchange with incoming rich liquor during passage out of the generator portion of the heat exchanger.

LOW TEMPERATURE PROCESSES

INTRODUCTION

The refrigerant capacity per volume pumped of a refrigerant is largely a function of boiling point, the lower boiling refrigerants generally offering the greater capacity at a given evaporator temperature. This factor to a great extent influences the design of refrigeration equipment and affects capacity, power requirements, size and cost of the unit.

Another important factor related to boiling point of the refrigerant is minimum cooling temperature desired during the refrigeration cycle, the lower boiling refrigerants being used to achieve the lower refrigeration temperatures. For these reasons, a large number of refrigerants of different boiling temperature and capacity are required to permit flexibility of design and the art is continually faced with the problem of providing new refrigerants as the need arises for new capacities and types of installations.

The lower aliphatic hydrocarbons when substituted by fluorine and chlorine are well-known to have potential as refrigerants. Many of these fluoro-chloro hydrocarbons exhibit certain desired properties including lower toxicity and nonflammability which have resulted in extensive use of such compounds in a large number of refrigeration applications. Trichlorofluoromethane and dichlorodifluoromethane are two of the most commonly available chlorine-fluorine hydrocarbon refrigerants available today. There is a recognized need for refrigerants with boiling point temperatures between the relatively high boiling point temperature of trichlorofluoromethane, 23.78°C at atmospheric pressure, and the relatively low boiling point temperature of dichlorodifluoromethane, -29.8°C at atmospheric pressure, in order to have available refrigerants of good performance in varying capacities.

Several fluoro-chloro hydrocarbons have boiling points in this range but suffer from other deficiencies such as flammability, poor stability or poor thermodynamic performance. Some examples of these types of refrigerants are tetrafluorodichloroethane, fluorodichloromethane, difluorochloroethane and fluorochloromethane.

92

It is also possible to achieve the desired boiling point by mixing two refrigerants with boiling points above and below the desired one. In this case, e.g., mixtures of trichlorofluoromethane and dichlorodifluoromethane could be used.

REFRIGERANT PROCESSES

Dual Flash Economizer Refrigeration System

A process described by *C.M. Anderson and L.H. Leonard; U.S. Patent 4,144,717; March 20, 1979; assigned to Carrier Corporation* involves the provision of a flash economizer within a single stage vapor compression refrigeration system. Therein the condenser is connected to a compressor, the condenser condensing the gaseous refrigerant received from the compressor to a liquid refrigerant. A flash economizer receives liquid refrigerant from the condenser and flashes that refrigerant such that part of the refrigerant changes state to a gas absorbing heat from the remaining liquid refrigerant. The liquid refrigerant then travels to the evaporator where it changes state from a liquid to a gas absorbing heat from the fluid to be cooled.

The gaseous refrigerant from the evaporator is then transported to the compressor where it is recompressed to start the cycle again. The flashed gas from the step of flash economizing is recompressed in a second compressor. The recompressed gas is then condensed in an economizer-condenser to the liquid state. The liquid refrigerant is flashed through an orifice into the flash economizer from which the liquid refrigerant is allowed to travel to the evaporator and the gaseous refrigerant is again conducted to the second compressor.

Referring to Figure 3.1a, a schematic drawing of a vapor compression refrigeration system, it can be seen that a dual channel compressor 10 is provided having two separate centrifugal compressors 11 and 17 located on a single axis driven by an electric motor 33. A primary compressor 11 has increased temperature and pressure refrigerant gas exiting therefrom at outlet 14 into line 20. From line 20, the gaseous refrigerant enters condenser 22 wherein it changes state to a liquid refrigerant. Liquid refrigerant is collected in the bottom of the condenser and then transported through line 24 to flash economizer 28.

In the flash economizer liquid refrigerant is flashed through nozzles 26 such that part of the refrigerant changes state to a gas, absorbing heat from the remaining liquid refrigerant. Liquid refrigerant collects at the bottom of the flash economizer shown as reservoir 30. Therefrom via line 32 liquid refrigerant passes to expansion control device 34 wherein the pressure of the liquid refrigerant is dropped. From the expansion control device the liquid refrigerant travels to chiller 36 wherein the liquid changes state to a gas, absorbing heat from the fluid to be cooled as it passes through the chiller. Line 40 then conducts the gaseous refrigerant from the chiller to inlet 12, inlet to compressor 11 wherein the gaseous refrigerant is recompressed to begin the refrigeration cycle again.

Within chiller 36 is located a coil 38 through which refrigerant flows. Water or other fluid to be cooled enters the chiller through line 64 and then typically floods over the coils in heat exchange relationship therewith. The now cooled water exists through the line 66 to the enclosure to be cooled. Connected to the flash economizer is line 50 which conducts the gaseous refrigerant to inlet 16 of the compressor 17.

Figure 3.1: Refrigeration System

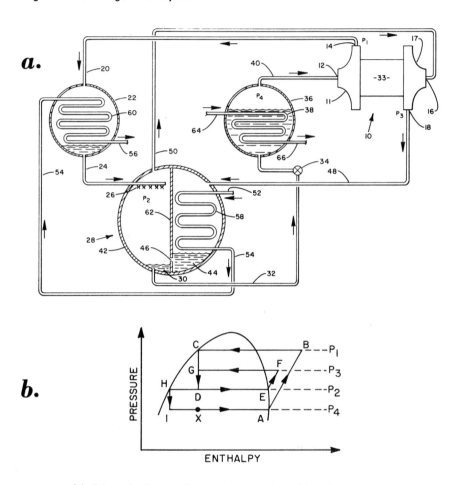

(a) Schematic diagram of a vapor compression refrigeration system
(b) Pressure-enthalpy graph showing the refrigeration cycle

Source: U.S. Patent 4,144,717

Both compressor **11** and compressor **17** are driven by electric motor **33**. Therein compressor **17** increases the temperature and pressure of the flashed refrigerant gas and delivers the recompressed gas to outlet **18**. This recompressed gas travels through line **48** to economizer-condenser **42** wherein the recompressed gas is recondensed into a liquid. The liquid is collected in reservoir **44** such that it may be flashed from the economizer-condenser to the flash economizer through orifice **46**.

The flashed refrigerant from the orifice travels upward and is conducted through line **50** back to the second compressor. The liquid refrigerant from the orifice is collected in reservoir **30** and it travels to the chiller **36**. Entering condensing water travels through line **52** through coils **58** of economizer-condenser **42** through lines **54** through condenser **22** and through coil **60** to exiting condenser water line **56**. The condensing water picks up heat in the economizer-condenser and then picks up additional heat in the main condenser **22**.

Compressor **11** increases the pressure of the gaseous refrigerant to P_1. Thereafter the pressure of the refrigerant is decreased in the flash economizer to P_2. The second compressor increases the pressure of the flashed gaseous refrigerant from P_2 to P_3, the economizer-condensing refrigerant at the P_3 pressure. From the reservoir of liquid refrigerant **44**, liquid refrigerant at pressure P_3 is then flashed to the lower pressure P_2 through orifice **46**. Expansion control device **34** allows the pressure to drop from P_2 to P_4 for cycling through the chiller **36**. Refrigerant enters inlet **12** at P_4 and is thereafter increased to P_1 by compressor **11**.

Figure 3.1b is a graph of pressure vs enthalpy for a typical refrigerant such as R-11 which is used within this system. Starting at Point **A** it can be seen that the pressure and enthalpy of the refrigerant is increased from Point **A** to Point **B**, the change in pressure and enthalpy due to compressor **11**. From Point **B** to Point **C** represents the change in enthalpy in condenser **22** as the gaseous refrigerant changes state to a liquid refrigerant. Thereafter in the flash economizer the refrigerant travels from Point **C** to Point **D**, representing the pressure decrease as the refrigerant is flashed.

From Point **D** the liquid refrigerant is cooled to Point **H** and the gaseous refrigerant travels to Point **E** absorbing heat from the now cooled liquid refrigerant. The distance from Point **E** to Point **F** represents the increase in enthalpy and pressure as the gaseous refrigerant is compressed in the second compressor. The distance from Point **F** to Point **G** represents the recondensing of the recompressed refrigerant in the economizer-condenser. The distance from Point **G** to Point **D** represents the decrease in pressure as the liquid refrigerant is flashed through orifice **46** from the economizer-condenser to the flash economizer.

The distance from Point **H** to Point **I** represents the pressure drop through expansion control device **34** and the distance from Point **I** back to the original Point **A** represents the change in enthalpy that occurs in the chiller when heat is absorbed from the liquid to be cooled. As can be seen in Figure 3.1a and Figure 3.1b, P_1, P_2, P_3 and P_4 are indicated on both showing the respective pressure relationships.

In a pressure-enthalpy diagram the left portion of the curve indicates the pressure-enthalpy line at which the liquid refrigerant is 100% saturated and the right side of the curve indicates pressure-enthalpy line when gaseous refrigerant is 100% saturated. The area between the two lines indicates a two phase mixture of liquid and vapor.

In order to obtain the most cooling work from a given amount of refrigerant it is desirable to cool the refrigerant as close as possible to the left side of the curve such that when the refrigerant is flashed in the chiller as much heat as possible, proportional to the distance from **I** to **A**, is absorbed from the refrigerant to be cooled. Without the flash economizer, it is obvious that the heat available to be absorbed by the refrigerant is proportional to that distance represented in the graph from **X** to **A**, **X** being that point to which the refrigerant would travel from point **C** if the pressure were dropped to P_4 in one step. By the provision of the flash economizer the refrigerant is cooled to Point **H** allowing the heat to be absorbed from the refrigerant to be cooled to be increased to the distance indicated by the line from **I** to **A**. This increase in the length from distance **XA** to distance **IA** represents an overall efficiency increase in the amount of heat that may be absorbed in the refrigeration system.

For optimization of this dual economized refrigeration system, the entering condensing water is circulated first through the economizer-condenser and then through the main condenser **22**. The economizer-condenser operates at a temperature considerably lower than the main condenser and consequently the cooling water is advantageously used by circulating first through the economizer-condenser and then through the main condenser. Of course, additional condensing water may be supplied to the main condenser to meet the load thereon.

Heat Transfer Surface for Nucleate Boiling

A process described by *C.E. Albertson; U.S. Patent 4,018,264; April 19, 1977; assigned to Borg-Warner Corporation* involves nucleate boiling or ebullition in pool boiling applications by the use of a heat transfer surface having dendrites or nodules electroplated onto the substrate. The nodules are formed by plating at high current densities, and may be further electroplated at lower current densities to strengthen and enlarge them.

Example 1: A ¾" copper tube having a wall thickness of ³/₁₆" was sanded, cleaned by etching 15 seconds in 50% HNO_3 at room temperature, rinsed, and then immersed in a sulfuric acid solution of a proprietary copper plating composition known as Cubath No. 2. This composition is believed to contain a copper salt, such as copper sulfate and additives such as stabilizers and brighteners. The tube was electrically connected to a source of direct current such that it functioned as the cathode; and an annular, consumable copper anode was placed around the tube so that it was uniformly spaced from the surface of the tube. A current density of 1,000 A/ft² was applied for about 20 seconds with gentle solution agitation. The current density was then reduced to 50 A/ft² and plating continued for 1½ to 2 hours to coat the nodules with a strong, dense layer of copper.

Following the electrodeposition of the final layer of copper, the boiling heat transfer was further enhanced by rolling the tube between three rolls of a sheet metal bending machine to partially compact the nodules to closer proximity to one another and to strengthen them by work hardening and mechanical interlocking.

The tube was tested in a heat transfer test cell with refrigerant R-12 at 37 psig. Figure 3.2 represents a plot of heat flux density (Btu/hr-ft²) vs the temperature differential between the refrigerant and the tube wall. The nodularized tube represented by plot **A** was clearly superior to the heat transfer efficiency of a standard finned tube (¾" o.d., 26 fins/linear inch). The latter is shown in plot **B** in Figure 3.2. Some temperature differential hyteresis was observed in generating the data shown in plot **A**, so the curve represents an average of the temperature differential values as the heat flux density was increased and then decreased.

Figure 3.2: Performance Data For Nucleate Boiling Heat Transfer Surface

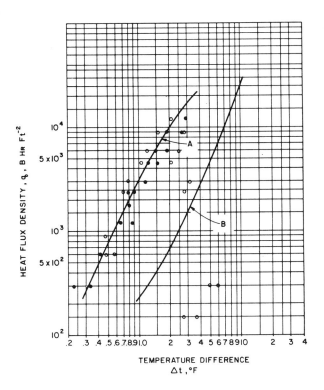

Source: U.S. Patent 4,018,264

Example 2: Instead of the concentric anode described in Example 1, the tubes may be rotated while adjacent one or more flat plate anodes of a more standard (and economical) design.

A ¾ " o.d. copper tube with an overall length of 8" was mounted on a device which slowly rotated it in the bath while being plated. Electrical contact was made to the tube by a copper plate bolted to one leg of a Teflon support structure. This copper plate had a cylindrical center section which extended halfway through the leg of the mount. This section butted the copper tube which rotated against it continuously making electrical contact. The sample was rotated at 11 rpm by a low speed motor bolted to the top of the Teflon mount. An O-ring transferred power between pulley wheels.

The lower wheel was attached to a Teflon axle the other end of which was shaped to fit snugly into the copper tube. A pin could be put through a small hole in the end of the copper tube and into the Teflon to insure that no slippage occurred. Electrical contact at the other end of the tube was insured by a spring.

Two 5" x 11" phosphorized copper anodes ¼" thick were placed in the electro-
lyte and arranged vertically, spaced about 4½" apart. A Clinton Plater (Model
109 CP) with a power supply capable of 0 to 100 A and 0 to 15 V was used.
A simple acid copper plating bath was used, containing 52.2 g/l sulfuric acid
and 210 g/l $CuSO_4;5H_2O$. Plating was initiated by supplying 100 A (750 A/ft^2)
for one minute. Power was then reduced to a level of 5 A (38 A/ft^2) and plating continued for 1 hour. The plated tube showed good dendrite formation,
especially near the ends of the tubes.

Suction Line and Capillary Tube Assembly

A process described by *R.B. Gelbard and R.M. Schreck; U.S. Patent 4,147,037;
April 3, 1979; assigned to General Electric Company* relates generally to heat
exchangers and, more particularly, to a refrigeration system suction line and
capillary tube assembly providing efficient heat exchange between the cool
gaseous refrigerant conveyed by the suction line and the warm liquid refrigerant conveyed by the capillary tube.

Referring to Figure 3.3a, a closed-circuit refrigeration system **10** includes a compressor **12**, a condenser **14** for cooling and condensing hot compressed gaseous
refrigerant received from the compressor to warm liquid refrigerant, and an
evaporator **16** within which liquid refrigerant vaporizes to produce cooling.

Figure 3.3: Heat Exchange for Refrigeration System

a.

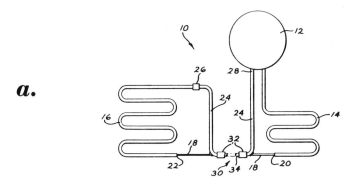

b.

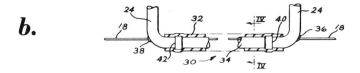

(continued)

Figure 3.3: (continued)

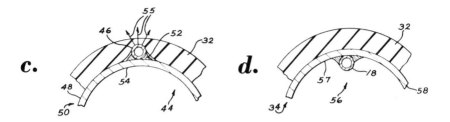

c.   **d.**

(a)   Schematic representation of a refrigeration system
(b)   Enlarged view of a portion of the refrigeration system of Figure
      3.3a, showing the suction line and capillary tube assembly
(c)   Greatly enlarged cross-sectional view of a prior art suction line
      and capillary tube assembly wherein the capillary tube is
      soldered to the outside of the suction line
(d)   Greatly enlarged partial section taken along line IV-IV of Figure
      3.3b, showing an assembly according to this process wherein the
      capillary tube is soldered to the inside surface of the suction line

Source: U.S. Patent 4,147,037

In the refrigeration system **10**, a flow-restricting copper capillary tube **18** conveys the warm liquid refrigerant from the outlet **20** of the condenser **14** to the inlet **22** of the evaporator **16**. Additionally, a suction line **24** conducts cool gaseous refrigerant from the outlet **26** of the evaporator back to the inlet **28** of the compressor **12**.

In order to provide heat exchange between the warm liquid refrigerant conveyed by the capillary tube and the cool gaseous refrigerant conveyed by the suction line, thereby to improve the efficiency of the refrigeration system, a suction line and capillary tube assembly **30** is formed by bringing the suction line and capillary tube together, in a manner hereinafter described in detail, for a portion of their respective lengths.

As is known to those skilled in the art of refrigeration, when applied to a household refrigerator, the suction line and capillary tube assembly is usually four to six feet in length and extends between the compressor and condenser, typically physically located together in a compartment in the uninsulated portion of the refrigerator, and the evaporator which is physically located in the insulated portion of the refrigerator.

For convenience of illustration, most of the relatively long midportion of the suction line and capillary tube assembly is omitted, as indicated by the broken lines. The major portions of both the capillary tube and the suction line are included within the suction line and capillary tube assembly, the remaining portions of the capillary tube and the suction line shown outside the suction line and capillary tube assembly being relatively smaller portions.

The suction line and capillary tube assembly is usually four to six feet in length.

This length is typically that which is necessary to physically reach between the compressor 12 and condenser 14 at one end, and the evaporator 16 at the other end. Additionally, this length provides sufficient beneficial heat exchange for improved thermodynamic efficiency. For shorter suction line and capillary tube assembly lengths, less heat exchange occurs. Eventually, a lower limit is reached beyond which, as a practical matter, the cost of providing a heat exchange structure is not worth the minimal efficiency advantage gained. Approximately two feet is considered to be the lower practical limit. Accordingly, the heat exchange assemblies of the process are at least two feet in length.

Depending upon the particular cabinet construction, the suction line and capillary tube assembly 30 may pass through the wall insulation space between the inner liner and outer case of the refrigerator cabinet. The space refrigerated by the evaporator is of course defined by the inner liner. Portions of the suction line and capillary tube assembly, particularly where it passes completely outside of the refrigerator cabinet, may be covered by a layer of thermal insulation material 32.

Referring to Figure 3.3b, an enlarged view of the suction line and capillary tube assembly 30 is shown. Figure 3.3b is intended to show only general details of construction of the suction line and capillary tube assembly without showing the specific details for promoting heat exchange between the capillary tube 18 and the gaseous refrigerant carried by the suction line 24.

From Figure 3.3b, it will be apparent that the capillary tube lies generally within a midsection 34 of the suction line. To permit the capillary tube to pass from outside to the interior of the suction line midsection and back outside again, the capillary tube is simply inserted at connections 36 and 38 through suitable apertures in the suction line. Either conventional silver soldering or brazing is employed to seal the connections. It will be understood that the connections are exemplary only, as numerous devices are known in the art for passing a capillary tube from the exterior to the interior of a capillary tube. As an example of a particular variation, the connection at 38 may be associated with the entry portion of a well-known single-entry evaporator.

To make the connections between the capillary tube and the suction line requires that the portion of the suction line at which the connections are made be copper or steel so that a metallurgical bond by silver soldering or brazing can be accomplished. To permit the use of a low-cost aluminum suction line over most of the length of the suction line and capillary tube assembly, the midsection may be formed of aluminum and suitably adhesively bonded at connections 40 and 42 to copper extensions of the suction line leading to the compressor inlet and the evaporator outlet, respectively.

Although other connections are possible, the adhesive type connections 40 and 42 provide a low-cost, high integrity method of connecting dissimilar metals that cannot be brazed. Further, the connections 40 and 42 are sufficiently localized that they may be reliably protected from atmospheric moisture by a heat shrinkable plastic sleeve or a plastic coating, if needed, in the specific application.

Referring to Figure 3.3c, there is illustrated a greatly enlarged fragmented cross-sectional view of a prior art suction line and capillary tube assembly 44 wherein a copper capillary tube **46**, similar to the capillary tube **18** of Figure 3.3a and Figure 3.3b, is longitudinally soldered to the exterior surface **48** of a copper suction line **50**. The solder is designated **52**. The thermal insulation material **32** surrounds appropriate portions of the assembly **44**. Due to the direct bond between the capillary tube **46** and the suction line, and resultant good thermal contact therebetween, the entire suction line, particularly the interior surface **54**, provides efficient heat exchange with the cold gaseous refrigerant passing through the suction line. However, as represented by arrows **55**, not all the heat of the capillary tube **46** flows to the cool gaseous refrigerant. Some of the heat escapes through the insulation into the surrounding ambient.

This heat leakage, since the ambient is generally warmer than the cool suction line gas, results in the capillary tube **46** not being cooled as much as it might be. Additionally, in those refrigerator cabinets where the suction line and capillary tube assembly passes in part through the insulation space between the inner refrigerator liner and the outer cabinet, some of the heat loss represented by the arrows may flow through the inner liner into the refrigerated food storage space, thus placing an additional unnecessary heat load on the refrigeration system.

Referring to Figure 3.3d, a sectional view taken along line IV–IV of Figure 3.3b illustrates a suction line and capillary tube assembly **56** contemplated by a first specific embodiment of the process. In Figure 3.3d, the midsection **34** of the suction line **24** is formed of copper and the capillary tube **18** is soldered to the interior surface **57** of the suction line wall **58** along substantially the entire length of the suction line and capillary tube assembly **30**. In operation, the assembly provides good heat exchange between the warm capillary tube and the gaseous refrigerant carried by the suction line due to effective employment of the entire interior surface of the suction line midsection. Additionally, heat loss from the capillary tube directly to the ambient, unlike in the prior art Figure 3.3c assembly, is eliminated. This results in increased efficiency due to elimination of the two effects mentioned above with reference to Figure 3.3c.

Cryogenic Plants

According to a process described by *W.M. Small; U.S. Patent 4,050,506; September 27, 1977; assigned to Phillips Petroleum Company* in a cryogenic plant utilizing a liquid sparged into a cold vapor in heat exchange relationship with hotter vapor, variations in the load in terms of the volume of hot vapor are compensated for by a stepwise complete closing of a uniformly-spaced-apart fraction of the cold vapor passageways.

The apparatus of this process is applicable in any cryogenic heat exchange operation and is of particular utility in connection with a liquefied natural gas plant. Referring to Figure 3.4a, there is shown an elongated heat exchanger **2** having a core comprised of a stack of elongated longitudinally extending platelike passages. As can be seen from Figure 3.4b, there are alternately spaced first and second fluid passages **4** and **6** respectively, only a single passage **4** being directly visible in Figure 3.4a, passage **6** being shown by cutout. Each passage is formed by interposing fin material **44** (see Figure 3.4d) between two spaced metallic plates **16** (Figure 3.4b or Figure 3.4c). The formation of such passages is known to those skilled in the art. The composite of these platelike passages may then be brazed together as an integral unit.

Figure 3.4: Cryogenic Plant

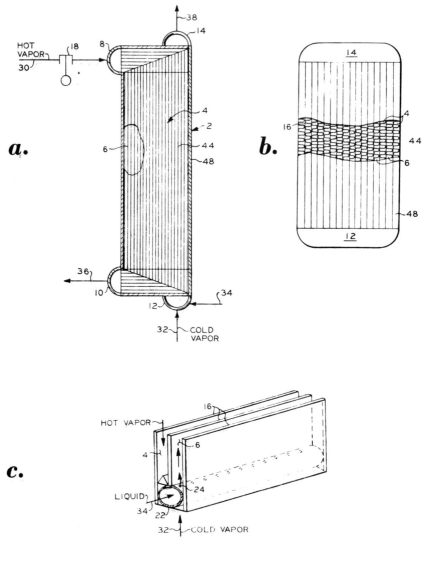

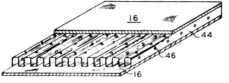

(continued)

Figure 3.4: (continued)

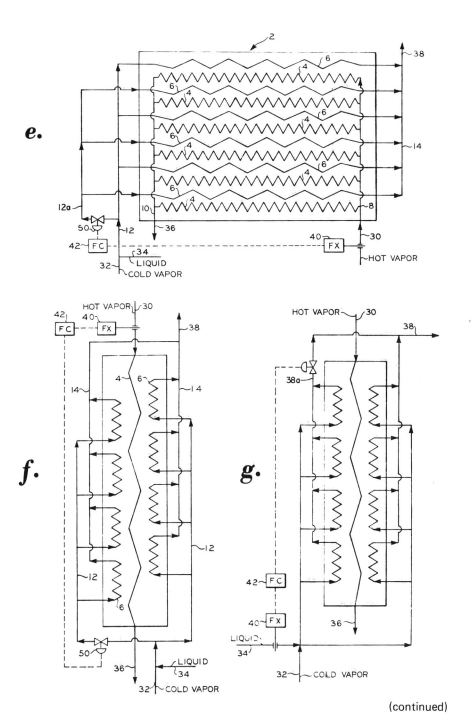

(continued)

Figure 3.4: (continued)

h.

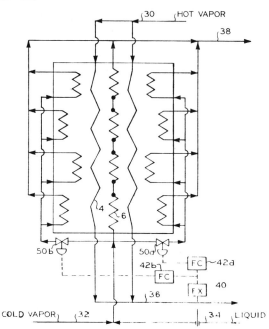

(a) Side elevation with the side plate removed showing one vertical
passageway and header system in diagrammatic form
(b) View taken at right angles to that of Figure 3.4a with a portion of
the end bars cut away so as to show alternate passageways **4** and **6**
(c) Detailed perspective view of one sparger means suitable for introducing
liquid into the vapor
(d) Perspective view of a portion of perforated fin material used in the
heat exchanger platelike passages
(e) Schematic representation of a heat exchanger having a turndown
ratio of ½
(f) Schematic representation similar to Figure 3.4e utilizing another
conventional shorthand (schematic) method of depicting the
passageways
(g) View similar to Figure 3.4f showing the flow control valve placed
downstream of the heat exchanger.
(h) View similar to Figure 3.4f and Figure 3.4g showing a system capable
of a turndown ratio in increments of ⅓

Source: U.S. Patent 4,050,506

Communicating with first fluid passages **4** are first inlet and outlet headers **8** and
10 respectively. Communicating with second fluid passages **6** are second inlet and
outlet headers **12** and **14**, respectively. Hot vapor carried by a line **30** is conveyed by means of gas compressor **18** to first inlet header and thence through
first fluid passageways **4**, collected by first outlet header **10** and removed by
hot vapor outlet line **36** (i.e., means to remove the thus cooled fluid). Cold
vapor is conveyed by line **32** to second inlet header **12** and thence through

second fluid passages 6, collected by second outlet header 14, and removed via refrigerant outlet line 38. Liquid from liquid refrigerant inlet line 34 is sparged into the vapor at the entrance to the second fluid passages by means such as that shown in detail in Figure 3.4c. Other conventional sparging means can also be used. Once distributed within a passage, the heat exchange fluid moves longitudinally through and around longitudinally extending perforated fin material 44 (see Figure 3.4d) to the opposite end of the passage. The fin material can be formed from corrugated sheet or otherwise fabricated metal such as solid or perforated aluminum as with apertures 46 which are shown in detail in Figure 3.4d. These apertures may comprise from 10% to 20% of the total sheet metal surface. The fluid material is confined within the heat passageway by end bars 48.

Referring specifically to Figure 3.4c, there is shown a preferred means for sparging the liquid into the cold vapor comprising conduit 22 positioned in each of the second fluid passages 6 communicating with second inlet header 12 and lines 32 and 34 as shown in Figure 3.4a. Each of the conduits has a plurality of fluid exit openings 24 formed along its length and opening into the respective fluid passageways for passing and distributing a liquid from the conduit into the second fluid passages.

Referring to Figure 3.4e, there is shown a heat exchanger 2 in more diagrammatic form employing the turndown control of this process. As can be seen, hot vapor enters via line 30 and is distributed by means of header 8 to alternate passageways 4. Header 10 collects the resultant fluid and passes same from the heat exchanger via line 36. Cold vapor enters via line 32 and liquid via line 34 and the two phase stream is distributed by header 12 to the other alternate passageways 6, the warmed refrigerant exiting by header 14 and line 38, flow controller 42 operates valve 50 one half of header 12 in response to a measured rate of flow of the hot vapor.

This operation is carried out by linear flow transmitter 40 which transmits a signal representative of the flow rate in conduit 30. When the rate of flow is reduced to a preset level or lower, flow controller 42 closes valve 50. This shuts off line 12a to one-half of the second inlet header system 12, thus shutting off the cold vapor and liquid being sparged into alternate second passageways 6. As can be seen, every other passageway is shut off, this being every fourth passageway since half of the passageways are hot vapor passageways. It is essential that the fraction of the refrigerant passageways shut off be substantially uniformly spaced apart. In this way the entire heat exchanger core is always used to its maximum efficiency. This avoids an entire section being unused since even the passageways shut off still conduct heat as a result of their proximity to the active passageways.

Thus, the sequence is a hot vapor passageway 4, a closed off cold vapor passageway 6, a hot vapor passageway 4, and an open cold vapor passageway 6, etc. Similarly, a turn down ratio in increments of successive one-thirds can be effected by shutting off every third refrigerant passageway and a turn down ratio in increments of successive one-fourths can be achieved by completely shutting off every fourth refrigerant passageway.

Referring to Figure 3.4f, there is shown a schematic representation of a heat exchanger identical to that of Figure 3.4e. Figure 3.4e is duplicated in order to introduce another shorthand form for depicting the hot vapor passageways **4** and the alternating, parallel, countercurrent cold vapor passageways **6**. As is apparent, this does not represent the actual spaced relationship of the passageways but is a conventional shorthand form depicting passageways in a complex multipass heat exchanger. For instance, passageway **4** depicted in Figure 3.4f is actually a plurality of passageways having passageways **6** sandwiched therebetween.

Figure 3.4g is a schematic representation of a heat exchanger similar to Figure 3.4f, except that the flow transmitter measures the flow rate of cold liquid and in response to a decreased flow, the controller shuts off a refrigerant exit line **38a** rather than an inlet line, thereby both the measurement and control loci are different from Figure 3.4e and Figure 3.4f.

In order to have a more accurate measurement, it is essential to measure a reliably single phase flow rate. Accordingly, the preferred measured stream is that as shown in Figure 3.4e and Figure 3.4f, where the flow rate of hot vapor is measured. However, in certain plants such as liquefied natural gas plants where the liquid is produced by compressing and cooling hot vapor, the flow rate of liquid is proportional to or at least correlatable with the incoming flow rate of hot vapor.

Figure 3.4h is similar to Figure 3.4g except that there are provided two flow controllers **42a** and **42b** operating valves **50a** and **50b**, respectively, so as to shut off completely uniformly-spaced-apart refrigerant feed lines in increments of $\frac{1}{3}$ of the total number in the heat exchanger. Also, in this case, the refrigerant liquid flow rate coming into the heat exchanger is utilized to regulate when the stepdown is to occur. Thus, when the flow rate of liquid is reduced to a preset level of less than $\frac{2}{3}$ but more than $\frac{1}{3}$ of the normally expected flow rate, flow controller **42a**, in response to the comparison of its signal from linear flow transmitter **40** with the $\frac{2}{3}$ flow rate set point thereto, shuts off valve **50a**, thus completely shutting off the uniformly-spaced-apart $\frac{1}{3}$ fraction of the refrigerant lines.

When the liquid flow rate is further reduced to a second preset level (less than $\frac{1}{3}$ normal flow), flow controller **42b** shuts off valve **50b**, thus shutting off a second bank of refrigerant lines leaving only a single uniformly-spaced-apart $\frac{1}{3}$ fraction of the refrigerant lines in operation.

Raw Natural Gas Processing

J.E. Schneider; U.S. Patent 4,042,011; August 16, 1977; assigned to Phillips Petroleum Company describes a process and apparatus for controlling the flow of a refrigerant in countercurrent heat exchange with a stream being cooled.

Although the process is applicable broadly, it finds particular utility in processing raw natural gas. In the gathering of such raw natural gas it is frequently expedient to process relatively small volumes, for example, in the neighborhood of 1 to 25 million standard cubic feet per day (MMscfd) which is about 0.03–0.80 m³/s in the field. Such processing may often include steps of phase separation, compression, and heat exchange.

According to the process, a first fluid stream is cooled by passage through two heat exchange zones in sequence in countercurrent flow with a second fluid stream, a part of the second fluid stream is bypassed around the second heat exchange zone and passed through the first heat exchange zone and the relative amounts of the second fluid stream passing through the second heat exchange zone and being bypassed around the second heat exchange zone are controlled in response to the temperature of the first stream between the heat exchange zones.

The part of the second fluid stream which passes through the second heat exchange zone can be recovered separately or can be used to provide additional refrigeration by passage through all or a part of the first heat exchange zone. This can be accomplished by passage through separate heat exchange passageways or the second part of the second fluid stream can be combined with the bypassed portion either before or after the bypassed portion passes through the first heat exchange zone or combined at an intermediate point in the first heat exchange zone.

As noted, the process finds particular utility in processing raw natural gas and, in a specific example, provides the steps of compression and air cooling of the incoming raw gas, further cooling by gas-gas heat exchange to a temperature safely above the hydrate temperature, phase separation to remove any condensed water and hydrocarbon, and injection of a hydrate inhibitor into the gas before or as it enters a second gas-gas exchanger where the incoming gas is further cooled. Additional hydrate inhibitor can be injected into the gas after it issues from the second gas-gas heat exchanger and the gas passed through a refrigerated third exchanger where the incoming gas is further cooled for condensation of C_{2+} hydrocarbons and/or water. Gas and condensates from the refrigerated exchanger pass through a hydrate inhibitor separator from which liquid hydrocarbons pass to further processing and water-rich hydrate inhibitor passes to regeneration facilities.

Cold residue gas from the hydrate inhibitor separator flows via a three-way control valve to the two gas-gas heat exchangers and thence to a pipeline or other use. The three-way control valve provides for a bypass around the second gas-gas heat exchanger for use in the first gas-gas heat exchanger. As indicated above, the gas which flows through the second heat exchanger can be used to provide additional refrigeration by passage through all or a part of separate passageways in the first heat exchange zone or can be combined with the bypass gas either before or after it enters the first exchanger or at an intermediate point.

By controlling the three-way valve in response to the temperature of the incoming gas between the two exchangers, it is possible to control the system in such a way that the maximum amount of water is condensed from the raw gas in the first gas-gas exchanger and removed from the system via the subsequent phase separator before the injection of hydrate inhibitor for subsequent cooling to lower the gas below its hydrate temperature. In this manner the use of hydrate inhibitor is minimized thus reducing the load on the hydrate inhibitor regeneration unit. In addition, full utilization of the cool residue gas is always obtained since all of the gas flows through one or both of the two exchangers at all times.

Dual Thermal Storage Vessels and Refrigeration System

J.E. Randall; U.S. Patent 4,037,650; July 26, 1977; assigned to National Research Development Corporation, England describes a thermal storage apparatus comprising two thermal storage vessels, each thermal storage vessel containing a thermal storage medium, one medium being capable of being maintained at a temperature in excess of a preselected temperature, and the other medium being capable of being maintained at a temperature below the preselected temperature. A refrigeration system is provided for transferring heat from the storage medium at the lower temperature to the storage medium at the higher temperature.

By providing a refrigeration system, heat which is absorbed in the first or cold storage vessel when a room is being cooled by the apparatus can be transferred to the second or hot storage vessel for use at a convenient time for heating the room. By using both a hot storage vessel and a cold storage vessel and a refrigeration system, the refrigeration system can be operated continuously, even though heating and cooling are required only intermittently.

In Figure 3.5a, the thermal storage apparatus comprises a first thermal storage vessel **1** within which are located containers **8b**. The containers contain a freezable liquid which may be water or a suitable crystallizable salt hydrate, or mixture of hydrates, which crystallizes at a suitably low temperature or temperatures. For the sake of clarity, it will be assumed that the containers contain water and are sealed.

The storage vessel also has an air inlet **9** and an air outlet **10**. As shown, both the inlet and the outlet are at the upper end of the storage vessel. Between the inlet and the outlet, a baffle **11** is provided. This baffle may be advantageously integrally formed with a cover **12** for the storage vessel. Within the storage vessel, spaces are provided between the containers to allow air to pass therearound. A fan **13** is provided to draw air through the storage vessel, the baffle preventing air from flowing directly from the inlet to the outlet without passing over the containers.

Also located within the storage vessel is an evaporator **3** forming part of a refrigeration circuit. A refrigerant fluid flows through the evaporator. The evaporator is in series with a compressor **5** and heat exchanger **6**. This heat exchanger is located within a second storage vessel **14**. The second storage vessel **14** is, in essence, extremely similar to the storage vessel **1**. The containers **8a** within the storage vessel **14** contain a crystallizable salt hydrate. Alternatively, these containers **8a** may be filled with a crystallizable salt hydrate plus water.

As with the storage vessel **1**, a fan **15** is provided to draw air through the storage vessel **14**. In the storage vessel **14**, it will be noticed that the air inlet **16** and the air outlet **17** are both located at the bottom of the vessel. A baffle **18**, which may be integrally formed with a base plate **19** is provided between the inlet and the outlet. The purpose of this baffle is to prevent air passing directly from the inlet to the outlet without passing over the containers **8a**.

The operation of the illustrated apparatus will now be described. The storage vessel **1** is maintained at below the desired room temperature and so may be regarded as a cold store while the storage vessel **14** is maintained at above the desired room temperature and so may be regarded as a hot store.

(continued)

Figure 3.5: Thermal Storage Apparatus

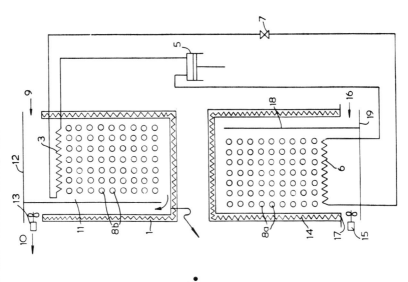

a.

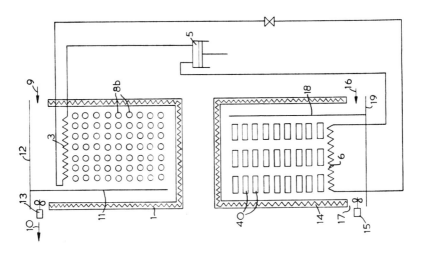

b.

Figure 3.5: (continued)

c.

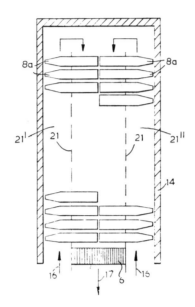

(a)(b) Schematic views of a thermal storage apparatus
(c) Schematic longitudinal section through a storage vessel
forming part of a thermal storage apparatus

Source: U.S. Patent 4,037,650

Assuming for the moment that the compressor **5** is operating, the refrigerant
fluid flows through the evaporator **3** in the storage vessel **1**. In so doing, the
refrigerant fluid extracts heat from the freezable liquid contained in the containers **8b** located in the storage vessel. This causes the freezable liquid to freeze
and the refrigerant fluid to be heated. The heated refrigerant fluid passes through
the compressor and then through the heat exchanger **6** located in the storage
vessel **14**.

Passage of the refrigerant through the heat exchanger causes the crystallizable
salt hydrates contained in the containers **8a** in the storage vessel **14** to become
heated. The crystallizable hydrates mentioned hereinbefore all have one common characteristic. This characteristic is that they lose some of their water
of crystallization when heated, and, on cooling, they recrystallize with the evolution of considerable quantities of heat. Thus, when heated by the refrigerant
fluid, the crystallizable salt hydrates will lose at least some of their water of
crystallization.

A pressure reduction device **7** is located, in the refrigerant circuit, between the
external heat exchanger **6** and the evaporator. The compressor may be run in
dependence upon load conditions. Thus, it can be run when electricity is cheap,
for example, at night, and then be switched off. Alternatively, it may be run

continuously. For the sake of example, it will be assumed that the compressor **5** has been run until substantially all of the freezable liquid contained in the containers **8b** within the storage vessel **1** have frozen and all of the crystallizable salt hydrate contained in the container **8a** located in the storage vessel **14** have been heated until they are at a temperature which is above their crystallization temperature.

The thermal storage apparatus of the process can now be used for supplying either heated air or cooled air to a room. Suppose, for example, that it is desired to supply heat to the room. This is done by switching on the fan **15** in the storage vessel **14** to draw air over the heated containers **8a** and to supply it to the room. The air flow over the containers **8a** means that the temperature of the contents of these containers **8a** will commence to fall. At a certain temperature, the salt hydrates will begin to recrystallize. As previously mentioned, when such recrystallization occurs, relatively large quantities of heat are evolved. The temperature of the containers **8a** will remain constant, or fall slowly, until substantially all of the hydrate has recrystallized. Accordingly, heat may be supplied to the room for a relatively lengthy period of time.

If on the other hand, it is desired to cool the room, then the fan **13** in the storage vessel **1** is switched on instead of switching on fan **15**. The fan **13** causes air to flow over the containers **8b** within the storage vessel **1**. In so doing, the air will be cooled, and the fan further directs air to the room to be cooled.

The heat given up by the air in passing over the containers **8b** will be absorbed by the freezable liquid. In due course, the freezable liquid will commence to melt. If the freezable liquid is water, the latent heat of fusion of ice is absorbed by the water thus contributing to the cooling of the air passing thereover and/or enabling a larger quantity of air to be cooled to the desired temperature than before. The large quantity of air to be cooled may, for example, be extracted from a room which has been heated by the sun. The ice may also be melted by use of an additional heating element.

The arrangement shown in Figure 3.5b is virtually identical to that shown in Figure 3.5a. The sole difference is that the hot store **14** contains not a crystallizable salt hydrate but refractory bricks **40**. These bricks are stacked one upon the other in rows. It will be readily apparent that while refractory bricks have a high thermal storage capacity, only the heat stored in the bricks can be extracted. Unlike the crystallizable salt hydrate, it is obviously not possible to obtain any latent heat of fusion or heat of crystallization from these bricks. Nevertheless, under certain circumstances the use of refractory bricks may be more advantageous than utilizing a crystallizable salt hydrate. Thus, if the apparatus is not required to have an extremely high heat storage density, it is easier to manufacture an apparatus using bricks rather than containers of a crystallizable salt hydrate. Moreover, refractory bricks may be heated to a higher temperature by an auxiliary heating element.

It will also be recalled that the storage vessel **14** is a hot store, heat being obtained from this store by cooling the contents of the containers **8a** from an elevated temperature to their recrystallization temperature. It is therefore extremely desirable to ensure that substantially uniform cooling of the containers throughout the hot store takes place. It has been found that if a storage vessel has containers located therein, and air is passed over these containers, the containers near the inlet are cooled more rapidly than those near the outlet. This is obviously undesirable since less heat is obtainable, in practice, from the containers near the outlet.

Figure 3.5c shows a design of a storage vessel which is intended to overcome this disadvantage. For ease of understanding and for the sake of clarity, Figure 3.5c will be taken as illustrating the storage vessel **14** shown in Figure 3.5a. However, the design is equally applicable to the storage vessel **1** shown in that Figure. Thus referring to Figure 3.5c, it will be seen that storage vessel **14** is cylindrical and is open at one end. The walls of the storage vessel are made of a thermally insulating material. In the central region of the open end of the storage vessel **14**, there is shown the external heat exchanger **6**. The annular space around the heat exchanger constitutes the air inlet **16**. The side walls of the heat exchanger have extension portions **21** which project into the storage vessel **14**. These extension portions constitute a baffle.

Within the interior of the storage vessel **14**, a plurality of containers **8a** are provided. These containers extend substantially from the center of the storage vessel to the peripheral wall. The containers pass through the extension portions into an annual channel **21'** defined between wall **14** and extension portion **21**. It will be assumed that the contents of the containers are at a temperature above their crystallization temperature. In use, a fan causes air to pass into the inlet **16** and then flow over those outer parts of the containers present in annular channel **21'**.

When it has heated the region of the closed end of the storage vessel **14**, the air passes over the end of the extension portion and (as shown) flows down the central passage **21"** defined by the extension portion **21**. After passing over the inner portions of the containers contained in passage **21"**, the air leaves the storage vessel through the outlet **17**, over the heat exchanger. Such an arrangement encourages substantially uniform cooling of the contents of the containers throughout the storage vessel. When the air enters channel **21'** via inlet **16**, the nearer the containers to the inlet, the more efficiently are their outer portions cooled by the flow through channel **21'**. However, it must be remembered that while the air is flowing through channel **21'**, heat transfer will be taking place within each individual container, and the cooling effect on the outer ends of the containers is transmitted along each individual container by conduction and/or convection to their inner ends and vice versa.

Obviously, the more container portions the air has passed over, the hotter the air will become and the less cooling effect it will have on subsequent containers. Hence, because the order in which the containers are passed by the air flow in channel **21'** is reversed in channel **21"**, it will be appreciated that the nearer a container is to the open end of the vessel **14**, the less efficient will be the cooling effect on its inner portion. It will be appreciated that there is a tendency for the different cooling effects at different portions of the container to tend to cancel one another out and in practice, it is found that the net heating effect produced by each container is more or less the same throughout the vessel. This represents the most efficient heating situation for the system as a whole.

T. Yamada, S. Mori, K. Kato, Y. Arai and K. Sakitani; U.S. Patent 4,051,888; October 4, 1977; assigned to Daikin Kogyo Co., Ltd., Japan describe a low temperature energy carrying apparatus and method adapted for central cooling which comprises a hydrate circulating system for a liquid hydrate agent and hydrate crystals incorporated with a conventionally designed water circulating system. In the hydrate circulating system, the hydrate crystals are fed to the users side together with cooled water and decomposed therein into hydrate agent and cooled water absorbing the latent heat of decomposition.

Therefore, the temperature of the cooled water during transport to the users side is stabilized and in addition to the sensible heat of the cooled water the latent heat of the hydrate crystals can be utilized in the users side, thereby improving cooling effect in the users side. Also, the hydrate circulating system is adapted to use a hydrate agent in the liquid phase area which allows the operation of the system to be stabilized.

REFRIGERANT FLUIDS

Trichlorotrifluoroethane and Carbon Dioxide

A process described by *D.H. Nail; U.S. Patent 3,915,875; October 28, 1975; assigned to McDonnell Douglas Corporation* relates to a low temperature fluid, and is particularly concerned with the provision of a refrigerant or low temperature heat transfer fluid or functional fluid comprising as the essential component, 1,1,2-trichloro-1,2,2-trifluoroethane, also referred to herein as trichlorotrifluoroethane, and marketed as Freon TF or Freon 113.

It has been found, that although 1,1,2-trichloro-1,2,2-trifluoroethane freezes at a temperature of −31°F, 1,1,2-trichloro-1,2,2-trifluoroethane saturated with carbon dioxide is a clear, low viscosity liquid at temperatures as low as −108°F to −112°F, and at even lower temperature, at ambient pressure. Such trichlorotrifluoroethane fluid saturated with carbon dioxide, has such advantageous low temperature properties, with or without the presence of excess solid carbon dioxide (i.e., dry ice).

It has been found from experience that a true depression of the freezing point of the trichlorotrifluoroethane occurs when this fluid in liquid form contains or is saturated with carbon dioxide at the temperature of use, rather than a supercooling of the liquid. Thus, for example, the above trichlorotrifluoroethane has been frozen solid, and solid carbon dioxide (dry ice) placed on top of the trichlorotrifluoroethane. Within a few minutes the trichlorotrifluoroethane, though still well below its melting point, melts and a mixture of liquid 1,1,2-trichloro-1,2,2-trifluoroethane saturated with carbon dioxide and containing suspended particles or chunks of solid dry ice, results.

The freezing point of liquid 1,1,2-trichloro-1,2,2-trifluoroethane saturated with carbon dioxide gas has been found to be approximately −122°F. The viscosity of such trichlorotrifluoroethane saturated with CO_2, e.g. at −112°F, is substantially less than the viscosity of water at room temperature.

If the trichlorotrifluoroethane fluid saturated with carbon dioxide is subjected to pressure, as by pressurizing with CO_2, a fluid is thus provided which is useful as a refrigerant, heat transfer fluid, or functional fluid at even lower temperatures than the above-noted approximately −122°F freezing point for such CO_2-saturated fluid at ambient pressure.

An additional advantage is that the trichlorotrifluoroethane and CO_2 components of the liquid are nonflammable, and hence, there is no danger of reaction with condensed liquid air or oxygen at extremely low temperatures. The following examples further illustrate the process.

Example 1: 390 g of 1,1,2-trichloro-1,2,2-trifluoroethane was slowly poured over 285 g of dry ice. There resulted a liquid consisting of 1,1,2-trichloro-1,2,2-trifluoroethane saturated with carbon dioxide and containing several pieces of dry ice. This liquid was prepared in a 2-liter Dewar, and was then poured into a beaker and the total volume of the liquid was found to be 350 ml, of which 5% to 10% of the total was solid dry ice. The temperature of the liquid was then found to be $-110°F$.

Example 2: A container containing 66 g of 1,1,2-trichloro-1,2,2-trifluoroethane was placed in a cooling bath consisting of liquid 1,1,2-trichloro-1,2,2-trifluoroethane saturated at ambient pressure with carbon dioxide and containing solid crushed dry ice. The trichlorotrifluoroethane in the container froze solid. 12 g of solid CO_2 was then placed in the container and the container with its contents was placed back in the 1,1,2-trichloro-1,2,2-trifluoroethane-dry ice bath.

At this point, the container held solid 1,1,2-trichloro-1,2,2-trifluoroethane at the bottom and solid CO_2 powder at the top. After 15 minutes, the container was withdrawn and the interface of the solid trichlorotrifluoroethane and solid dry ice was observed to be liquid. Bath temperature was $-112°F$.

After 2 hours in the cold bath, the contents at the bottom of the container were still solid, as were the contents in the top thereof, and there was approximately 15 ml of liquid at the interface. The container was then withdrawn from the cold bath and thus was supplied heat from the ambient room temperature atmosphere. In a few minutes the entire container was in liquid form except for a few pieces of dry ice, and rapidly evolving CO_2 gas.

This example shows that no super-cooling of the 1,1,2-trichloro-1,2,2-trifluoroethane occurs upon saturation thereof with carbon dioxide, but rather an actual melting point suppression of the trichlorotrifluoroethane.

1-Chloro-2,2,2-Trifluoroethane and Hydrocarbons

According to a process described by *K.P. Murphy, R.F. Stahl and S.R. Orfeo; U.S. Patent 4,101,436; July 18, 1978; assigned to Allied Chemical Corporation* constant boiling mixtures have been discovered which consist essentially of 1-chloro-2,2,2-trifluoroethane and a hydrocarbon selected from the group consisting of isopentane, n-pentane, n-butane, isobutane and 2,2-dimethylpropane. The compositions are as follows:

Table 1

Mixture No.	Component A (mol %)*	Component B (mol %)*	Boiling Point (°C)**
1	1-chloro-2,2,2-trifluoroethane (88)	isopentane (12)	4

(continued)

Table 1: (continued)

Mixture No.	Component A (mol %)*	Component B (mol %)*	Boiling Point (°C)**
2	1-chloro-2,2,2-trifluoroethane (96)	n-pentane (4)	5
3	1-chloro-2,2,2-trifluoroethane (39)	n-butane (61)	−5
4	1-chloro-2,2,2-trifluoroethane (55)	2,2-dimethyl-propane (45)	1

* at 20°C
** at 760 mm Hg

The azeotropic compositions of the process all have boiling points lower than those of their individual components. From the properties of the components alone, the reduction in the boiling point temperature and azeotropic characteristics in the mixtures are not expected.

The azeotropic mixtures provide increased refrigeration capacity over the components and represent new refrigeration mixtures especially useful in systems using centrifugal and rotary compressors. The use of the azeotropic mixtures eliminates the problem of segregation and handling in the operation of the system because of the behavior of azeotropic mixtures essentially as a single component. The azeotropic mixtures are substantially nonflammable.

Example 1: The azeotropes were determined in the following manner. Phase studies were made wherein the compositions of the various binary mixtures were varied and the vapor pressures were measured at a temperature of 20.0°C. In all cases azeotropic compositions at 20°C were obtained at the maximum pressure as reported in the above Table. The azeotrope of 1-chloro-2,2,2-trifluoroethane and isobutane was verified but its precise composition was not determined.

All the azeotropes have boiling points lower than the individual components and thus afford higher refrigeration capacity for the azeotropes than the individual components and new refrigerating capacity levels.

An evaluation of the refrigeration properties of the 1-chloro-2,2,2-trifluoroethane-isopentane azeotrope of the process and its fluorocarbon component is shown in the following Table. Isopentane alone is not suitable as a refrigerant in view of its flammability.

Table 2: Comparison of Refrigeration Performance

	1-Chloro-2,2,2-trifluoroethane	Azeotropic Composition*
Evaporator pressure, psia	13.75	14.34
Condenser pressure, psia	53.35	54.36
Evaporator temperature, °F	40	40

(continued)

Table 2: (continued)

	1-Chloro-2,2,2-trifluoroethane	Azeotropic Composition*
Condenser temperature, °F	110	110
Discharge temperature, °F	116	110
Net refrigeration effect, Btu/lb	69.9	72.78
Coefficient of performance	6.13	6.04
Displacement ft³/min/ton	9.02	8.64
Compression ratio	3.88	3.79

*Consisting of 88 mol % 1-chloro-2,2,2-trifluoroethane and 12 mol % isopentane.

By net refrigeration effect (NRE) is intended to mean the change in enthalpy of the refrigerant in the evaporator or, in other words, the heat removed by the refrigerant in the evaporator. By coefficient of performance (COP) is intended to mean the ratio of the NRE to the compressor work. It is a measure of the efficiency of the refrigerant. The azeotrope exhibits a 4.4% increase in capacity over 1-chloro-2,2,2-trifluoroethane.

Additives such as lubricants, corrosion inhibitors and others may be added to the compositions for a variety of purposes provided they do not have an adverse influence on the compositions for their intended applications.

Trifluoropropene

A process described by *A.J. Butler; U.S. Patent 3,884,828; May 20, 1975; assigned to Dow Corning Corporation* relates to a refrigeration system in which the refrigerant undergoes a change from the liquid to the vapor state wherein the improvement comprises employing trifluoropropene as 1% to 100% by weight of the refrigerant.

Trifluoropropene can be used as the sole refrigerant in a system or it can be used in combination with other refrigerants. By way of illustration, trifluoropropene can be used in admixture with fluorodichloromethane, difluorochloromethane, fluorotrichloromethane, difluorodichloromethane, 1,2,2-trifluoro-1,1,2-trichloroethane and 1,1,2,2-tetrafluoro-1,2-dichloroethane.

Trifluoropropene has a boiling point of –22°C, and a vapor pressure of 60 psig at 25°C. Trifluoropropene is flammable, and acute studies indicate it to be relatively nontoxic. All percents are by weight unless otherwise specified.

Example: When the following refrigerants are used in a refrigeration system in which the refrigerant undergoes a change from the liquid to the vapor state, such as the refrigeration systems found in refrigerators and freezers, good cooling is obtained.

(A)	100% trifluoropropene
(B)	90% trifluoropropene
	10% difluorodichloromethane

(C)	80% trifluoropropene
	20% 1,2,2-trifluoro-1,1,2-trichloroethane
(D)	50% trifluoropropene
	50% fluorotrichloromethane
(E)	45% trifluoropropene
	45% fluorotrichloromethane
	10% isobutane
(F)	1% trifluoropropene
	99% fluorotrichloromethane

Difluoromethyl Trifluoromethyl Ether and Dimethyl Ether

W.M. Hutchinson; U.S. Patent 3,922,228; November 25, 1975; assigned to Phillips Petroleum Company has found that difluoromethyl trifluoromethyl ether forms with dimethyl ether substantially constant-boiling admixtures, and, indeed, at least two disparate azeotropes.

The following data are presented to illustrate the process and to show the determination of the azeotropic admixtures.

The azeotropic compositions were obtained by distilling a mixture of difluoromethyl trifluoromethyl ether and dimethyl ether and obtaining a constant-boiling, constant-composition mixture. The distillation unit comprised an Ace Glass Company No. 9219 concentric tube fractionating column 13 inches long, about 10 mm internal diameter; silvered vacuum jacket; rated 40 theoretical plates at 80 cc/hr boil-up rate, a vacuum jacketed head with magnetic take-off fitted with a copper-constantan thermocouple and Thermoelectric Minimite potentiometer giving temperature readings within 0.2°C accuracy; a graduated receiver, and a 25 cc conical kettle heated by a mantle and wrapped in glass wool.

The condenser was cold-finger type cooled by dry ice and trichlorofluoromethane. The receiver was cooled to prevent reboiling. The pressure was atmospheric, and the pressure was measured frequently with a calibrated aneroid barometer graduated in mm Hg. The overhead products (both liquid phases when two appeared) or the kettle product of Example 2 were analyzed with an Aerograph 1520 gas chromatograph having a disc integrator on its recorder.

Example 1: Run 1 — A mixture of 10.4 weight per cent difluoromethyl trifluoromethyl ether and 89.6 weight per cent dimethyl ether was charged to the kettle and heated to its atmospheric boiling point and distilled at a high reflux ratio of at least 5:1 and up to 40:1. A maximum boiling azeotrope was obtained as the constant-boiling, constant-composition mixture consisting essentially of 17.7 weight per cent difluoromethyl trifluoromethyl ether and 82.3 weight per cent dimethyl ether at 742.8 mm Hg pressure, boiling point –23.4°C.

Run 2 — To approach this azeotrope from the other side, a mixture of 26.2 weight per cent difluoromethyl trifluoromethyl ether and 73.8 weight per cent dimethyl ether was charged to the kettle and heated to its atmospheric boiling point and distilled at a 20:1 reflux ratio. A maximum boiling azeotrope was obtained as the constant-boiling, constant-composition mixture consisting essentially of 17.2 weight per cent difluoromethyl trifluoromethyl ether and 82.8 weight per cent dimethyl ether at 742 mm Hg pressure. The azeotropic boiling point was –23.4°C.

The existence of an azeotrope thus was established by two approaches. The feed composition of Run 1, 10.4 weight per cent difluoromethyl trifluoromethyl ether upon distillation became enriched in this component until reaching a concentration of 17.7 weight per cent at which concentration it distilled completely without change in composition. In Run 2, the feed composition of 26.2 weight per cent difluoromethyl trifluoromethyl ether became depleted in this component upon distillation until reaching approximately the same concentration of 17.2 weight per cent as in Run 1.

Example 2: Run 3 — A mixture of 80.8 weight per cent difluoromethyl trifluoromethyl ether and 19.2 weight per cent dimethyl ether was charged to the kettle and heated to its atmospheric boiling point and distilled at a 10:1 reflux ratio. A maximum boiling azeotrope was obtained in the kettle consisting essentially of 65.7 weight per cent difluoromethyl trifluoromethyl ether and 34.3 weight per cent dimethyl ether at 741.4 mm Hg pressure, boiling point –22.0°C.

Run 4: To approach the azeotrope from the other side a mixture of 62.0 weight per cent difluoromethyl trifluoromethyl ether and 38.0 weight per cent dimethyl ether was charged to the kettle and heated to its atmospheric boiling point and distilled at a 20:1 reflux ratio. A maximum boiling azeotrope was obtained as the constant-boiling, constant-composition mixture consisting essentially of 65.2 weight per cent difluoromethyl trifluoromethyl ether and 34.8 weight per cent dimethyl ether at 740.8 mm Hg pressure, boiling point –22.0°C.

Summary of Binary Maximum Boiling Azeotropes

	CHF_2OCF_3 (wt %)	CH_3OCH_3 (wt %)	BP (°C)	Pressure (mm Hg)
Example 1				
Run 1	17.7	82.3	−23.4	742.8
Run 2	17.2	82.8	−23.4	742.0
Average	17.4	82.6	−23.4	742.4
Example 2				
Run 1	65.7	34.3	−22.0	741.4
Run 2	65.2	34.8	−22.0	740.8
Average	65.4	34.6	−22.0	741.1

The above table shows a summary of the runs of the several examples to illustrate the azeotropes of the process. Each of the two azeotropes was demonstrated by approaching the azeotropic boiling point by compositions starting on each side.

The azeotrope containing the major amount of difluoromethyl trifluoromethyl ether and minor amount of dimethyl ether is useful as a refrigerant in particular. The azeotrope containing the minor amount of difluoromethyl trifluoromethyl ether and major amount of dimethyl ether is particularly useful as a degreasing solvent for degreasing of wool. The azeotropes have other specialty applications such as specialty refrigeration, vapor phase degreasing, flux solvents, aerosol propellants, cleaning solvents for various purposes such as garments and the like, and a variety of other uses.

Azeotropic Compositions Containing 1-Chloro-2,2,2-Trifluoroethane

K.P. Murphy, R.F. Stahl and S.R. Orfeo; U.S. Patent 4,057,974; November 15,

1977; assigned to Allied Chemical Corporation describe azeotropic or constant boiling mixtures which consist essentially of 22 mol % of 1-chloro-2,2,2-trifluoro-ethane and 78 mol % of octafluorocyclobutane at 20.0°C.

The azeotropic composition of the process has a boiling point of –8°C at atmospheric pressure (760 mm Hg). 1-chloro-2,2,2-trifluoroethane has a boiling point of 6.1°C at atmospheric pressure and octafluorocyclobutane has a boiling point of –6°C. The azeotropic mixtures exhibit marked reduction in boiling point temperature as compared with the boiling temperatures of the components. From the properties of the components alone, the marked reduction in the boiling point temperature and azeotropic characteristics in the mixtures are not expected.

The azeotropic mixtures provide substantially increased refrigeration capacity over the components and represent new refrigeration mixtures especially useful in systems using centrifugal and rotary compressors. The use of the azeotropic mixtures eliminates the problem of segregation and handling in the operation of the system because of the behavior of azeotropic mixtures essentially as a single component. The azeotropic mixtures are substantially nonflammable.

Example: A phase study was made on 1-chloro-2,2,2-trifluoromethane (BP 6.1°C 760 mm) and octafluorocyclobutane (BP –6°C 760 mm) wherein the composition was varied and the vapor pressures were measured at a temperature of 20.0°C. An azeotropic composition of 20°C was obtained at the maximum pressure and was as follows: 1-chloro-2,2,2-trifluoroethane: 22 mol %; and octafluorocyclobutane: 78 mol %.

The lower boiling point of the azeotrope compared to its components, affords increased refrigerating capacity over both components and a new level of refrigerating capacity. Additives such as lubricants, corrosion inhibitors and others may be added to the compositions for a variety of purposes provided they do not have an adverse influence on the compositions for their intended applications.

In addition to refrigerant applications, the constant compositions of the process are also useful as heat transfer media, gaseous dielectrics, expansion agents such as for polyolefins and polyurethanes, working fluids in power cycles, solvents and as aerosol propellants which may be particularly environmentally acceptable.

K.P. Murphy, R.F. Stahl and S.R. Orfeo; U.S. Patent 4,057,973; November 15, 1977; and U.S. Patent 4,055,054; October 25, 1977; both assigned to Allied Chemical Corporation also describe constant boiling mixtures of 1-chloro-2,2,2-trifluoroethane and 2-chloroheptafluoropropane, as well as mixtures of dichloro-monofluoromethane and 1-chloro-2,2,2-trifluoroethane which are useful as refrigerants, heat transfer media, gaseous dielectrics, expansion agents, aerosol propellants, working fluids in a power cycle and solvents.

In related effort, *K.P. Murphy; U.S. Patent 3,901,817; August 26, 1975; assigned to Allied Chemical Corporation* has discovered that monochlorotrifluoromethane ($CClF_3$) BP –114.6°F, and methyl fluoride (CH_3F), BP –109°F, in certain proportions form an azeotropic mixture or essentially azeotropic mixtures, all of which boil at a temperature lower than the lower boiling $CClF_3$ component and which exhibit only negligible fractionation on boiling under refrigeration conditions.

The azeotropic mixtures have boiling points of approximately –122.5°F or below, and represent a marked reduction as compared with the boiling temperature of the lower boiling $CClF_3$ component (–114.6°F). These compositions
also provide substantially increased refrigeration capacity and represent new refrigerant compositions useful in obtaining high-capacity, low-temperature refrigeration.

The azeotropic mixtures exhibit a number of desired properties for refrigeration
purposes such as higher refrigeration capacity than either of the components,
higher efficiency, negligible flammability, low toxicity and others. Moreover,
the azeotropic mixtures exhibit higher solubility of the refrigerant gas in refrigerant oils, as compared to monochlorotrifluoromethane alone, thereby producing
better oil circulation, and also exhibit higher water solubility than that of monochlorotrifluoromethane alone.

A further property of the mixtures is that the compression ratios of the mixtures are lower than those of the individual mixture components alone. This
has practicable significance in terms of higher volumetric efficiency of the
compressor and longer compressor life.

The true azeotropic mixture of the process consists of 58 mol % monochlorotrifluoromethane and 42 mol % methyl fluoride at atmospheric pressure (14.7
psia) and has a normal boiling point of –122.5°F. The true azeotropic composition will, of course, vary with the pressure.

Azeotropic Compositions Containing 1,1,2-Trichlorotrifluoroethane

According to a process described by *K.P. Murphy and R.F. Stahl; U.S. Patent
4,054,036; October 18, 1977; assigned to Allied Chemical Corporation* azeotropic or constant boiling mixtures have been discovered which consist essentially of 73.0 wt % of 1,1,2-trichlorotrifluoroethane and 27.0 wt % of cis-
1,1,2,2-tetrafluorocyclobutane at 20.0°C.

The azeotropic composition has a boiling point of 43.0°C at atmospheric pressure (760 mm Hg). 1,1,2-trichlorotrifluoroethane has a boiling point of 47.6°C
at atmospheric pressure and cis-1,1,2,2-tetrafluorocyclobutane has a boiling
point of 50.4°C at atmospheric pressure. The azeotropic mixtures exhibit
marked reduction in boiling point temperature as compared with the boiling
temperatures of the components. From the properties of the components
alone, the marked reduction in the boiling point temperature and azeotropic
characteristics in the mixtures are not expected.

The azeotropic mixtures provide substantially increased refrigeration capacity
over the components and represent new refrigeration mixtures especially useful
in systems using centrifugal and rotary compressors. The use of the azeotropic
mixtures eliminates the problem of segregation and handling in the operation of
the system because of the behavior of azeotropic mixtures essentially as a single
component. The azeotropic mixtures are substantially nonflammable.

Example 1: A phase study was made on 1,1,2-trichlorotrifluoroethane (BP
47.6°C 760 mm) and cis-1,1,2,2-tetrafluorocyclobutane (BP 50.4°C 760 mm)
wherein the composition was varied and the vapor pressures were measured at
a temperature of 20.0°C. An azeotropic composition at 20°C was obtained at

the maximum pressure and was as follows: 1,1,2-trichlorotrifluoroethane, 73.0 wt % and cis-1,1,2,2-tetrafluorocyclobutane, 27.0 wt %.

Example 2: An evaluation of the refrigeration properties of the azeotropic mixtures of the process and its components are shown in the following table.

Comparison of Refrigeration Performance

	1,1,2-Tri-chlorotri-fluoroethane	cis-1,1,2,2-Tetra-fluorocyclobutane	Azeotropic Composition*
Evaporator pressure, psia	2.70	2.23	3.14
Condenser pressure, psia	12.75	11.47	14.86
Evaporator temperature, °F	40	40	40
Condenser temperature, °F	110	110	110
Discharge temperature, °F	110	110	110
Net refrigeration effect, Btu/lb	51.8	82.5	59.5
Coefficient of performance	5.98	6.26	6.05
Displacement, ft^3/min/ton	41.4	44.9	35.0
Compression ratio	4.73	5.15	4.74

*Consisting of 73.0 wt percent 1,1,2-trichlorotrifluoroethane/27.0 wt percent cis-1,1,2,2-tetrafluorocyclobutane.

By net refrigeration effect (NRE) is intended to mean the change in enthalpy of the refrigerant in the evaporator, or, in other words, the heat removed by the refrigerant in the evaporator.

By coefficient of performance (COP) is intended to mean the ratio of the NRE to the compressor work. It is a measure of the efficiency of the refrigerant. The azeotropic composition exhibits an 18% increase in capacity and a 1.2% increase in efficiency over 1,1,2-trichlorotrifluoroethane and a 28% increase in capacity over cis-1,1,2,2-tetrafluorocyclobutane.

In related work *K.P. Murphy and R.F. Stahl; U.S. Patent 4,055,049; October 25, 1977; assigned to Allied Chemical Corporation* have found constant boiling mixtures consisting essentially of 43 wt % of 1,2-difluoroethane and 57 wt % 1,1,2-trichlorotrifluoroethane at 760 mm Hg.

Example: Equal molecular quantities of 1,2-difluoroethane (BP 29.6°C 760 mm) and 1,1,2-trichlorotrifluoroethane (BP 47.6°C 760 mm) were charged to a still equipped with a fractionating column. This mixture was heated to reflux and then distilled. A fraction boiling at 24.9°C at 760 mm pressure was collected. Redistillation of this fraction showed no change in boiling point or composition. This fraction was analyzed by gas liquid chromatography and found to possess the following composition: 1,2-difluoroethane, 43 wt %, and 1,1,2-trichlorotrifluoroethane, 57 wt %.

Intermetallic Compound for Cryogenic Refrigerants

K. Andres and P.H. Schmidt; U.S. Patent 4,028,905; June 14, 1977; assigned to Bell Telephone Laboratories, Incorporated have found that $PrNi_5$ is capable of producing refrigeration to lower temperatures than other known materials of the same class by the process of adiabatic demagnetization. This material is convenient to use, in that it is relatively chemically stable in air and easily solderable to the metal wires or foils incorporated in the cooling pill needed for an adiabatic demagnetization refrigerator. This material has been used to produce temperatures in the millidegree Kelvin range. It has a relatively large cooling entropy and, since the specific heat peak is well below one millidegree Kelvin, it shows promise of utility as a refrigerant to temperatures below one millidegree. The use of oriented crystals with their hexagonal axes perpendicular to the applied magnetic field has been shown to be particularly advantageous. Suitable single phase PrNi$_5$ bodies have been produced by annealing bulk melted samples, by Czochralski growth and by an extrusion method.

The bodies of PrNi$_5$ suitable for use as an adiabatic demagnetization refrigerant have been produced by several methods. These methods have started by the melting of stoichiometric amounts of constituent elements (Pr and Ni) in an inert atmosphere, for example, in an arc furnace. The melting temperature of the compound is approximately 1450°C. If the molten mass is allowed to solidify in the furnace hearth a polycrystalline body is produced. Such bodies have not been found to be suitable for use, as thus formed, because they have been found to consist of more than the desired one crystal phase.

The polyphase nature of such bodies is evidenced by a marked increase in magnetic susceptibility at temperatures of the order of 4.2° Kelvin. However, it has been found that the annealing of such polycrystalline bodies at temperatures of the order of 1000°C to 1100°C for some extended time, for example, two days, in inert atmosphere (e.g., argon) or, preferably, in vacuum, produces a body which is sufficiently single phase.

Single crystals of PrNi$_5$ have also been produced by Czochralski growth from an arc melted liquid. Such crystals tend to grow with the hexagonal axis parallel to the direction of crystal growth. Crystals thus grown have shown no discernible evidence of the presence of a second phase. Suitable rods have also been grown by an extrusion process.

Branched Chain Alkylbenzene Lubricant

S.A. Olund; U.S. Patent 4,046,533; September 6, 1977; assigned to Chevron Research Company describes a working fluid consisting essentially of a refrigerant and a chemically inert, wax-free lubricant. The refrigerant is a halo-substituted hydrocarbon having 1 to 3 carbons and at least 40% by weight of fluorine. The lubricant is a mixture of monosubstituted branched chain alkylbenzenes having an average molecular weight of from 300 to 470, the alkyl groups of the alkylbenzenes being at least 60% by weight polypropylene and having an average of at least one branch for every 5 carbon atoms.

The branched chain monoalkylbenzenes have from 16 to 28 carbon atoms in the alkyl group and are excellent lubricants for use in admixture with the highly fluorinated halogenated hydrocarbon refrigerants in sealed compressor-type refrigeration apparatus. The resulting mixture is a homogeneous refrigeration working fluid, compatible even at the low temperature existing in the evaporator of a refrigeration apparatus.

Alkylbenzenes for this use are prepared by alkylating benzene with an alkylating agent in the presence of a catalyst. Typical alkylating agents are the branched chain olefins or branched chain halides, preferably chlorides. The preferred method of preparation is by the HF-catalyzed reaction of benzene with a branched chain olefin.

Corrosion Inhibitors for Alcohol-Lithium Bromide-Zinc Bromide Fluids

R.H. Krueger; U.S. Patent 4,019,992; April 26, 1977; assigned to Borg-Warner Corporation has found that concentrated solutions of lithium bromide and zinc bromide in methanol are inhibited in their corrosion of ferrous metals by the addition of a synergistic combination of ethylenediaminetetraacetic acid (EDTA) and arsenic trioxide.

More particularly, methanol solutions containing from 48 to 75% by weight of a 2:1 (mol ratio) mixture of lithium bromide and zinc bromide to which is added from 0.2 to 1% by weight of a 1:1 mixture of EDTA and arsenic trioxide are markedly inhibited toward corrosion of ferrous metals.

Example: A test procedure was devised to provide an accelerated corrosion test for evaluation purposes. Metal coupons measuring $3\frac{1}{4}"$ x $\frac{3}{4}"$ x $\frac{1}{8}"$ were first polished with No. 500 emery paper, then degreased in methanol and air dried. The metal samples were weighed, then placed in a 200 ml round bottom flask containing 100 ml of a methanol-zinc bromide-lithium bromide solution containing 32.5% by weight LiBr, 42.5% by weight $ZnBr_2$ and 25% by weight methanol such that the samples were half immersed. The mixture was heated at reflux (340°F) for 4 days, then cooled. The metal samples were then cleaned, dried and reweighed to determine the metal loss. The weight loss data in mils per year for mixtures with and without added corrosion inhibitors are given in the table.

Example No.	Metal (Steel)	Inhibitor	Inhibitor (% by wt)	Metal Loss (mpy)
1*	1018	None	–	17.8
2	1018	As_2O_3	0.4	5.4
3	1018	EDTA (Na salt)	0.4	5.9
4	1018	As_2O_3	0.2	
		EDTA (Na salt)	0.2	3.5
5	1018	Sb_2O_3	0.4	30.4
6	1018	$(CH_3)_4NBr$	0.4	19.0

*Control

Thus it will be apparent that the rapid corrosion of 1018 mild steel in alcohol-LiBr-$ZnBr_2$ solutions is retarded by about $\frac{2}{3}$ by the addition of either As_2O_3 or EDTA. When the two are used together, however, a synergistic improvement occurs, and the corrosion rate is lowered to approximately 20% of the uninhibited system.

That the inhibition of corrosion for these alcohol-based working fluids is unexpected is shown by comparative tests with corrosion inhibitors commonly employed with aqueous systems, as in Example 5, wherein the corrosion rate is doubled by the addition of antimony oxide and in Example 6, where the corrosion rate is somewhat increased by the addition of tetramethylammonium bromide. Thus it will be seen that the behavior of corrosion inhibitors in alcohol-based systems cannot be extrapolated from data for aqueous systems.

Reduction of Alkalinity in Aqueous Lithium Bromide Solutions

According to a process described by *H.W. Sibley; U.S. Patent 3,968,045; July 6, 1976; assigned to Carrier Corporation* carbon dioxide gas is bled into an absorption refrigeration system to reduce the alkaline normality of an aqueous lithium bromide solution. The carbon dioxide reacts with lithium hydroxide contained in the solution to form a carbonate thereby reducing the alkaline normality of the solution.

The following stoichiometric relationship describes the hydroxide normality reduction that takes place with the addition of carbon dioxide into such a solution:

$$2\,LiOH \ + \ CO_2 \ \longrightarrow \ Li_2CO_3\downarrow \ + H_2O$$

Apparatus is described to accomplish an alkalinity adjustment of solution normality in a refrigeration system.

Example 1: An alkalinity adjustment was conducted on 200 gal of a 54% lithium bromide solution inhibited with approximately 0.2% by weight lithium hydroxide and a secondary inhibitor of lithium chromate whereby the solution had an alkaline normality of 0.03N by bleeding CO_2 into the solution at a flow rate of 7 cfh until 0.86 lb of CO_2 had been introduced into the system. The corrected solution alkalinity was sampled and found to have a desired normality of 0.005N.

Example 2: An absorption refrigeration machine having 600 gal of a 54% lithium bromide solution containing 0.2% by weight lithium hydroxide and being inhibited with lithium chromate at an alkaline normality of 0.1 N was treated with an excess of CO_2. It was determined that an addition of 102 lb of CO_2 would be needed to reduce the normality down to 0.005 N. An excess of 10 lb of CO_2 was bled into the system resulting in unreacted CO_2 being present in the solution.

The excess caused the solution to temporarily become acidic due to the formation of carbonic acid changing the yellow chromate ion to the orange dichromate ion. However, the ionization potential of the acid was such as to permit the hydrogen ion to react with the lithium carbonate to form free CO_2 and lithium hydroxide. As a result, the solution self-corrected to bring the normality back to the desired level and the orange dichromate ion reverted back to the more desirable yellow chromate ion.

THERMAL ENERGY STORAGE
AND TRANSPORT

INTRODUCTION

Present concern for the conservation of energy has led investigators to focus upon more effective ways of using available energy sources. The effective storage of energy has also become of principal concern. Investigators have concluded that the ability to store and retrieve heat, as well as collect and dissipate heat, would contribute much to energy conservation. Air-to-air heat pumps have been devised to accomplish this end, but their effectiveness depends somewhat on the ambient conditions. It is well known, for example, that as outdoor temperature declines, efficiency of the heat pump drops off as it maintains constant indoor comfort conditions. Similarly, when operating in the cooling mode, the efficiency of the heat pump diminishes with increasing outdoor temperature. Therefore, there is a distinct advantage in being able to produce and store heat during periods of maximum heat pump efficiency, i.e., during periods when the heat pump will operate at lower efficiency, i.e., during the cooler portions of the night. Conversely, when the heat pump is operating in the cooling mode, the ability to store "coolness" during periods of maximum cooling efficiency will also provide a distinct advantage. It is to this end that the heat storage and heat sink materials are described in this chapter.

In this connection, the late Dr. Farrington Daniels, noted physical chemist and past president of the Solar Energy Society, presents an excellent morphological survey of thermal energy storage concepts in his book, *Direct Use of the Sun's Energy*, Yale University, 1964. Dr. Daniels divides thermal energy storage concepts into three basic categories: (1) sensible heat, storage by heat capacity; (2) physical changes, particularly heats of fission and/or vaporization; (3) reversible chemical reaction.

When economic analysis is made of these three categories, it becomes apparent that the two major considerations are: the inventory cost of the heat storage medium and container and, secondly, the engineering complexity of inputting and extracting the heat. Most analyses tend to favor sensible heat storage as the most economically and operationally attractive system. For temperatures up to about 200°F, water is by far the best medium. To quote Dr. Daniels, "Water has about the highest heat capacity per kilogram per liter or per dollar of any ordinary material." The next lowest price heat storage medium is gravel or crushed rock, which is available at a cost of a few dollars per ton and is suitable for storage of heat at temperatures up to at least 1500°F, the upper limit being determined by the fluid which flows through the rock bed to input or extract heat.

CHEMICAL REACTION PROCESSES

Use of Nuclear Reactor Heat to Dissociate Sulfur Trioxide

The utilization of nuclear reactors as sources of electrical energy in petrochemical complexes does not present any particular difficulties but problems arise in the use of nuclear reactors as sources of process heat.

For economical, ecological and safety reasons, very high power nuclear plants are usually sited at relatively long distances, about several hundred kilometers, from the main centers of utilization. Whereas the conveyance of the electrical energy over these distances is not particularly difficult, the conveyance of the thermal energy is altogether a different matter.

G. Cocuzza and G. Beghi; U.S. Patent 4,091,864; May 30, 1978; assigned to Societa Italiana Resine S.I.R. SpA, Italy describe an energy conversion, conveyance and utilization system, particularly for providing process heat in industrial plants, which makes use of a source of heat such as a nuclear reactor to dissociate sulfur trioxide according to the reaction: $2SO_3 \rightleftharpoons 2SO_2 + O_2$; the dissociation products are conveyed through a pipeline to a remote utilization station where the heat of recombination is utilized. The resulting sulfur trioxide is returned through the pipeline to the reactor site. The pipeline incorporates separate pipes in which the sulfur dioxide and sulfur trioxide are conducted in liquefied form, surrounded by a duct in which the gaseous oxygen flows.

The process is based on the equilibrium reaction: $2SO_3 \rightleftharpoons 2SO_2 + O_2 - 47$ kcal, which has the following advantages:

(a) The thermal levels are in conformity with the necessary requirements: at about 850°C the sulfur trioxide is mostly dissociated and this dissociation occurs within a temperature range sufficiently wide to facilitate thermal coupling with the gaseous cooling circuit of a nuclear reactor of the HTGR type. Furthermore, the temperature range within which the sulfur dioxide and the oxygen recombine allows the generation of high-pressure superheated steam.

(b) Coproduction of oxygen is possible, in which case the oxygen required for the oxidation of the sulfur dioxide in the utilization plant can conveniently be supplied from the atmosphere.

(c) Sulfur dioxide and sulfur trioxide are easily liquefied, an extremely important point as it allows their conveyance in the liquid state, with consequent reduction in costs as compared with processes involving the conveyance of considerable quantities of gas, such as that, in the following example, which is based on the equilibrium reaction: $CH_4 + H_2O \rightleftharpoons CO + 3H_2$.

Furthermore, the easy liquefaction of the sulfur dioxide and sulfur trioxide allows their storage in the liquid state, which makes it possible to cope with the possible load fluctuations without any change in the rate of operation of the nuclear reactor.

(d) The cost of the chemical substances used is low.

(e) The risk of pollution due to the sulfur dioxide and trioxide may be reduced, especially if specific conveyance devices are used.

Phosgene Conversion

According to a process described by *H.S. Spacil; U.S. Patent 3,967,676; July 6, 1976; assigned to General Electric Company* at a heat source, such as a nuclear reactor, phosgene is reacted (i.e., at about 700°C) to form a mixture of carbon monoxide and chlorine. This mixture of gases is cooled by heat exchange with incoming cold phosgene and is pumped through a first pipeline at ambient temperature to an energy use area. At the energy use end of the first pipeline the gas mixture is heated to a temperature (i.e., near, but below 550°C) in the presence of activated charcoal or platinum as a catalyst. The CO and Cl_2 react exothermically to form $COCl_2$ (phosgene) with the liberation of 26 kcal per mol of phosgene formed. The heat evolved from this reaction is released across a heat exchanger for the boiling of water and superheating of the resultant steam to a temperature in the 400° to 500°C range for use as process heat or the generation of electricity. The phosgene produced is cooled and then returned to the heat source by a second pipeline for repetition of the closed loop process.

Ammonia Conversion

C.G. Miller; U.S. Patent 4,044,821; August 30, 1977; assigned to NASA describes a system and process in which a working substance such as a chemical compound is decomposed into simpler substances at relatively low temperature and low pressure. The low temperature heat which is absorbed is effectively stored in the form of chemical potential in the simpler substances. The simpler substances are subsequently recombined into the original compound, generally at higher pressure. During the recombination heat of recombination is released which raises the environment in which the recombination takes place to a substantially higher temperature. A substantial portion of the released heat of recombination is extracted to perform useful work, such as for example, to produce vapor at high temperatures for driving a turbine in an electrical generator. The working substance is then led back to be decomposed by the absorption of low temperature heat and the process is repeated. The system can be thought of as a temperature converter, or low-to-high temperature energy conversion system.

One example of a low-to-high temperature energy conversion system includes a decomposition chamber in which ammonia (NH_3) is decomposed into hydrogen and nitrogen by absorbing heat of decomposition from a low temperature, e.g., 300°C energy source. The separated hydrogen and nitrogen are then supplied to a recombination chamber wherein they recombine to produce ammonia. The recombination process is associated with a significant increase in temperature, used to increase the temperature of a fluid to temperatures about 500°C.

Methane and Carbon Dioxide Reactions

According to a process described by *R.H. Wentorf, Jr.; U.S. Patent 3,958,625; May 25, 1976; assigned to General Electric Company* using a heat source, such as a nuclear reactor, methane and carbon dioxide are reacted at about 800° to 900°C to form a mixture of carbon monoxide and hydrogen. This reaction absorbs about 62 kcal/g mol of CO formed. The mixture of gases produced is cooled by heat exchange with incoming cold CH_4 and CO_2 and is then pumped through a first pipeline at ambient temperature to an energy use area. At the energy use end of the first pipeline the gas mixture is heated to about 350° to 500°C in the presence of steam and a catalyst. The CO and H_2 react exothermically to form CH_4 and CO_2 and thereby liberate about 61 kcal/mol of CO consumed. The heat evolved from this reaction is released across a heat exchanger for use as process heat or for conversion to electricity. Water is condensed and separated from the mixture of gaseous reactants and the dried, cooled CH_4/CO_2 mixture is returned to the heat source end by a second pipeline for repetition of the closed loop process.

Conversion and Reforming of Methane

F. Hilberath and J. Teggers; U.S. Patent 4,109,701; August 29, 1978; assigned to Rheinische Braunkohlenwerke AG, Germany describe a process of conveying heat which involves applying heat energy for the catalytic steam-reforming of methane; conveying the cooled gases obtained thereby to the desired place of energy consumption; catalytically reacting the gases to again form methane while utilizing the heat released thereby; and optionally, recycling the reformed methane back to the steam-reforming unit.

In accordance with the process the above reaction sequence is carried out in such a manner that water is substantially completely removed from the gas obtained from the steam-reforming of methane. Carbon dioxide present in the gas is partially or wholly separated and the subsequent catalytic reaction to form methane (methanization reaction) is carried out under elevated pressure of between 10 to 100 atmospheres gauge and at least partially at temperatures of between 400° to 650°C. By maintaining these conditions, it becomes possible to carry out the entire process considerably more economically, despite the increased use of apparatus, and to increase the heat utilization generally by 10% or more as compared to a procedure which includes the presence of water vapor in customary amounts.

Vaporized Reactants

B. Ulano; U.S. Patent 4,018,263; April 19, 1977; describes a process which relates to vapor phase reactions and more particularly to a method and apparatus for generating and storing vaporized reactants.

Many processes involve the reaction of one or more reactants while in the vapor phase. Such vapor phase processes are utilized for example in the textile industry for imparting certain desirable characteristics to the textile thus treated. An example of such a reaction involves the treatment of cellulosic fiber with a vapor phase mixture of formic acid and formaldehyde to render the cellulosic fiber shrink resistant and crease resistant.

This process provides an improved method and apparatus for vaporizing and storing reactants whereby a reserve of vaporized reactants is maintained in a pressurized condition and the reactant is driven by the force of its own pressure into the reaction chamber as required. In this manner a substantially uniform supply of vaporized reactant is provided, undesirable fluctuation in reactant concentrations in the reaction chamber avoided and the production rate of the vapor phase process is substantially increased.

More particularly, the method comprises introducing a vaporizable reactant in liquid or solid form into a fluid heat exchange medium under confined conditions wherein the reactant is vaporized and entrained in the heat exchange medium. The medium and entrained vaporized reactant are led into a first chamber and maintained therein under nonconfined conditions whereby at least a portion of the entrained vaporized reactant rises from the medium and is led into a storage vessel which is also maintained at a sufficiently high temperature to keep the reactant in the vaporized form. The medium is then returned to the heat exchanger for vaporizing contact with additional reactant.

The vaporized reactant is maintained in the storage vessel under pressure and at sufficient temperature to maintain it in the vaporized state. Upon demand, the pressurized and vaporized reactant is led from the storage vessel to the reaction chamber for vapor phase reaction. In a preferred case, pressure in the storage chamber is generated by vaporized reactant contained therein and a drop in pressure in the storage chamber is compensated for by increasing the amount of medium containing entrained vaporized reactant flowing into the first chamber so that the rate of replenishment of vaporized reactant in the storage chamber is increased to replace the used reactant and to return the storage chamber to its desired storage pressure.

The generator comprises generally an expansion chamber into which heated medium containing entrained vaporized reactant is introduced and the entrained reactant allowed to escape from the heated medium. A storage tank communicates with the expansion chamber by means of one or more risers through which the vaporized reactant leaving the expansion chamber is led into the storage vessel.

A conventional heat exchanger is connected to the expansion chamber for fluid communication therebetween and conventional pump means are provided for pumping heated media containing entrained vaporized reactant into the expansion chamber and for returning media from which at least a portion of the entrained vaporized reactant has been removed back to the heat exchanger. The generator is suitably insulated as are the lines leading to and from the generator to reduce thermal losses.

Metal Hydrides

According to a process described by *A.W. McClaine; U.S. Patent 4,039,023; August 2, 1977; assigned to the U.S. Secretary of the Navy* heat transfer and

thus temperature control of an environment is achieved by a method and apparatus which comprises withdrawing hydrogen from a first hydride reaction system
in a two-phase equilibrium at a certain temperature and pressure thereby causing
a decomposition of a portion of the hydride. Heat is then added to the reaction
system to compensate for the loss of heat caused by the endothermic hydride
decomposition reaction, the withdrawn hydrogen transferred to a second hydride
reaction system in a two-phase equilibrium at a higher temperature and pressure,
thereby causing a formation of hydride and release of heat. The heat created by
the exothermic hydride formation reaction is removed.

Example: Figure 4.1 shows the experimental apparatus used. The apparatus
comprised a constant temperature bath **210** filled with water in which container
121 containing lanthanum pentanickel was immersed. The container was connected to a switching valve **114** which was connected to T-union **161** which was
connected to switching valve **118** and switching valve **241**. Valve **118** was connected to regulator **201** which was connected to hydrogen supply **221**. Valve
241 was connected to wet test meter **261**.

Figure 4.1: Heat Transfer Using Metal Hydride

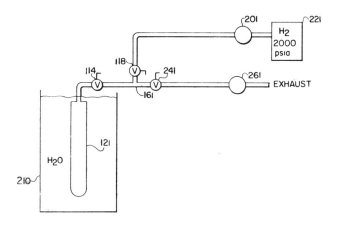

Source: U.S. Patent 4,039,023

Hydrogen was introduced to the hydride for given periods at given pressures and
then the hydride was totally desorbed to determine how much hydrogen had
been absorbed. This was repeated until the hydride hydrogen composition had
reached the high side of the two-phase region. To determine desorption rates
hydrogen was released from the hydride at atmospheric pressure. Hydrogen
evolved was measured by a wet test meter while being timed. These measurements were made at fixed bath temperatures.

The results of the experiments were that hydrogen was desorbed at atmospheric
pressure and 70°F, evolving 0.3 cubic feet of hydrogen in 6 minutes and that

hydrogen was absorbed at 100 psi and 140°F absorbing 0.3 cubic feet of hydrogen in 6 minutes. Thus if a compressor were inserted between two containers of lanthanum pentanickel hydride which would compress hydrogen 100 psi, then heat could be transferred from a body of water at 70°F to a body of water at 140°F.

J.G. Cottingham; U.S. Patent 4,044,819; August 30, 1977; assigned to the U.S. Energy Research and Development Administration describes a method and apparatus for the use of hydrides to exhaust heat from one temperature source and deliver the thermal energy extracted for use at a higher temperature, thereby acting as a heat pump. For this purpose there are employed a pair of hydridable metal compounds having different characteristics working together in a closed pressure system employing a high-temperature source to upgrade the heat supplied from a low-temperature source.

Heating Pack

C.S. Krupa; U.S. Patent 3,980,070; September 14, 1976; assigned to Scotty Manufacturing Company describes a heating pack containing a granular chemical composition. The granular composition includes a suitable ferrous metal and an oxidizing agent. When water is added to this chemical composition, an exothermic reaction results which produces heat useful as a hand warmer. The particular combination of chemical elements provides an improved chemical mix which results in faster heating of the composition, a lower level of objectionable odor and reduced weight through a reduction in the amount of iron filings which are necessary.

The chemical composition is contained in an inner bag which is encloseable in a second outer bag. The inner bag comprises a rectangular bottom portion and a tapered flap portion which has a liquid fill opening. A barrier-producing envelope sewn in the flap portion of the inner bag prevents spillage of the granular material while allowing water to be added through openings in the envelope. The openings are effectively closed to the passage of granular material by means of a closure rivet and folding of the flap portion. This two bag container in combination with the improved chemical composition provides a heating pack which is economical to produce and use and which produces a relatively fast heating cycle. When the chemical composition is allowed to dry, the exothermic reaction is stopped and the heating pack can be recycled repeatedly by adding water to the chemical composition for each use.

The preferred chemical composition comprises an oxidizing component and a ferrous metal component. The ferrous metal component preferably consists essentially of cast iron filings. About 5 to 10 parts by weight of the oxidizing component are combined (e.g. dry mixed) with each 100 parts by weight of the ferrous component. The oxidizing component preferably comprises by weight: 20-50% cupric carbonate; 20-60% suitable water-soluble metal halide salt (e.g. NaCl); 15-40% citric acid; and 6-12% alkali metal chlorate (e.g. $KClO_3$).

FUSION PROCESSES

Discharging Apparatus for Thermal Accumulator

Storage accumulators for the storage of latent heat energy continue to gain in importance. At a high heat flux density, the thermal discharge of these accumu-lations encounters substantial difficulties because crusts of thermally depleted storage substance form on the surface of the heat sink portions of the accumu-lators. As their thickness builds up, these crusts increasingly inhibit heat trans-mission so that the full rate of heat withdrawal from the accumulators can only be achieved at the very first instant.

A process which is directed towards overcoming this problem is described by *N. Laing; U.S. Patent 4,064,931; December 27, 1977*. The process provides, inside an accumulator container, a heat sink component, which rotates or else is equipped with a stripping device rotating around it. In this way, the storage substance, which is continuously thermally discharged by the heat transfer sur-faces of the heat sink and builds up as a crust, is thrown off or stripped off these surfaces so that the favorable initial condition of a stationary accumulator endures. By this means, it becomes possible to accomplish the thermal discharge of the accumulator at high rates of heat flow. The heat sink component is ad-vantageously arranged at the center of the accumulator container. A fluid vortex is then formed by which the solid portions of the storage substance of higher specific weight are thrown towards the periphery of the accumulator container and become deposited there. Containers with rotational symmetry are especially favorable. The thermal charging of the accumulator takes place through a heated bottom.

The same apparatus can be used for the thermal charging of accumulators in which the density of the storage substance is smaller in the crystalline phase than in the liquid phase. In particular, this situation is true for ice. Any device which generates angular motion can be used for the thermal charging of such accumulators.

The process is described with the aid of figures. Figure 4.2a shows the accumu-lator container 1 surrounded with an insulating layer 2. A heat sink component 4 is situated at the center of the container. A fluid heat carrier 5 flows through the pipeline 6 into the heat sink body and emerges again through the pipeline 7. Stripping blades 8 are arranged around this thermal discharge heat exchanger which are rotated via a coupling 10 by a motor 9. The stripping blades, simi-larly to the blades of a windscreen wiper, are in contact with the surface of the heat sink component and thus continuously strip off the continuously deposited layer of crystallized storage substance.

Simultaneously, these stripping blades generate a fluid vortex, whereby the solid particles 11 in the storage substance are thrown outward and are deposited along the layer surfaces 12 so that not until almost the entire storage substance is thermally depleted does the front face of the solid storage substance portion finally stop the rotation of the stripping blades. The left-hand part of the figure shows the apparatus in the thermally charged state (the storage substance is en-tirely melted), while, on the right-hand side, a partly discharged accumulator is illustrated.

With some storage substances, the crust adheres to the stripping blades. In this case, the process provides that the blades slide over the surface of the heat sink component, which is shaped as a solid of revolution, under a preload, so that, by virtue of the mechanical friction, the blades are subject to a small temperature rise relative to the melt.

Figure 4.2: Thermal Storage Accumulator

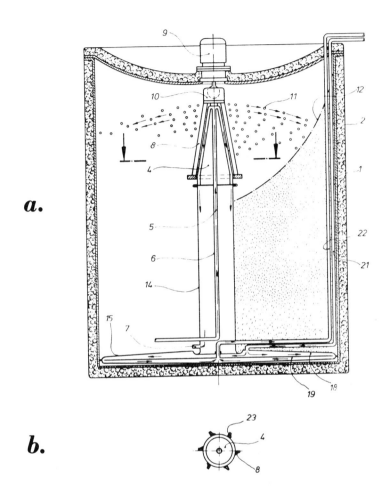

a.

b.

(a) Cross section of a thermal storage accumulator
(b) Cross section of the center portion of the accumulator of
     Figure 4.2a.

Source: U.S. Patent 4,064,931

Another method consists in conducting an electric current via a liquid heat carrier through the stripping blades **8** for the purpose of artificial heating. The heat
sink component **4** is mounted on a column **14** so that the component **10**, together
with the stripping blades, can be withdrawn upwards. The column is connected to
a heat exchanger **15**, which has two chambers **18** and **19** intercommunicating at
the periphery. Heat carrier fluid for melting the storage substance enters through

the pipeline **21**. The cooled heat carrier emerges again through the pipeline **22**.

Figure 4.2b shows a cross section through the stripping blades **8** and through the heat sink component **4**. The stripping blades **8** are made hollow. A fluid heat carrier for heating the blades can be conducted through the duct **23**.

Heat Engine Using Solidification and Melting Operations

C. Cheng and S.-W. Cheng; U.S. Patent 3,953,973; May 4, 1976 describe a heat engine or a heat pump in which the working medium used is subjected alternately to solidification and melting operations. A working medium so used is referred to as an S/L type medium. In this process using a nonaqueous S/L medium, the medium is melted under a high temperature T_H and a high pressure P_H to absorb heat and is solidified under a low temperature T_L and a low pressure P_L to release heat. Since the nonaqueous medium expands as it is solidified under the low pressure, the system does work on its surroundings.

Since water expands on solidification and contracts on melting, in this heat engine with an aqueous medium, the medium is melted under a high temperature and a low pressure to absorb heat and is solidified under a low temperature and a high pressure to release heat. Since the aqueous medium expands under the high pressure as it is solidified and contracts under the low pressure as it is melted, the system does work on the surroundings. The operation of a heat pump is just the reverse operation of a heat engine. The engine comprises a multiplicity of longitudinal conduits which are connected through a first check valve to a high pressure zone and through a second check valve to a low pressure zone.

Referring to a heat engine utilizing a nonaqueous S/L medium, medium liquid in an amount equivalent to volume expansion associated with the high pressure melting operation is discharged to the high pressure zone through the first check valve from each conduit during a melting step and an equivalent amount of medium enters each conduit during a solidification step through the second check valve. The medium discharged under the high pressure may be depressurized through a hydraulic motor to do work and becomes low pressure medium.

For a heat engine utilizing an aqueous S/L medium, medium liquid in an amount equivalent to volume expansion associated with high pressure solidification is discharged to the high pressure zone through the first check valve from each conduit during a solidification step and an equivalent amount of medium enters each conduit during a melting step through the second check valve.

The melting point of an abnormal substance, such as water, bismuth and gallium, is lowered as the applied pressure is increased. Such an abnormal substance expands as it is solidified and contracts as it is melted. In this discussion S/L type working mediums are classified into nonaqueous mediums and aqueous mediums which are respectively used to mean the normal substances and abnormal substances. Melting point of normal substance increases as applied pressure is increased.

Figure 4.3a illustrates the principle of operation of the heat engine utilizing a nonaqueous and normal S/L type medium. Figure 4.3b illustrates the phase diagram (P-T) of the normal working medium, density of solid being higher than that of liquid. It shows the triple point **1**, the vaporization line **1-2**, the melting line **1-3**, and the sublimation line **4-1**. Referring to points **5** and **6**, the medium is to be subjected to a cyclic operation comprised of a high pressure P_H high temperature

Figure 4.3: Heat Engine Utilizing Solidification and Melting Operations

An Engine Utilizing an S/L Type Working Medium

a.

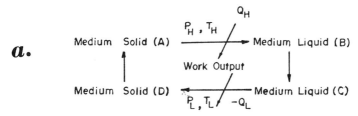

(a)   Illustrates the principle of operation of heat
engine utilizing a nonaqueous S/L type
medium.

b.

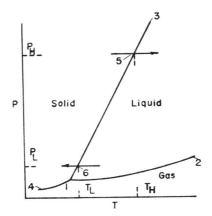

(b)   Illustrates the principle of the heat engine
utilizing a nonaqueous S/L type medium
on a P-T diagram

c.

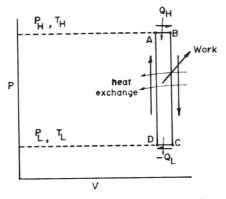

(c)   Illustrates the principle of the heat engine
utilizing a nonaqueous S/L type medium
on a P-V diagram

(continued)

Figure 4.3: (continued)

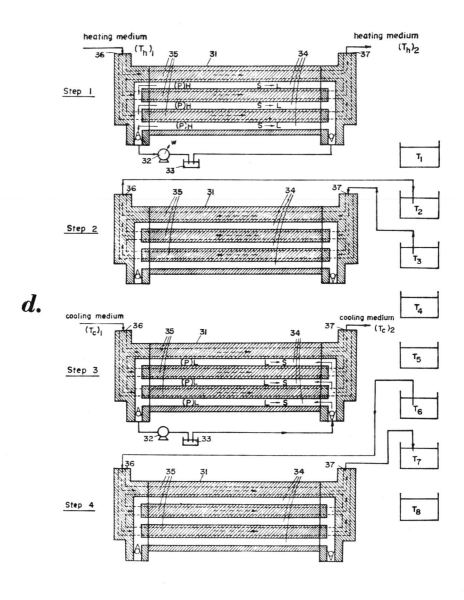

(d)   Processing steps

(continued)

Figure 4.3: (continued)

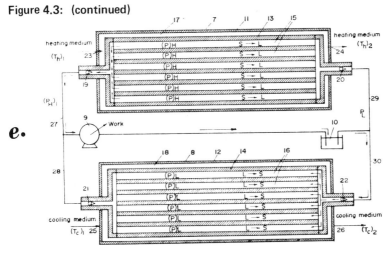

(e) Construction and operation of the heat engine

Source: U.S. Patent 3,953,973

T_H melting step and a low pressure P_L low temperature T_L solidification step. Figure 4.3c shows a P-V diagram of the medium and shows the volume changes at points **5** and **6** and along line **5-6** of Figure 4.3b.

Referring to Figures 4.3a, b and c, the medium in the heat engine undergoes a cyclic operation which comprises four steps: **A-B, B-C, C-D** and **D-A**. During the first step, the medium solid (**A**) is melted to become medium liquid (**B**) under a high pressure and a high temperature by absorbing heat Q_H. The volume expands during this step. During the second step, the medium liquid (**B**) is depressurized and cooled to P_L and T_L and becomes medium liquid (**C**). During the third step, the medium liquid (**C**) is solidified by releasing heat ($-Q_L$) under a low pressure and a low temperature and becomes medium solid (**D**). The medium contracts during this step. The medium solid (**D**) is then pressurized from P_L to P_H and heated from T_L to T_H and becomes medium solid (**A**).

The next cycle is then initiated. The engine does work by expanding under the high pressure and contracting under the low pressure. At least a part of the heat released in cooling the medium liquid from T_H to T_L, from state **B** to state **C**, may be utilized in heating the medium solid from T_L to T_H, from state **D** to state **A**.

When an abnormal substance such as an aqueous solution is used as working medium, the process description given above has to be modified. For such a medium, the volume expands as it is solidified and the volume contracts as it is melted and the melting line in Figure 4.3b has a negative slope. The medium therefore melts at a lower temperature under a higher applied pressure. In such case, the medium in the heat engine undergoes a cyclic operation which comprises four steps: **A-B, B-C, C-D** and **D-A**. During the first step, the medium solid is melted under a low pressure P_L and under a high temperature T_H to become medium liquid (**B**) by absorbing heat Q_H. The volume of the medium contracts during this step. During the second step, the medium liquid (**B**) is pressurized to P_H and cooled to T_L and becomes medium liquid (C). During the third step, the medium liquid (C) is solidified by releasing heat ($-Q_L$) under a high pressure and a low temperature and becomes medium solid (D). The medium expands during this step. The medium solid (D) is then depressurized from P_H to P_L and heated from T_L to T_H and becomes medium solid (A). The next cycle is then initiated. This engine also does work by expanding under the high pressure P_H and contracting under the low pressure P_L. Again, at least a part of the heat released in cooling the medium liquid from state B to state C may be utilized in heating the medium solid from state D to state A.

Figure 4.3e shows two heat engine units 7 and 8, a hydraulic motor or a turbine 9 and a medium storage tank 10. The two heat engine units 7 and 8 respectively consisting of outer vessels 11 and 12, multivoid metal blocks 13 and 14 containing multitude of conduits 15 and 16 which contain an S/L type working medium and are connected to high pressure lines 27 and 28 through check valves 19 and 21 respectively and are connected to low pressure lines 29 and 30 through check valves 20 and 22 respectively. The medium in each heat engine unit undergoes the four steps described earlier. As shown, a heating medium is passed through the first unit from 23 to 24, to cause a high temperature and high pressure melting of the medium.

A volume of medium in equivalent to the volume expansion associated with the melting operation is discharged through the check valve 19 and the high pressure line 27 and depressurized through the hydraulic motor to do work. The depressurized medium liquid is stored in the medium tank. A cooling medium is shown to be passed through the second unit from 25 to 26 to cause a low temperature low pressure solidification of the medium. A volume of medium in equivalent to the volume contraction associated with the solidification operation enters from the medium tank through the low pressure line 30 and through the check valve 22 and into the conduits of the unit. The functions of the two units alternate. The second step (B-C) and the fourth step (D-A) are not illustrated in Figure 4.3e but are illustrated in Figure 4.3d.

Figure 4.3d illustrates steps A-B, B-C, C-D and D-A described earlier that are conducted in a multivoid heat engine. The figures illustrate a multivoid heat engine unit 31, a hydraulic motor 32, a medium storage tank 33 and tanks containing intermediate heat exchange mediums at intermediate temperatures T_1 through T_8 between T_H and T_L. The heat engine unit contains two sets of conduits 34 and 35. The first set of conduits contains a normal S/L type working medium. The second set of conduits are to be used for passing a heating medium, a cooling medium and intermediate heat exchange mediums. During the first step, step A-B, a high temperature $(T_h)_1$ heating medium enters the conduits 35 from entrance 36 and exits at the exit 37 at $(T_h)_2$.

The medium in the conduits 34 is melted under a high pressure P_H and a high temperature T_H. A quantity of medium liquid is discharged at the high pressure and depressurized through the hydraulic motor and stored in the medium tank. Tanks T_1 through T_8 store intermediate heat exchange mediums at temperatures T_1 through T_8 which are intermediate temperatures between T_H and T_L. During the second step, step B-C, intermediate heat exchange medium at T_2 is passed through the unit and heated to T_1 and stored in tank T_1, then intermediate heat exchange medium at T_2 is passed through the unit and heated to T_1 and stored in tank T_1, then intermediate heat exchange medium at T_3 is passed through the unit and heated to T_2 and stored in tank T_2 etc., till heat exchange medium at T_8 is passed through the unit and is heated to T_7 and stored in tank T_7. By this time, the medium in the unit is cooled down to T_L and the second step is completed.

During the third step, step C-D, a cooling medium enters at $(T_c)_1$ at 36 and leaves at $(T_c)_2$ at 37 to solidify the medium at a low temperature T_L and low pressure P_L. A quantity of low pressure medium enters the conduits 34 to compensate for the volume shrinkage associated with the solidification of the medium. During step 4, step D-A, intermediate heat exchange mediums at T_1 through T_7 are successively passed through the unit to heat medium in the unit from T_L to T_H and respectively become mediums at T_2 through T_8 and are respectively stored in tanks T_2 through T_8. The operations described are repeated.

Flow of Heat Transfer Fluid over Molten Salt

According to a process described by *N.D. Greene; U.S. Patent 4,109,702; Aug. 29, 1978* energy is stored by heating a salt to a temperature above its latent heat of fusion to convert the salt to a liquid state. Heat is retrieved by moving a heat transfer fluid that is immiscible with the salt and has a density less than that of the salt over the top surface of the liquid salt at such a velocity that the upper layer of the salt is emulsified with the heat transfer fluid to crystalize the salt in the upper layer and exothermally surrender heat from the salt to the heat transfer fluid. The crystalized salt gravitates from the top surface thereby maintaining the top surface in a liquid state.

The process is fully operative with both normal salts and those having a supercooled liquid phase. Use of supercooled liquid phase salts allows heat to be stored at ambient or room temperatures that are lower than the latent heat of fusion of the salt.

By using a heat transfer fluid that is immiscible with the salt and of a lower density than the salt, an emulsion between the two can be formed to increase the heat transfer area. At the same time, a sufficient demarcation between the heat transfer fluid and the salt is defined to prevent significant entrainment of salt crystals in the heat transfer fluid outside of the emulsion layer. A higher heat transfer rate is obtained because of the larger heat transfer area between the salt and the heat transfer fluid incident to their direct contact with one another in the emulsion.

The heat transfer fluid is moved across the salt in such a pattern as to create a large shear rate between them. Two preferred patterns are a vortex pattern and an outward radial pattern. Although the heat transfer fluid is moved across the top surface of the salt at such a velocity that the upper layer of the liquid salt is emulsified with the heat transfer fluid to increase the heat transfer area, this velocity must not be so high that crystalized salt is entrained in the heat transfer fluid outside of the emulsion layer. If the salt crystals become so entrained they may coat and clog whatever means, if any, are used to move the heat transfer fluid out of the container containing the salt to a heat exchanger.

In one example both the heat transfer fluid and the salt are enclosed in the same container and a coolant is moved through tubing that makes contact with the heat transfer fluid to remove heat from the heat transfer fluid through the tubing to the coolant. Because the heat transfer fluid is enclosed in the container, a larger emulsion layer can be tolerated since there is no danger of salt crystals being circulated outside of the container; and as a result a greater heat transfer area and crystalization rate can be achieved.

Lithium Perchlorate Trihydrate

R.J. Ruka and R.G. Charles; U.S. Patent 4,057,101; November 8, 1977; assigned to Westinghouse Electric Corporation have found that lithium perchlorate trihydrate, ammonium aluminum disulfate dodecahydrate, and oxalic acid dihydrate are excellent heat sink materials. It has also been found that the corresponding deuterates are excellent heat sink materials, that their transition temperatures are different, and that a continuous range of transition temperatures can be obtained using deuterate-hydrate mixtures.

Oxalic acid dihydrate has the highest known experimentally determined heat of fusion per unit weight of any material between $2°$ and $100°C$, as far as is known. Lithium perchlorate trihydrate and ammonium aluminum disulfate dodecahydrate both have high heats of fusion in the $90°$ to $95°C$ range, where few other suitable materials are available. Though the resolidifying characteristics of ammonium aluminum disulfate dodecahydrate have not yet been thoroughly explored, both oxalic acid dihydrate and lithium perchlorate trihydrate have been found to resolidify with very little tendency to supercool.

The fusible materials of this process have the general formulae $LiClO_4 \cdot 3R_2O$, $NH_4Al(SO_4)_2 \cdot 12R_2O$, and $(COOH)_2 \cdot 2R_2O$, where each R is independently selected from 0 to 100% hydrogen and 0 to 100% deuterium. The following table gives the properties of compounds which have been tested:

Compound	Melting Temperature ($°C$)	...Heat of Fusion... cal/g	cal/cm^3
$NH_4Al(SO_4)_2 \cdot 12H_2O$	93.5	64	105
$LiClO_4 \cdot 3H_2O$	95	67	123
$(COOH)_2 \cdot 2H_2O$	101.5	91–94	150–155
$(COOH)_2 \cdot 2D_2O$	94	83	–

The above hydrates can be purchased commercially, but they and the dideuterate were prepared by heating the corresponding anhydrous compound with stoichiometric amounts of water or deuterium oxide until dissolution. Mixtures of deuterates and hydrates, e.g., $(COOH)_2 \cdot \frac{3}{2}H_2O \cdot \frac{1}{2}D_2O$, can be made in the same way to achieve intermediate melting temperatures. Lithium perchlorate trihydrate and oxalic acid dihydrate have been found to freeze with little tendency to supercool or form the anhydrous compound. The other materials have not been as yet fully tested. Oxalic acid dihydrate is readily available at a low price, can be transferred in the presence of air and water vapor at room temperature, and requires no special handling. Lithium perchlorate trihydrate is an oxidizing agent and may explode in the presence of oxidizable materials. Isolated, however, it is very stable even at temperatures above its melting point.

Both lithium perchlorate trihydrate and aluminum ammonium disulfate dodeca-hydrate are relatively inexpensive and can be handled conveniently in air. The heat sink can be used in combination with a radiator to protect electronic circuitry or other equipment from temporary overheating.

Clathrate Forming Compounds

L. Leifer; U.S. Patent 3,976,584; August 24, 1976; assigned to Michigan Technological University describes a thermal energy storage material comprising a substantially solid clathrate which has a melting point above 32°F, is stable at atmospheric temperature and pressure and has a relatively high specific heat and heat of fusion. The thermal energy storage material is formed by dissolving a clathrate forming compound in water and is characterized by the water being structured in a manner so that the distance between the oxygen atoms is less than 2.87 A.

The thermal storage material is placed in heat exchange relationship with a region to be thermally conditioned or with the air being introduced into the region. When the temperature of the thermally conditioned region is higher than the melting point of the thermal storage material, such as during the hot hours of the day, heat is absorbed from the region or the air passing over the material. As the thermal storage material absorbs heat, its temperature will eventually rise to its melting point. Because of its high heat of fusion, the thermal storage material can absorb a large quantity of heat per unit mass. When the temperature of the region or the air passing over the thermal storage material is below the melting point of the material, the heat stored during melting is released to the region or air as the material cools to its freezing point, thereby maintaining the temperature of the thermally conditioned region within a relatively narrow range.

By selection of an appropriate solid clathrate or mixtures, a room or similar region can be maintained substantially constant in this manner. The power required for cooling and/or heating required for comfort control can be substantially reduced and even eliminated in some circumstances depending on the ambient climate condition.

The clathrate forming compounds used in this process can be represented by the formula: $(R_4Y)^+(X)^-$ where each R is selected from the group consisting of alkyl, alkenyl, alkynyl, cycloalkyl, cycloalkenyl and aryl radicals and combinations thereof, such as alkaryl, aralkyl, and the like, containing 1 through 8 carbon atoms; Y is nitrogen, phosphorus, or sulfur; X is chloride, fluoride, or bromide when R contains 1 through 8 carbon atoms, iodide when R contains 1 through 5 carbon atoms, or another anion, such as butyrate.

The clathrate forming compounds are represented by the formula $R_4'NX$ where R' is a straight chain alkyl radical containing 1 through 4 carbon atoms or combinations thereof and X is fluoride, chloride, or bromide. When dissolved in water this particular class of clathrate forming compounds has been found to have a particularly good ability to form the desired substantially solid clathrate. Within this specific class, tetra-n-butyl ammonium fluoride is the most preferred.

To form the thermal energy storage material by this process, one or more of the above starting materials is dissolved in water to form a substantially solid clathrate. This dissolution is preferably carried out at substantially atmospheric temperature and pressure; however, higher temperatures and pressures can be used if desired to increase the rate of dissolution and, hence, the rate of solid clathrate formation.

The resulting clathrate has a melting point higher than $32°F$, which of course varies depending upon the specific clathrate forming compound used. For example, tetra-n-butyl ammonium fluoride forms a clathrate, $(C_4H_9)_4NF\cdot32.8H_2O$, which has a melting point of about $77°F$; tetra-n-butyl ammonium chloride forms a clathrate, $(C_4H_9)_4NCl\cdot33H_2O$, which has a melting point of about $50°F$, and tetra-n-butyl ammonium butyrate forms a clathrate which has a melting point of about $59°F$.

The clathrate is then separated from the solution in any suitable manner. For instance, if the solution temperature is above the melting point of the clathrate formed, the solution is cooled to a temperature below the melting point of the clathrate but above $32°F$ causing the clathrate to precipitate. The excess water is then removed, such as by decantation and/or evaporation.

The resultant clathrates used as the thermal energy storage material have latent heats of fusion and specific heats approaching that of water and have melting points higher than $32°F$. As mentioned above, the oxygen to oxygen distance of the water in the hydrogen-bonded cage of the clathrate is less than that for normal water. As, for example, the O to O distance for the clathrate, $(n-C_4H_9)_4NF\cdot32.8H_2O$ formed from tetra-n-butyl ammonium fluoride is 2.67 A. The thermal energy storage material, much like water, is capable of absorbing or storing reasonably large quantities of heat upon "melting" or changing in phase from a solid to a liquid state and then releasing this absorbed or stored heat upon "freezing" or changing in phase from a liquid to a solid state.

Dual-Temperature Thermal Energy Storage Composition

H. Spauschus and L. Loeb; U.S. Patent 4,100,092; July 11, 1978; assigned to General Electric Company describe a combination thermal energy storage material capable of functioning in two discrete temperature ranges and comprised of two constituents, one of which has a phase change in the low temperature range between about $35°F$ and $55°F$, and the other of which has a phase change in the high temperature range of between $90°F$ and $130°F$. Both temperature ranges are selected to correspond with the operable and desirable ranges for a heat pump operation. The two constituents are selected to be nonreactive with one another and in the mixture maintain independent phase change characteristics.

However, the mixture itself alters in some cases, depending upon concentration, the melting points of each member of the pair so that the transition temperatures of the phase changes in the mixture are somewhat different and somewhat lower than the phase changes characteristic of the independent constituents and are dependent upon the concentration of the constituents in the mixture. The pairs of materials selected can, therefore, function in the heat pump environment to provide a heat sink into which heat may be dissipated for air cooling purposes as well as a heat source which provides heat for heating purposes. Preferred pairs of materials include a hydrocarbon oil paired with a wax fraction; a C_8 fatty acid paired with a C_{14}-C_{18} fatty acid; a C_{10}-C_{12} alcohol with a C_{14}-C_{18} fatty acid; and a C_{10}-C_{12} alcohol with a wax fraction. Preferred weight percent of the components are also selected to conform to the high and low transition temperature ranges used as parameters.

The process thus provides a combination heat storage and heat sink material wherein the same formulation can function either as a heat source or, in the alternative, as a heat sink. The process utilizes the same material to provide auxiliary heat supplied during heating mode operation of a heat pump and, alternatively, as a heat sink to receive heat during air-cooling periods.

Aqueous Solution of Potassium Fluoride

A process described by *J. Schroder; U.S. Patent 4,091,863; May 30, 1978; assigned to U.S. Philips Corporation* relates to a method of reversibly storing latent heat in a heat storage medium comprising a liquid phase part and a solid phase part wherein heat is stored by melting the solid phase part of the heat storage medium and wherein the liquid phase part is subjected to supercooling.

The process is characterized in that the heat storage medium is charged by being conducted past a heat exchanger having a temperature above the melting point of the heat storage medium and is discharged by the temperature of the heat exchanger being kept at a sufficiently low value below this melting point and the flow rate of the heat storage medium near the heat exchanger being maintained at a sufficiently high value so that substantially no crystallization of the heat storage medium occurs near the heat exchanger and that substantially no crystals are deposited on the heat exchanger. The liquid phase part of the heat storage medium which is supercooled and supersaturated is subsequently conducted past a location where a crystal nucleating material is present or past the heat storage medium solidified at this location so that the supersaturated part of the heat storage medium is solidified and separated and the remaining liquid heat storage medium is returned to the heat exchanger.

In the method, the usually detrimental supercooling tendency of the heat storage material is used in a positive sense for storing heat in and extracting heat from this heat storage medium in a simple manner, without carrier materials spread through the entire volume of the heat storage medium or dispersions of seeds being required.

In a preferred example, the surface of the heat exchanger is briefly heated above the melting point of the heat storage medium in the case of crystal formation on the heat exchanger. This can be effected by reversing a heat pump in which the heat exchanger is included.

An aqueous solution of 44 to 48% by weight KF is preferably used as the heat storage medium. $Na_2SO_4 \cdot 10H_2O$ is also very suitable.

Regenerative Nucleating Agents

A process described by *D.D. Chadha; U.S. Patent 3,956,153; May 11, 1976; assigned to Bio-Medical Sciences, Inc.* pertains to a method of minimizing the phenomenon of undercooling in the liquid-solid phase transformation of thermally responsive materials and to special compositions demonstrating such properties.

A variety of industrial products and processes utilize transformation of various materials from liquid to solid phase or solid to liquid phase. For example, molten masses have been employed to achieve heat storage. The reversible absorption of heat in passing from a solid phase into a liquid phase or into another solid phase is employed, for example, in accumulators or reservoirs in refrigerated containers or vehicles. Phase transformations are utilized in a number of temperature measuring devices, the devices being designed so that a phase transformation occurs at a given and preselected temperature so as to indicate the attainment of at least that temperature. Hair curlers and solar heating devices also utilize the enthalpy of two different phases.

In all of these applications, the phenomenon of undercooling in the passage of the thermally responsive material from the liquid phase to the solid phase is encountered. Thus it is known that while such thermally responsive materials may present a distinct melting point in passing from the solid phase to the liquid phase solidification of the liquid phase through the removal of heat often occurs at a temperature below the melting point. This phenomenon, which is known as undercooling, is particularly evident in the case of extremely pure materials. It is desirable for a number of reasons that the temperature of solidification be as close to the melting point as possible and that the phenomenon of undercooling be minimized.

There is described in French Patent No. 2,010,241 a means of minimizing undercooling by providing crystal seeds within the heat-responsive material. These are generally of a fibrous crystal structure and are typified by such materials as silicates, aluminates, light metal ferrites, oxides and the like.

The process utilizes a totally different type of nucleating substance. While the nucleating substance of the process shares with the prior art nucleating substances the characteristic of having a crystal structure similar to the thermally responsive material, that is at least one a/b, a/c or c/b unit cell ratio and a space group substantially identical with the heat-responsive material, the process represents a distinct departure from the prior art approach in that the nucleating substance is intentionally soluble, in the heat-responsive material and increasingly so at temperatures above the temperature at which the phase transformation occurs.

Moreover, the nucleating substance will have a melting point significantly higher than the temperature at which the phase transformation occurs. The magnitude of the difference between the melting point of the heat-responsive material and the nucleating substance can vary widely, depending primarily upon the temperatures to which the heat-responsive material will normally be subjected, it being desirable that the nucleating substance melt above such 'normally encountered temperatures.'

In operation, a nucleating substance satisfying the above criteria performs as follows. An amount of the nucleating agent in excess of that which will dissolve in the thermally responsive material at the phase transformation temperature is added to the thermally responsive material, the latter being in the liquid phase. The amount of excess is not critical and indeed it is generally desirable to add only small excess. The absolute percent composition will of course depend upon the particular heat-responsive material and nucleating substance, primarily the solubility of the latter in the former, but generally there will be from about 0.1 to 50% by weight of the nucleating substance in the thermally responsive material.

In view of the fact that the nucleating substance is only sparingly soluble in the heat-responsive material, it has little effect on depressing the melting point of the material. Any small amount of melting point depression which occurs is readily calculated and easily nullified through adjustment of the composition of the thermally responsive material or the selection of other chemicals.

Example 1: A series of solid solutions of o-bromonitrobenzene and o-chloro-nitrobenzene having weight percent compositions ratios ranging from 56.2:43.8 to 96.0:4.0 respectively were prepared as described in Belgian Patent 770,290; These demonstrated graduated melting points ranging from 96.0° (35.55°C) to 104.8°F (40.44°C). Such solutions containing no other components demonstrate an undercooling ($\Delta T_{uc} = T_m - T_s$) of about 34°F; the temperature of resolidification of the liquid phase is approximately 34°F below the temperature at which the solid phase melts. Thus the liquid phase must be cooled to approximately 62° to 63°F in order to achieve resolidification.

These solid solutions are heated to 140°F and saturated with anthraquinone which was previously purified by recrystallization from toluene. The saturated solutions are then cooled and demonstrate a ΔT_{uc} of only 11°F, i.e., it instantly resolidifies at temperatures of from 82° to 83°F.

Example 2: A series of tests were conducted to determine the time required for recrystallization of a solid solution of 56.2% o-bromonitrobenzene and 43.8% o-chloronitrobenzene containing a nucleating substance upon being subjected to room temperature (73°F) after storage at elevated temperatures. The nucleating substance is anthraquinone as set forth in Example 1.

| Sample No. | Saturation Temperature (°F) | Aging Conditions | | | Resolidif-ication Time (min) |
		Temp (°F)	RH (%)	Time (days)	
1	130	120	*	30	<5
2	140	120	50	21	<5
3	125	120	0	16	<5
4	125	120	*	16	<5
5	125	120	50	16	<5

*Ambient

A sixth sample saturated at 125°C was repeatedly heated to 120°F and cooled to 73°F at ambient humidity for 50 cycles. The sample continued to require less than 5 minutes to resolidify upon being subjected to the lower temperature.

Samples in actual thermometers in all cases achieved resolidification within 15 minutes of being subjected to room temperatures (70° to 73°F). Only in the case of aging the materials at temperatures above the temperature of saturation was slower resolidification observed, presumably because all the anthroquinone was then dissolved and a supersaturated solution was obtained upon cooling.

Suppression of Ice Formation in Bodies of Water

J.S. Best; U.S. Patent 3,932,997; January 20, 1976; assigned to The Dow Chemical Company describes a method for suppressing the formation of ice in natural or manmade bodies of water such as lakes, rivers, sounds, straits, bays, shipping canals, shipping locks, and harbors. The method is particularly useful in areas where ice breakers or the like cannot be efficiently operated.

The method provides for the direct replacement of heat lost to the atmosphere from a body of water during cold seasonal days when the air temperature is below that necessary for the formation of an ice cover on the body of water. This replacement of heat results in maintaining the upper layer of water in the body of water just above freezing, thus suppressing the formation of an ice cover. By relating heat losses to differences in temperature between air and water, the rate of transfer of heat between surfaces of various bodies of water and the atmosphere can be established with a fair degree of accuracy. It has been found that this rate may be taken at about 95 Btu, transferred per day per square foot of water surface per degree fahrenheit, and is independent of the surface character of the body of water in question, that is, the surface of rapids, the surface of lakes and the surface of smooth sections of canals or rivers all have about the same cooling coefficient or ratio of heat transfer.

The process takes advantage of low-level thermal energy sources by using liquid compositions to store heat near the body of water in which it is desired to suppress the formation of an ice cover. The compositions freeze substantially above $0°C$ ($32°F$) and have relatively large latent heats of fusion associated with their melting and freezing phase change, which heat can be transferred to the surface layer of the body of water by heat transfer means. Heat transfer apparatus may be utilized which is of a permanent type located in retaining side walls and/or in the top layer of water adjacent the sides of a body of water to prevent natural bridging of an ice cover on the same or may be a traveling heat exchanger disposed in a barge, scow or the like which moves along a desired path in a body of water.

When a traveling heat exchanger is used such as a barge or scow, it is necessary to utilize intermediate compositions which have freezing temperatures substantially above $0°C$ ($32°F$) but below that of the compositions which are used to store heat near a body of water. By utilizing the release of the relatively large latent heats of fusion of the compositions when they freeze coupled with the temperature differential driving forces from the freezing temperature of the storage compositions to the freezing temperature of the intermediate compositions, if used, and then to the water which is just above freezing, a large heat transfer can be accomplished from the heat storage compositions to a body of water.

Many compositions useful in the process are described in U.S. Patent 3,834,456. The compositions described are single phase solutions comprising water and at least one organic component which are mixed in such proportions that a homogeneous, crystalline, icelike solid hydrate is formed upon freezing. The aqueous organic compositions have latent heats of fusion in the order of about 140 Btu per pound. Examples of compositions found in the above-noted patent which are particularly useful are mixtures comprising one mol of pinacol and six mols of water which composition freezes at about 45°C (113°F), and one mol of butane-2,3-diol and six mols of water which composition freezes at about 14°C (57°F). Another beneficial property of many of the organic compositions described is a relatively low percent expansion or contraction of the same during freezing which significantly reduces mechanical stress on their encapsulating structures.

The thermal energy source utilized to regenerate or melt the heat storage compositions can be provided in numerous ways as, for example, from condenser water of steam power plants, nuclear energy power reactors, solar energy, exhaust gases or cooling water of diesel engines or the like and direct heating with fossil fuels. The thermal energy sources may also be used in combination as, for example, where solar energy, used as the primary source to regenerate or melt the storage compositions, is supplemented with fossil fuel heat during peak loads when the available solar energy is inadequate.

In addition, warm bottom waters may also be pumped to the surface and used in combination with fixed position or traveling heat exchangers to relieve the strain on other thermal energy sources. In some instances, it may even be practical to use bottom water as a primary thermal energy source to regenerate or melt the heat storage compositions of the fixed storage facilities or the traveling heat exchanger as, for example, where bottom water from a large industrial holding pond is used to melt the heat storage composition in a barge heat exchanger. Of course, the extent bottom water is utilized in combination with other thermal energy sources is governed by its temperature and location relative to the area where an ice cover is being controlled.

OTHER PROCESSES

Rock Thermal Storage System

R.C. Mitchell, J. Friedman, R.J. Holl and C.R. Easton; U.S. Patent 4,124,061; Nov. 7, 1978; assigned to Rockwell International Corp. describe a thermal storage unit comprising a bed of particulate solid material, a liquid situated in heat-exchanging relation with the bed and cooperating with the bed to define a liquid-solid system containing a thermocline, and means for introducing liquid into and extracting liquid from the liquid-solid system.

Thus, it is known that water (and other liquids) may produce a thermocline, that is, the hot and cold water may be made to separate into layers having a fairly distinct boundary, which rises or lowers within the container as water is added or withdrawn, so that the temperature of the water being drawn off can be substantially constant until the thermocline is reached and, at that point, will drop sharply to the temperature of the unheated water. This phenomenon is familiar in domestic hot water heaters.

The thermal storage system of the process permits the thermocline principle to be employed with rock thermal storage systems and, consequently at considerably higher temperatures than have been economically feasible with the prior art systems.

The advantages of the process are preferably attained by controlling such factors as bed material and particle size, fluid velocity, void fraction and method of fluid distribution so as to produce a thermocline in the bed of crushed rock or the like and, thus, to obtain the thermocline principle with storage temperatures up to at least 1500°F.

Heat Sink Encapsulated in Polymeric Resinous Matrix

A process described by *J.S. Best and W.J. McMillan; U.S. Patent 4,003,426; Jan. 18, 1977; assigned to The Dow Chemical Co.* provides a heat or thermal energy storage structure comprising a crosslinked polymeric resinous matrix having a plurality of substantially unconnected small closed cavities and a heat sink material encapsulated within the cavities. The storage structure is characterized in that the heat sink material forms an essentially stable dispersion in the uncured polymeric resinous matrix when mixed therewith before the matrix is crosslinked. Depending on the service requirements to be met, a heat sink material having a specific melting point can be selected to take advantage of the relatively large latent heat capacity at its melting and freezing phase change for supplying heat to or removing heat from the storage structure. Beneficially, the structure can be used in cooperation with low level heat or thermal energy collector means and heat transfer means to provide a space heating or cooling apparatus, the storage structure can also effectively provide a secondary function of a building component in a building construction such as at least one wall of a building.

The heat or thermal energy storage structure which can be utilized in space heating or cooling apparatus can be formed with a matrix of any crosslinked or thermoset polymeric resinous material provided the heat sink material will form an essentially stable dispersion in the uncured polymeric resinous matrix before the matrix is crosslinked. Particularly suited crosslinked polymeric ressinous matrixes are selected from the group consisting essentially of polyesters, polyvinyl esters, and epoxies. Beneficially, the crosslinked polymeric resinous matrix comprises as low as 25 wt % but preferably 35 wt % or more of the heat or thermal energy storage structure.

The heat sink materials useful in forming a heat or thermal energy storage structure should be selected to take advantage of their relatively large latent heat capacity at their melting and freezing phase change which supplies heat to or removes heat from the heat or thermal energy storage structure. The heat sink materials are characterized in that they should form an essentially stable dispersion in the uncured polymeric resinous matrix when mixed therewith before the matrix is crosslinked. Depending on the service requirements to be met, the heat sink materials forming the heat or thermal energy storage structure useful in space heating or cooling apparatus should beneficially melt between 5° and about 100°C and should not supercool more than about 5°C below the melt temperature of the heat sink material. The heat sink material can comprise up to about 65 wt % of the heat or thermal energy storage structure.

A wide variety of heat sink materials may be employed in the heat or thermal energy storage structure in accordance with the present process. Particularly suited as heat sink materials are inorganic hydrates such as barium hydroxide octahydrate, zinc nitrate hexahydrate, strontium bromide hexahydrate, calcium bromide hexahydrate, calcium chloride hexahydrate, ferric bromide hexahydrate and the like. Other inorganic hydrates which have poor supercooling characteristics such as sodium sulfate decahydrate and calcium nitrate tetrahydrate can be used as heat sink materials provided they are nucleated.

Water and Magnesium Chloride

L. Greiner; U.S. Patent 4,126,016; November 21, 1978 describes a heating and cooling apparatus which comprises two vessels connected by a conduit, the first containing a vaporizable liquid, for example, water, and the second containing a vapor absorptive chemical such as magnesium chloride. Liquid evaporates from the first vessel, and the vapor from this evaporation passing through the conduit is absorbed in the vapor absorptive chemical. This process is exothermic at the vapor absorptive container and endothermic at the vaporizing container, and is thus useful for cooling or heating other materials. The conduit connecting these vessels must be evacuated so that other gases do not interfere with the vapor process. By this means the pressure within the conduit can be determined solely by the vapor pressure of the evaporating liquid. The process permits the operation of a very simple valve which assures maintenance of the vacuum in the conduit before and during operation of the refrigeration cycle. One form of the process involves a valve and interconnect which allows the conduit to be formed as two pieces, one connected to each of the vessels. This interconnect and valve seals each of the conduit segments prior to interconnection and permits interconnection of the evacuated conduit sections without introducing any air to the conduit. Another example of the process provides an extremely simple and inexpensive valve for fluidly interconnecting the pair of vessels while maintaining the vacuum integrity, but in this case the vessels are manufactured and evacuated while mechanically joined by the conduit and remain mechanically connected prior to and during use.

CONTROL OF HEAT TRANSFER
IN BUILDING STRUCTURES

AIR CONDITIONING PROCESSES

Optimal Utilization of Daily Temperature Variations

According to a process described by *L.O. Andersson, E. Isfält and A. Rosell; U.S. Patent 4,124,062; November 7, 1978* there is a device for exploiting the daily variations of the outside air temperature for controlling the air temperature, for example, in a structure. In the process, the outside air is caused to flow continuously or discontinuously through channels in a heat-storing mass and thereby is caused by absorbing or giving off heat to change its temperature in accordance with the demand of heating and/or cooling variations during a day. The process is characterized in that the heat-storing mass is at least in part of the building structure surrounding the premises.

For example, it is possible by means of the process by utilizing the relatively low night temperature of the air to achieve the necessary cooling effect during the hot part of the day. It is further possible by utilizing a relatively high temperature of the outside air during the day to achieve the necessary warming effect during the colder part of the day. It is possible to construct such devices with simple means and a relatively low cost.

What is required in principle is an accumulator with a sufficient thermal efficiency and such other properties that both the giving off and the absorbing of heat can be controlled in agreement with the variations of the cooling requirement.

The accumulator can be thought of as comprising both a solid material and also a liquid, for example, water. Economically, however, it is most expedient if some part of the building construction proper can be used as an accumulator.

A calculation of the cooling requirement and the available thermal capacity shows that in a normal building intended for some of the common purposes the use of the floors of the building as the accumulator provides a possible solution. In order to obtain the desired transfer of heat it is necessary, however, to provide between the supply air and the concrete a contact surface which is considerably larger than those found in previous constructions. This can be achieved, for example, by providing the floor with a system of channels through which flows the ventilation air supplied to the room. It is necessary in this connection to choose such a ratio between the transfer surface and the air flow area that both the α values (air-concrete heat transfer number) of the required magnitude are achieved and the thermal course, the giving off and absorbing of heat, is given a suitable periodicity in proportion to the cooling requirement of the premises.

This course is influenced also by the arrangement of the air channels in the concrete slab and also by the size of the air flow.

An arrangement of the channels at a greater distance from a surface results inter alia in a greater delay and also in a reduction in the temperature variations of the surface in relation to the variations of the outside air. This is of particular importance beyond the effect on the accumulated heat amount, because the heat transfer between the surfaces of the concrete slab and the room represents a substantial part of the heat exchange.

The greatest amount of heat on the premises where the air treatment is to be applied often is constituted by sun radiation through glass surfaces in the front of the building. Absorption of heat is, therefore, necessary for completely different periods of time on fronts facing in different directions. The length of the air channels, their position in the concrete slab and the air flow, thus, must be so adjusted that absorption of heat takes place in the slab at that precise period of time when there is a cooling requirement.

In office and business buildings and also residential dwellings the amount of heat by sun radiation, measured, for example, per longitudinal meter of the front, lies within certain pretty well defined limits. The thickness of the floors and the air flow suitable for the treatment of the air likewise lie within relatively narrow limits. This means that the maximum requirement for transferring heat between air and concrete also shows relatively small variations in different plants. The deciding factor for heat transfer at a given flow of air is A x α, where A is the circumferential surface of the channels and α is the air-concrete heat transfer number.

In order for a cooling plant of the construction to work correctly, the highest value for the factor A x α must, therefore, lie within a certain definite range. If the requirement for heat transfer falls, then A x α can be reduced by adjusting the air flow.

The whole process, in mathematical terms, is very complicated, and it would have been practically impossible before the introduction of the data processing technique to conduct the calculations which are necessary for a safe function. Today, however, it is possible to construct a data processing program which provides a safe and easily understandable basis for dimensioning.

Graphs (Figure 5.1a and 5.1b) show the variations in the room temperature in a 3-module room in an assumed office building to which outside air is supplied, the temperature of which varies in accordance with a curve which is normal for one day during a hot period in a northern European climate. The supply air was assumed to flow through floor channels which were dimensioned with respect to the desired accumulation effect. Both the thickness of the floor and also the air flow were chosen according to normally occurring values.

Figure 5.1:   Temperature Control in Building Structures

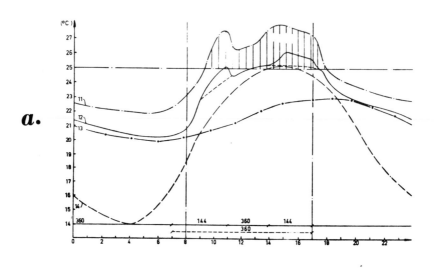

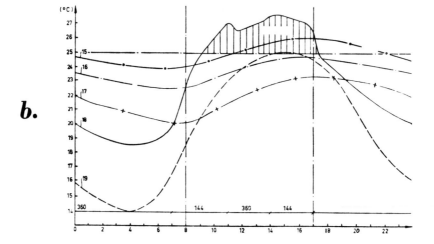

(continued)

Figure 5.1: (continued)

c.

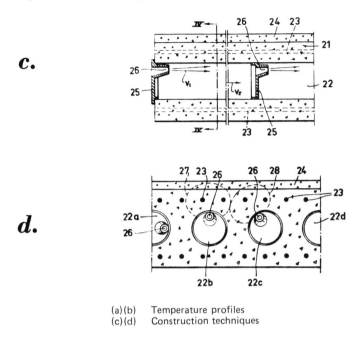

d.

(a)(b)   Temperature profiles
(c)(d)   Construction techniques

Source: U.S. Patent 4,124,062

These graphs as well as a series of similar graphs with different prerequisites were drawn up with the help of available data processing programs. They show unambiguously the possibility of controlling the temperature course in a satisfactory way.

The calculations were carried out with a data processing program which was developed by lecturer Brown of the Institute for Heating and Cooling Technology at the Royal Technical High School in Stockholm, in cooperation with AB Datasystem and with funds from the National Swedish Council for Building Research. The calculation model which forms the basis of the program is very complete.

Thus, Figures 5.1a and 5.1b show graphs of the temperature course in an office room with 3 modules (3 x 1.2 m), south front with circa 40% glass surface. The chosen outside conditions correspond to a period in July with clear weather and a maximum outside temperature of +24°C. The floor has a thickness of 0.3 m, and a blind is positioned between the window panes. Figures 5.1c and 5.1d, the latter a section IV–IV of Figure 5.1c, show in longitudinal/cross section an advantageous device according to the process for effecting the transfer of heat between channels in a floor, through which air is flowing, and the floor.

Curve **14** shows an assumed temperature course for the outside air which is blown into a floor according to the process. The temperature of the air flowing into the floor was assumed in this case to be equal to the temperature of the outside air with an addition of 1°C for the fan work. In this case, the temperature shown for zero time is circa 16°C. The temperature thereafter falls for 4 hours before it begins to rise again and reaches 25°C at circa 14 hours after which it falls again.

The temperature of the air blown into the room will, as a result of the heat-accumulating properties of the floor, vary according to curve **13**. During the night the floor will heat the air which is flowing through and during the day it will cool it. If the floor were not constructed in accordance with the process, i.e., without regard to the exchange of heat with the air flowing through, the temperature of the air blown into the room would have varied as the outside air, with the addition of 1°C according to curve **14** in Figure 5.1a or curve **19** in Figure 5.1b (**14** and **19** are the same curves).

The exchange of heat between the floor and the air flowing through was controlled by changing the air flow. For 7 hours the air flow into the room from the floor is 360 kg/hr, and between 7 hours and 11 hours it is 144 kg/hr, etc., in accordance with the scale above the hour axis.

Curve **12** shows how the temperature of the air in the room will vary. For 8 hours the temperature of the air in the room will be substantially equal to the temperature of the air which is blown in. After 8 hours the effect of sun radiation makes itself felt with a resulting increase in temperature. At 11 hours, the air flow increases to 360 kg/hr, the temperature of the air in the room falling as a result of the removal of heat from the floor which is thereby effected. The dot-dash section of curve **12** corresponds to the case where the air flow is constantly equal to 360 kg/hr during the whole day (and the night as well).

A substantial part of the cooling of the room during the day takes place by transmission of heat between the floor and the room. Curve **11** shows the temperature of the air in the room if the temperature of the air blown in is assumed to vary in accordance with curve **13**, and the floor is conventional, i.e., not adapted for heat accumulation. As the floor cannot be cooled in this case, its surface temperature will be higher in the morning, and as a result its cooling effect by radiation will be worse.

This latter behavior is shown by curves **15** and **17** in Figure 5.1b, where curve **15** shows the temperature of the surface of the ceiling at air temperatures according to curves **11** and **13** and curve **17** shows the temperature of the ceiling with a floor according to the process at air temperatures in accordance with curves **12** and **13**. With the floor according to the process, according to curve **17** the temperature of the ceiling surface will be circa 20°C at 8 hours as a result of the cooling during the night. The corresponding temperature for the ceiling with the floor which is not adapted to be cooled is circa 24°C in accordance with curve **15**. The difference in the temperature of the air in the room between curves **11** and **12** is thus caused by the difference in the ceiling temperature according to curves **15** and **17**.

The curves **16, 18** and **19** refer to a conventional air treatment system in which the external air is introduced directly into the room without the exchange of heat with an accumulator and the temperature of the air blown into the room in accordance with curve **19** is assumed to be 1°C higher than the temperature of the outside air. The temperature of the room air will in this case vary in accordance with curve **18** and the temperature of the surface of the ceiling in accordance with curve **16**. Here also the temperature of the surface of the ceiling is higher than the ceiling surface adapted to be cooled according to curve **17**.

In the device according to Figure 5.1c and 5.1d, the air is caused, during its passage through the channels **22**, to form one or more jets by means of throttle member **25**, the jets being directed against selected parts of the channel wall.

In order that a heat transfer of some magnitude can take place in conjunction between the jet and the wall, the jet must have such a speed that turbulence occurs between the jet and the channel wall. The device has the advantage that the jet and thus the transfer of heat can be concentrated on those places in the channel wall which have the greatest surrounding mass.

It is further possible by this means for the mass surrounding the channel walls to become ineffective in terms of heat storage by virtue of the fact that the air flow is reduced below the limit at which turbulence occurs between the jet and the channel wall. The cessation of turbulence leads to a drastic reduction in the transfer of heat to insignificant values, without this happening at the expense of the demand for adequate ventilation flow.

By means of the device according to Figures 5.1c and 5.1d it is also possible to use hollow floor arrangements which are already known without large-scale modifications.

The actual floor is designated by **21**. A number of parallel channels **22** extend through the floor in its longitudinal direction and can have a circular cross section as shown, for example, in Figure 5.1d. In Figure 5.1d, the channels are designated by **22a**, **22b**, **22c** and **22d**. The floor is further provided with suitable reinforcement **23** on which structural concrete **24** can also be poured. The inlet end of the channel **22** has a throttle member **25** with a nozzle aperture **26** as the only connection between the two sides of the throttle member **25**. The throttle member **25** is, for example, a rotatable unit inserted into the channel aperture. One or more further throttle members **25** with nozzle apertures **26** can be placed deeper inside the channel **22**.

The throttle member **25** with the nozzle aperture **26** is so used for the exchange of heat between air and floor **21** that an increased speed V_1, for example, circa 10 m/sec, is imparted to the air by means of the nozzle aperture **26**. The air jet from the nozzle aperture is directed against a suitable portion of the channel wall. In the channel **22b** in Figure 5.1d, for example, the jet is directed against the roof of the channel.

In this case, the mass is cooled and, respectively, heated on the inside in front of the dot-dash mark **27** by the air jet. In channel **22c** according to Figure 5.1d, in order to explain the process, the throttle member **25** has been rotated somewhat so that in this case it is substantially the mass within the marking **28** which will exchange heat with the air flowing through the floor.

In this way it is possible to concentrate the heat exchange at that point where most of the mass is situated. If, for example, by pouring on structural concrete **24** the mass above the channels becomes greater, it is possible by means of expediently directed nozzle apertures and adjustments of the air speed to concentrate the heat exchange on to this mass.

The transfer of heat between jets from the nozzle aperture will take place, as a result of the relatively high jet speed, with a heat transfer number which is many times greater than the heat transfer between the air further inwards in the channel where it can have, for example, a speed V_2 of circa 1 m/sec, as the heat transfer in the first case takes place at violent turbulence. If the jet speed is lowered sufficiently the turbulence stops and this results in a drastic reduction in the transfer of heat. This property of the device according to the process, can be exploited in such a way that, if it is desired to use the floor as an accumulator, the air flow is so increased that turbulence occurs between jet and channel wall.

If it is not desired that the air be influenced by and, respectively, influences the floor, the air flow is reduced below the value below what corresponds to turbulence between jet and wall. The flows required for good ventilation do not suffer from this as it is no problem to keep the air flow for initial turbulence between the limits recommended for good ventilation.

By varying the design, placement and number of the throttle members **25** it is easy to obtain the desired exchange of heat with the heat storing mass. At a relatively short channel a single throttle member in the channel may be sufficient, whereas at longer channels two or more throttle members distributed along the channel can be necessary so that the heat-storing mass surrounding the channel can be exploited to a sufficient degree.

At the floor in Figure 5.1c, the throttle member **25** at the inlet of the channel can also be imagined to be advanced by a short distance into the channel.

With an air conditioning plant as described here used in terms of channel dimension and channel arrangement, and controlled by a correctly placed control device, it is possible in a climate of North European type to keep the condition of the air in a room within conventional comfort limits without the installation having to include cooling machinery.

Temperature Control System

J. Faiczak; U.S. Patent 4,061,185; December 6, 1977; assigned to Canada Square Management Ltd., Canada describes a method and apparatus for controlling the temperature in a building having a heat transfer load, a temperature control circuit containing a temperature control fluid for transferring heat energy to or from the building load, and a heat energy storage reservoir for storing heat energy. Energy is transferred between the storage reservoir and the fluid in the control circuit at optimum times to reduce the external energy required to control the temperature of the building.

The amount of heat energy required to be transferred to or from the building load during a predetermined interval is predicted from load profiles. The amount of energy in the reservoir is measured to determine the amount of existing available energy to be transferred between the building load and the reservoir. A transfer unit is coupled to the control circuit to vary the heat energy in the field over the predetermined interval, so that the total energy available to be transferred is at least equal to the energy required to be transferred. Finally, the existing available energy to be transferred to or from the reservoir is dissipated by controlling the flow of fluid in the reservoir, so that the energy storage capacity is used at a rate proportional to the difference between the building heat transfer load and the rate of variation by the transfer unit. Heat energy or cooling capacity not immediately required is stored in the reservoir for later use during the predetermined interval.

Reverse Cycle Heat Pump Circuit

L.B. Chambless; U.S. Patent 4,057,977; November 15, 1977; assigned to General Electric Company describes a reverse cycle heat pump refrigeration system including multicircuited heat exchangers arranged to provide a series of refrigerant flow through the circuits for either heat exchanger when it is used as a condenser, and a parallel refrigerant flow through the circuits for either heat exchanger when it is used as an evaporator.

Referring to Figure 5.2, a conventional reversible cycle heat pump system is shown including a compressor **10** having a high pressure gas discharge connected by a conduit **12** to the intake of a reversing valve **14**. One reverse flow port **16** of valve **14** is connected by a conduit **18** to a line **19** of a heat exchanger **20** which when it is used as a condenser is the inlet line. Heat exchanger **20** is preferably located or arranged so that it is subjected to outdoor air and is referred to as the "outdoor coil." A second reverse flow port **22** of the valve **14** is connected by conduit **24** to a line **25** of a heat exchanger **26** which when it is used as a condenser is the inlet line. Heat exchanger **26** is disposed so that it is subjected to recirculated indoor air and is referred to as the "indoor coil."

The low pressure intake or suction port of compressor **10** is connected by a conduit **30** to the exhaust port **23** of valve **14**, which port is selectively connected with either reverse flow ports **16** and **22**. In the cooling position the valve **14** is arranged so that the high pressure discharge gas from conduit **12** is directed through port **16** and connected through conduit **18** to the outdoor coil **20** which in the cooling cycle is used as the condenser. The suction or intake low pressure gas returning to the compressor from heat exchanger **26** which is being used as the evaporator during the cooling cycle is through inlet line **25**, conduit **24**, reverse flow port **22**, suction line **30** and back to the compressor **10**. To complete the refrigeration cycle the heat exchangers **20** and **26** are interconnected by a conduit **28**.

Each of the heat exchangers **20** and **26** include a plurality of circuits **32** and **34** respectively. By the reversible heat pump refrigeration system of the process, a series refrigerant flow is provided through the circuits **32, 34** of either heat exchanger **20, 26** when it is used as a condenser, and a parallel refrigerant flow is provided through the circuits **32, 34** of either heat exchanger **20, 26** when it is used as an evaporator.

Each of the separate circuits **32** and **34** may comprise a conventional serpentine arranged tube **36** connected in series with the adjacent circuit of a heat exchanger by conduits or return bends **38**. In the cooling mode the high pressure discharge gas from compressor **10** is directed from port **16** of valve **14** through conduit **18** and inlet line **19** of outdoor heat exchanger **20** which is in this instance operating as the condenser. High pressure gas flows from inlet **19** through the series arranged circuits in heat exchanger **20** and is condensed to a high pressure liquid which flows into an outlet which in this flow direction is conduit **40**. From conduit **40** the high pressure gas passes through a one-way check valve **42** and into the conduit **28** interconnecting heat exchangers **20** and **26**.

It should be noted that the pressure drop through a heat exchanger operating as a condenser in a refrigeration system is generally more than that required for the heat exchanger operating as an evaporator. By this process, when a heat exchanger is operating as a condenser the circuits therethrough are arranged in series so that the condenser will have enough length through each circuit to provide subcooling for the liquid refrigerant. As the refrigerant gas condenses to a high pressure liquid, it becomes dense and needs less area and volume for a given mass.

Figure 5.2: Reverse Cycle Heat Pump Circuit

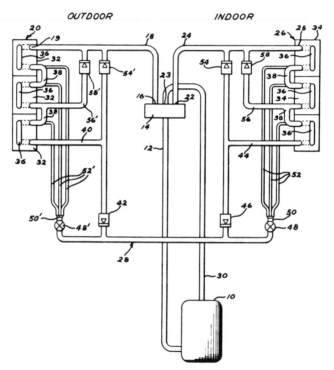

Source: U.S. Patent 4,057,977

In the heating mode the high pressure discharge gas from compressor **10** is directed from port **22** of valve **14** through conduit **24** and into the inlet line **25** of indoor heat exchanger **26** which is in this instance operating as the condenser. High pressure gas flows from inlet **25** through the series arranged circuits in heat exchanger **26** and is condensed to a high pressure liquid which flows into an outlet which in this flow direction is conduit **44**. From conduit **44** the high pressure gas pours through a one-way check valve **46** and into the conduit **28** interconnecting heat exchangers **20** and **26**. Accordingly the above arrangement provides a series refrigerant flow through the circuits for either heat exchanger **20** or **26** in the cooling and heating cycle when they are used as condensers. This series refrigerant flow arrangement of the heat exchangers **20, 26** when they are used as condensers allows an effective pressure drop that is sufficient to condense the high pressure gas into a liquid that may then pass effectively through the expansion or capillary portion of the system.

It should be noted that the pressure drop through a heat exchanger operating as an evaporator in a refrigeration system is generally less than that required of the heat exchanger operating as a condenser. By this process, when a heat exchanger is operating as an evaporator, the circuits therethrough are arranged in parallel so that the relative shorter length of each circuit keeps the pressure drop low while providing greater area and volume for the gas as boiling occurs at which time the liquid changes to a gas.

In the cooling mode high pressure liquid passing through valve **42** into conduit **28** continues through a thermostatic expansion valve **48** to a distributor **50**. At this time valve **46** is effective in preventing high pressure liquid from entering conduit **44** and heat exchanger **26** which is operating as an evaporator. Distribution conduits **52** are connected between the distributor **50** and each of the circuits **34** of indoor heat exchanger **26**. Accordingly the circuits **34** of the indoor heat exchanger **26** operating as the evaporator in the cooling mode are arranged in parallel so that a lower pressure drop is allowed therethrough relative to the series arranged circuits **32** in the outdoor heat exchanger **20** now operating as a condenser in the cooling mode.

In operation high pressure liquid flows into the thermostatic expansion valve **48** where expansion of the high pressure liquid takes place and the resultant low pressure liquid flows into each of the circuits **34** through conduits **52**. It should be understood that a capillary tube may be employed in place of valve **48** to provide the appropriate expansion of the high pressure liquid. One of the parallel circuits of heat exchanger **26** is completed from capillary **52** through a circuit **34** and into conduit **44**. Flow from conduit **44** then passes through a one-way check valve **54** and into conduit **24** which is connected to suction conduit **30** through the reverse flow port **22** of valve **14**. The presence of high pressure liquid on the other side of valve **46** at this time prevents passage of refrigerant from conduit **44** into conduit **28**.

A second parallel circuit of heat exchanger **26** is completed through a conduit **56** and a one-way check valve **58** and into conduit **24**. The third parallel circuit through heat exchanger **26** is completed directly into conduit **24** through inlet **25**. One-way check valves **54** and **58** are effective in preventing flow into conduits **56** and **44** when the heat exchanger **26** is being used as a condenser.

In the heating mode high pressure liquid passing through valve **46** into conduit **28** continues through a thermostatic expansion valve **48'** to a distributor **50'**. At this time valve **42** is effective in preventing high pressure liquid from entering conduit **40** and heat exchanger **20** which is now operating as an evaporator. Distribution conduits **52'** are connected between distributor **50'** and each of the circuits **32** of outdoor heat exchanger **20**. Accordingly the circuits **32** of the outdoor heat exchanger **20** operating as the evaporator in the heating mode are arranged in parallel so that a lower pressure drop is allowed therethrough relative to the series arranged circuits **34** in indoor heat exchanger **26** operating now as a condenser in the heating mode.

In operation high pressure liquid flows into the thermostatic expansion valve **48'** when expansion of the high pressure liquid takes place and the resultant low pressure liquid flows into each of the circuits **32** through conduits **52'**. Like heat exchanger **26** when it was operating as an evaporator, one of the parallel circuits of heat exchanger **20** is completed from capillary **52'** through a circuit **32** and into conduit **40**. Flow from conduit **40** then passes through a one-way check valve **54'** and into conduit **18** which is connected to suction line **30** through the reverse flow port **16** of valve **14**. The presence of high pressure liquid on the other side of valve **42** at this time prevents passage of refrigerant flow from conduit **40** into conduit **28**.

A second parallel circuit through heat exchanger **20** is completed through a conduit **56'**, a one-way valve **58'** and into conduit **18**. The third parallel circuit through heat exchanger **20** is completed directly into conduit **18**. Like valves **54** and **58**, valves **54'** and **58'** are effective in preventing flow into conduits **56'** and **40** when heat exchanger **20** is operating as a condenser.

All-Climate Heat Exchanger Unit

A process described by *W. R. Iriarte; U.S. Patent 4,064,932; December 27, 1977; assigned to Hughes Aircraft Company* relates to a heat pipe heat recovery unit for unidirectional, but reversible, and temperature controllable transfer of heat between at least two ducts.

The ever-increasing need to conserve energy resources in industrial and habitable environments has generated a large number of possible solutions. Such solutions generally involve the transfer of heat between exhaust and supply ducts in which the warmer fluid of one duct is extracted by means of heat pipes transferred to the fluid flowing through the other duct.

Two concepts are described in U.S. Patents 3,788,388 and 3,980,129. The latter patent is concerned with reducing frost build-up at the exit end of the exhaust-duct heat exchanger. The heat pipes are disposed horizontally within the heat exchanger which is rotated about an axis so that, when ice forms at the end of the warm air exhaust nearest the cold outside environment, the heat exchanger may be turned to 180° about its axis so that warm exhaust air impinges directly on the frosted heat pipes.

In U.S. Patent 3,788,388 heat transfer between ducts is bidirectional which is claimed to occur as a result of the liquid phase of the working fluid being normally free-standing along substantially the entire length of the heat pipes, with a substantial vapor space also substantially along the entire length of the heat pipes. Tilt is induced solely to limit the amount of heat transferred from the air stream in one duct to the air stream in the other duct by reducing the effective length of the heat pipes and, therefore, the overall efficiency of the thermal transfer unit.

The present process presents a solution to the heat recovery problem by improving upon the prior art. It comprises a heat exchanger whose heat pipes are in parallel and thermally coupling at least two ducts so that the fluid in the ducts flows through the heat pipes. The heat pipes are angularly movable about an axis extending through and normal to the ducts, with the important conditions that the heat pipes have a fixed angular offset from the axis about which they rotate. In addition, to operate at higher capacity, the heat pipes are inclined at an angle to a line normal to gravitational force in order to take advantage of gravity to return the working fluid to its evaporator end. Therefore, the heat input must be into the lower side of the heat pipes for proper operation.

When these heat exchangers are used in installations where all-climate operation is required, a simple reorientation of the inclination to gravity is effected by moving the heat pipes angularly about their axis. Therefore, the process permits orientation of the heat exchanger so that it can be operated in either heating (winter) or cooling (summer) conditions by 180° turn of the unit. Precise temperature control as well as defrost control is made by slight angular rotation of the heat exchanger unit such that a part of each air stream may be bypassed.

Defrost control is also possible by a complete 180° rotation, as in above-noted U.S. Patent 3,980,129; however, it is required that the heat pipes be turned another 180° to continue proper heat transfer, as a result of their angular inclination to gravity. Also as in U.S. Patent 3,980,129, by rotating the heat exchanger to 90°, a complete bypass is obtained so that no heat will be transferred.

Referring to Figures 5.3a and 5.3b, a ducting system includes a portion **10** which comprises a first duct **12** separated by a divider wall **13** from a second duct **14** through which, for example, supply air and exhaust air respectively flow. The ducting system may be secured to a habitable enclosure or industrial process. Extending between the two ducts are a plurality of heat pipes **16** which are formed in any manner known in the art. Heat pipes **16** are secured in conventional manners to a four-sided, generally rectangular frame **18** whose opposed sides **18a, 18b** and **18c, 18d** respectively extend generally parallel and normal to heat pipes **16**, so that the frame is open to the supply fluid in duct **12** and the exhaust fluid in duct **14**.

The heat pipes and their frame **18** are mounted on a plate **20**. Plate **20** is pivoted on shafts **22** or equivalent which are journaled in suitable bearings at **24** of ducts **14** and in a large circular bearing **23** in divider wall **13**. Shafts **22** are adapted to be angularly driven or moved by any suitable means **26** which is generally labelled "rotational or angular drive." Conventional seals **23b** and **32** extend about the peripheries **23a** and **27** of frame **18** and bearing **23** to ensure proper fluid sealing to the ducts and wall **13**.

Heat pipes **16** are operated at an angle θ of inclination to gravity. For heat ventilating and air conditioning of habitable enclosures, angle θ is preferably of at least 6°. For industrial or high temperature application, a larger angle θ, for example, 10° or greater, is preferred. In any case, the angle is dependent upon the requirements of the use in which the heat recovery unit is placed. Since shafts **22** and large bearing **23** are generally placed on a line which is normal to the direction of gravitational force, axis **28** of shafts **22** and bearing **23** therefore define such a normal to the direction of gravity. Therefore, as shown in Figure 5.3a, axis **30** of the heat pipes is also at angle θ from the axis about which plate **20** rotates. The angular disposition between axes **28** and **30** is fixed or permanent; however, because supporting plate **20** and angularly offset frame **18** and heat pipes **16** rotate about axis **28**, at 90°; the angle between the heat pipes axis

30 and a plane which is normal to the direction of gravity will be zero. It is intended, however, that heat pipes 16 not be operational when in the 90° position,
that is, 90° with respect to that shown in Figure 5.3a and as shown in Figure 5.3c.

Figure 5.3: All-Climate Heat Exchanger Unit

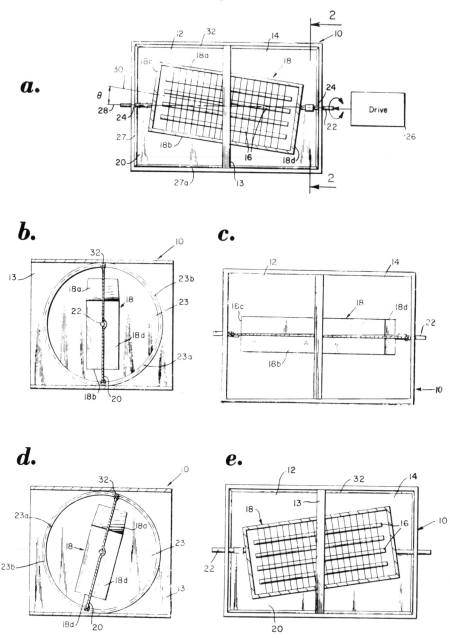

(continued)

Figure 5.3: (continued)

f.

(a)   Cross-sectional view of a pair of ducts
(b)   Cross-sectional view of the system depicted in Figure 5.3a
(c)   Another view similar to that shown in Figure 5.3a, but with the
        heat recovery unit turned 90° to that in the preceding figure
(d)   Another view similar to Figure 5.3b showing a slight angular move-
        ment of the heat recovery unit
(e)   View similar to that shown in Figures 5.3a and 5.3b, but turned
        180° from that illustrated
(f)   Second modification of the process showing accordion seals in one
        of the ducts

Source: U.S. Patent 4,064,932

This particular inclination angle θ of heat pipes **16** with respect to gravity permits gravity to be utilized for the return of the working fluid from the point
where it condenses to the point where it evaporates. In this arrangement, the
heat input must be into the lower side of the heat pipes for proper operation
thereof. In addition, such an angle of inclination to gravity ensures that the
heat pipes operate in a single direction, that is, with unidirectional heat transfer.

Therefore, as shown in Figures 5.3a and 5.3b, the fluid passing through duct **14**
is of higher temperature than that flowing through duct **12** so that heat transfer
is from duct **14** to duct **12**. Because duct **14** has been designated as an exhaust
duct, the configurations shown in Figures 5.3a and 5.3b show an operation of
the process during winter in which the hot or warmer air from the inhabitable
enclosure is transferred to incoming cold air passing through supply duct **12** in
order to preheat it and accordingly to conserve thermal energy.

If it should occur that the amount of heating is too great for the particular position shown, rotational or angular drive **26** may rotate the unit a few degrees so
that a part of the fluid streams passing through both ducts **12** and **14** will escape
over the edges and past seals **32** at edges **27a** of plate **20** which otherwise prevent
any air from escaping around the frame. This slight angular rotation decreases
the efficiency of heat transfer so that less than the full amount of heat capable
of being transferred from exhaust duct **14** to supply duct **12** occurs. Thus, by
varying the angle of inclination, a greater or lesser amount of heat can be controlled so as to obtain the precise temperature control.

Furthermore, in extreme cold where supply air will conduct sufficient heat away from the far end of the heat pipes in exhaust duct **14**, those heat pipes furthest spaced from the habitable enclosure will become frosted and prevent proper operation of the heat recovery unit such as described in above-noted U.S. Patent 3,980,129. To remove the frost, it is only necessary to slightly rotate the heat recovery unit in order to reduce the efficiency of thermal transfer of heat. The frost will therefore melt and be removed and, thereafter, the heat recovery unit may be returned to its normal angular rotation for maximum control.

It is possible, to fully rotate the unit 180°, as suggested in the abovementioned patent; however the frost is removed, and not like that taught by that patent, the heat recovery unit will have to be rotated another 180° back to its original position in order for it to operate.

Such 180° rotation, which was suggested above for purposes of defrost control, is used herein primarily for summer operation of the process, that is, as shown in Figure 5.3e, in which the lower side of the heat recovery unit is now in supply duct **12** so that the cooler air exhausting through duct **14** from the habitable enclosure will cool the incoming air passing through duct **12**. This is particularly effective if air conditioning or equivalent equipment is utilized. Rather than simply exhausting the cold air to the external environment and wasting the energy used in cooling such air, this cooler air is used to draw heat from air being supplied through duct **12**. Therefore, a simple change from summer to winter operation, or vice versa, can be obtained simply by rotating the heat recovery unit by 180°.

If it is desired to completely bypass the heat recovery unit so that it will not operate to transfer heat, it is rotated to a 90° position as shown in Figure 5.3c so that air or other fluid medium is completely bypassed around the thermal recovery unit. In certain cases, this 90° rotation is required in some industrial uses where the environment involves the use of high thermal energies. If such excess heat were allowed to exhaust onto pipes **16**, it would destroy or seriously degrade them. In such a case, it would be necessary to turn the heat exchanger to a 90° position as shown in Figure 5.3c.

In addition, as shown in Figure 5.3f, accordion pleated seals **40** may be attached to the heat pipes only in supply duct **12** so that there will be a certain amount of thermal conductivity by forcing air flow through only one side of the heat exchanger, that is, by drawing whatever heat might be placed on heat pipes **16** in exhaust duct **14**. Thus, possible damage by extremely hot blasts through exhaust duct **14** on the heat pipes is reduced.

According to a process described by *A. Basiulis; U.S. Patent 4,147,206; April 3, 1979; assigned to Hughes Aircraft Company* a selected number of heat pipes, located in the front rows of a plurality of otherwise operable heat pipes which are disposed between intake and exhaust ducts, have liquid-trap sections extending into a switching section. During normal operation, reservoirs in the switching section are dry and the plurality of heat pipes operate in a conventional manner. However, if some of the heat pipes in the exhaust duct become frosted over or otherwise too greatly cooled due to excessive cold in the intake duct, thermostatically or command-controlled valves or louvres cause the fluid stream in the exhaust duct to warm up or defrost the excessively cooled heat pipes therein. Prevention of excessive cooling is used to avoid frost build-up in air conditioning equipment, or solidification of solids and condensation of corrosive liquids.

Variable Terminal Control Box

According to a process developed by *J.L. Hufford; U.S. Patent 4,106,552; August 15, 1978; assigned to Armer Construction Company* an improved heating and air conditioning system and associated equipment are provided which utilize heat losses from other building facilities as well as heat collected from the sun for the heating of the temperature controlled areas.

Figure 5.4a shows a variable volume terminal box **10** contained in a rectangular housing **11** having a length approximately equal to its width and a height being approximately one-fourth the width. The box **10** is divided approximately in half lengthwise by a vertical partition **12** so that two parallel air flow channels are formed including a cooling channel **13** and a heating channel **14**.

The cooling channel **13** is provided at its intake end **15** with a connecting collar **16** which leads into an air valve **17**. The discharge end **18** opens into a discharge plenum **19** through a rectangular opening **21** at the center of which is mounted an air flow sensor **22**.

The heating channel **14** is provided at its inlet end **23** with an air filter **24**, and it houses midway lengthwise an electric heater unit **25**. A squirrel cage blower **26** is located at its outlet end **27**, the blower **26** exhausting through a rectangular opening **28** in end **18** into the discharge plenum **19** which is common to both channels **13** and **14**.

Attached to the outside of the heating channel **14** is a control box **29** which houses various pneumatic and electric control devices for the operation of valve **17**, heater unit **25** and blower **26**. Valve **17** is provided with a shroud **31** in the general form of a truncated cube having a circular opening **32** at one end opening into the inside end of connecting collar **16**. The truncation of shroud **31** forms an inclined rectangular opening **33** the lower boundary **34** of which runs horizontally across the lower portion of the end of shroud **31** opposite the opening **32**. The upper boundary **35** of shroud **31** runs parallel to the lower boundary **34** across the top of shroud **31**.

A rectangular vane of damper **36** having outer dimensions only slightly less than those of opening **33** is pivotally mounted within opening **33** by means of a pivot rod **37** which is attached to damper **36** along its horizontal center line. The ends of rod **37** extend beyond the edges of damper **36** and are journaled within holes **38** and **39** located near the edges of opening **33**. Hole **38** is located near the center of one inclined edge **42**. Thus, damper **36** may be rotated about rod **37** from a horizontal position, in which it offers essentially no resistance to air flow through valve **17** to an inclined position aligned with opening **33** where it may block air flow through channel **13**.

The rotational control of damper **36** is accomplished by means of pneumatic motor **43** which is mounted inside of channel **14** on the side of partition **12** opposite vane **36**. Motor **43** is a plunger-type motor having a substantially cylindrical shape with its axial plunger **44** extending outwardly thereof and then laterally thereof, as shown, into the slotted end of a lever arm **45**.

(continued)

Figure 5.4: Air Conditioning System

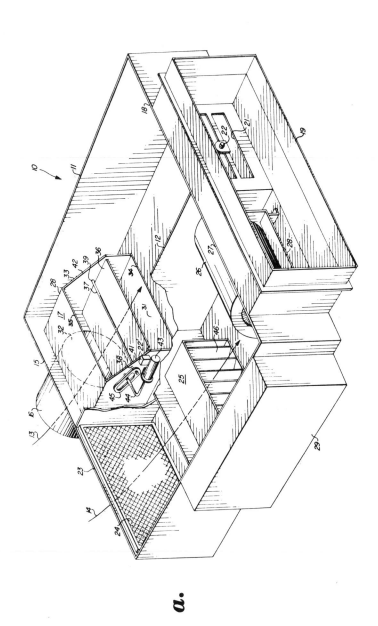

a.

Figure 5.4: (continued)

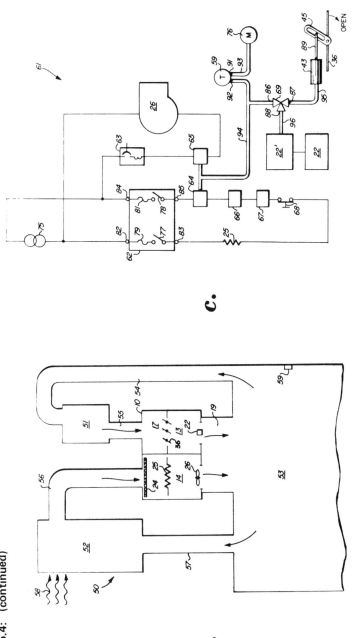

(a) Perspective view of the variable volume terminal box with the top cover removed
(b) Diagrammatic representation of the heating and air conditioning system
(c) Schematic diagram showing the arrangement of the electric and pneumatic control elements incorporated in the variable volume terminal box

Source: U.S. Patent 4,106,552

Lever arm **45** is attached laterally thereof to the end of rod **37**. The slotted opening of arm **45** is engaged by the lateral extension of plunger **44**. As plunger **44** is extended or retracted from the housing of motor **43**, arm **45** and damper **36** are rotated about pivot rod **37** as desired for the control of air flow through valve **17**.

The air flow sensor **22** is a limit detecting device which is employed to control motor **43** and valve **22'**. It is only activated when air flow through channel **13** exceeds a set level.

Air filter **24** may be any one of a number of commercially available types formed of a suitable material such as fiber glass.

Heater **25** is simply a metal housing with openings **46** front and rear to allow the passage of air therethrough and employs an internal resistance heating element diagrammatically shown in Figure 5.4b, which is electrically energized when it is desired to supplement the heat content of the air flowing through channel **14**.

While blower **26** is described and illustrated as a squirrel cage-type, the use of other types of fans or blowers may be employed.

The functional diagram of Figure 5.4b illustrates an example of the improved heating and air conditioning system **50** of the process incorporating the variable volume terminal box **10**. The system **50** comprises in addition to terminal box **10** an air conditioning unit **51** and heat collection and storage chamber **52** interconnected with a living, working or other space **53** in which temperature is to be controlled. Interconnecting these major elements or the system **50** are air handling pipes or ducts **54–57** and the plenum **19**.

Chamber **52** may be any space in which heat tends to accumulate such as an air space above false ceiling in which lighting fixtures are installed or an attic which collects heat from the sun. In some cases chamber **52** may be designed into a new building with the specific intention of accommodating the system **50** of this process. In any case, chamber **52** holds a volume of air which receives heat energy **58** from some energy source other than that produced directly as a heat source by public utilities through the combustion of fuels.

As shown in Figure 5.4b, system **10** incorporates two air flow circulation loops with the first loop beginning at air conditioning unit **51** and continuing through pipe **55**, chamber **13** and plenum **19** into space **53**, then returning to air conditioning unit **51** via pipe **54**. The second loop begins at chamber **52** and continues through pipe **56**, chamber **14**, plenum **19** and into space **53** and then through pipe **57** back into chamber **52**.

It should be noted that system **50** has two operating modes including a cooling mode and a heating mode. In the cooling mode, blower **26** is not energized so that no air is drawn from chamber **52** through channel **14**. A thermostat **59** monitors the temperature in space **53** and controls valve **17** as appropriate to regulate the air flow from air conditioning unit **51** thus regulating the temperature of space **53**. The air flow sensor **22** comes into operation only if the air flow through channel **13** exceeds a set amount and is particularly essential to the operation of a system in which there are more than one terminal box incorporated with each box utilized to control a separate space **53** and drawing cooled air from a single air conditioning unit **51**. In such an application sensor **22** prevents excessive air flow to any one space which would constitute a cause for discomfort in that space as well as an excessive load on the air conditioning unit which might reduce the effectiveness of system **10** in adequately providing the cooling requirements of other areas.

As the temperature in space **53** falls below a predetermined level valve **17** is closed and blower **26** is energized to draw warm air from chamber **52** through channel **14**. If the air from chamber **52** is not warm enough to hold the temperature in space **53** above a set lower limit, heater unit **25** is energized to introduce the additional required heat energy. Because heater unit **25** is only utilized when the "free" heat energy **58** proves inadequate, a saving in energy and heating costs is achieved in accordance with a primary object of the process.

The control of system **50** to effect the operating modes just described is accomplished by control system **61** of Figure 5.4c. Control system **61** incorporates some elements already mentioned and shown in Figures 5.4a and 5.4b, including fan **26**, thermostat **59**, air flow sensor **22**, motor **43**, lever arm **45**, damper **36** and heater unit **25**. Additionally, it includes a fused disconnect **62**, a fan disconnect **63**, pneumatic electric switches **64** and **65**, an air flow switch **66**, an automatic thermal cutout **67**, a manual reset thermal cutout **68**, and a pneumatic restrictor **69**. The control system **61** receives electric energy from a source of alternating current voltage **75** which is typically 120 V at 60 Hz; it is also energized pneumatically from a source of air pressure **76**.

Disconnect **62** comprises two manually operated contacts **77** and **78** and two fuses **79** and **81** with fuse **79** and contact **77** serially connected between a first line terminal **82** and a first load terminal **83**. Fuse **81** and contact **78** are serially connected between a second line terminal **84** and a second load terminal **85**. Fan disconnect **63** is a circuit breaker which is opened by excessive current and which may also be opened or closed manually.

Air flow switch **66** is any one of a variety of switches designed to close when placed in an air stream of a given minimum velocity. If the air velocity falls below the minimum level the switch **66** opens.

The thermal cutout **67** may be automatic or manual and opens when it senses a temperature above a given high level and recloses automatically if the temperature subsequently falls below a given lower level.

The manual reset thermal cutout **68** opens automatically if a given temperature is exceeded. It remains open until it is reset manually.

Thermostat **59** is a pneumatic-type which is connected in series with an air pressure line. It responds to temperature changes about a set level by producing a pressure drop or a pressure rise in the pneumatic line. Disconnect **63** is a manually operated switch controlling voltage to blower **26**.

Pneumatic-electric switches **64** and **65** have electric contacts which are opened or closed by pressure in a connected pneumatic control line. As pressure falls below a predetermined level, the switch closes, the level at which the closing occurs being determined by an adjustment of the switch.

Restrictor **69** has three ports **86, 87** and **88**. A pressure drop exists between ports **86** and **87** and this pressure drop increases as air is bled off through port **88** so that as an increasing amount of air is bled off through port **88** the pressure at port **87** becomes lower and lower.

Air flow sensor **22** operates an air flow operated valve **22′** which when air flow exceeds a predetermined level opens valve **22′** bleeding off air from the pneumatic line.

Damper motor **43** is operated by pneumatic pressure and as the pressure increases the axial plunger **89** of motor **43** extends; as pressure decreases plunger **89** is withdrawn inside the body of motor **43**.

As shown in Figure 5.4c, heater unit **25** is serially connected with fuse **79**, contact **77**, cutout **68**, cutout **67**, switch **66**, switch **64**, contact **78** and fuse **81** across source **75**. Cutout **62** serves as a means for manually disconnecting heater unit **25** and provides fuse protection against electrical shorts or failures. Air flow switch **66** opens if air flow through heater unit **25** is interrupted and thereby prevents damage to unit **25** by overheating. Cutouts **67** and **68** are redundant protective devices which are opened by excessive current. They are incorporated as safety features and are required to meet safety codes. Switch **64** automatically controls the energization of heater unit **25** and is operated by thermostat **59**.

Blower **26** is serially connected with cutout **63** and switch **65** across source **75**. Disconnect **63** serves as a means for manually disconnecting blower **26** from source **75** and switch **65** serves as a means by which blower **26** is automatically energized through the control of thermostat **59**.

Thermostat **59** has an input port **91** and output port **92**. Input port **91** is connected by pneumatic line **93** to pressure source **76**; output port **92** is connected by pneumatic line **94** to switches **64** and **65** and to port **86** of restrictor **69**. Port **87** of restrictor **69** is connected by pneumatic line **95** to damper motor **43**, and port **88** is connected by line **96** to sensor **22**.

Operation of systems **50** and **61** occurs as follows: Assuming that the temperature in space **53** is such that thermostat **59** calls for cooling and the pressure in line **94** is sufficiently high that switches **64** and **65** are held open so that heater unit **25** and blower **26** are not energized. There is, therefore, no air flow through heating channel **14**. The same relatively high pressure in line **95** causes damper motor **43** to extend plunger **89** and thereby to hold damper **36** in a relatively open position so that cooling air from the air conditioning unit is admitted through channel **13** into space **53** with warm air recirculating through duct **54** to unit **51**.

As the temperature in space **53** begins to fall, thermostat **59** responds by reducing the pressure in line **94**. A corresponding reduction in pressure is transmitted through restrictor **69** and line **95** to motor **43**. The reduced pressure to motor **43** causes plunger **89** to be retracted somewhat and damper **36** to be moved closer to a closed position so that a reduction in cooling air through channel **13** is effected as appropriate to regulate the temperature in space **53**.

Under certain conditions, a very high cooling demand as evidenced by a high temperature in space **53** will call for a high rate of air flow through channel **13**, but such a high rate of flow will cause creature discomfort in space **53** and may adversely affect the performance of other cooling channels connected to unit **10** (not shown in Figure 5.4b). When air flow through channel **13** exceeds the desired maximum level sensor **22** opens causing air to be bled off at port **88** of restrictor **69**. The attendant reduction in pressure at port **87** and line **95** causes plunger **89** of motor **43** to be withdrawn so that damper **36** is moved toward a closed position only to the degree necessary to limit air flow to the desired maximum level.

A drop in outside temperature as might be experienced, for example, during the late evening, will remove the requirement for cooling. The reduced temperature in space **53** as sensed by thermostat **59** results in a significant drop in pressure in line **94**. The reduced pressure in line **94** causes switch **65** to close energizing blower **26** just prior to the complete closing of damper **36** so that the total interruption of air flow in space **53** is prevented, but the source of air flow is now chamber **52** with its charge of warm air which had received thermal energy **58** during the warmer part of the day as from the sun or from the building lighting system.

The warm air from chamber **52** now warms space **53** as air is circulated from chamber **52** through duct **56**, and channel **14** into space **53** and returning through duct **57** to chamber **52**. As the outside temperature continues to fall, however, this source of heat becomes inadequate and thermostat **59**, sensing a still lower temperature in space **53** causes a still lower pressure in line **94** which causes switch **64** to close, thereby energizing heater unit **25**. Unit **25** is then cycled on and off by thermostat **59** and switch **64** as appropriate to regulate the temperature in space **53** during the ensuing heating cycle.

A complete and effective heating and cooling system is thus provided wherein the variable volume terminal box permits the controlled utilization of collected thermal energy from the sun and the lighting system as a first source of heating energy with provision as well for supplementing the first source by means of utility-supplied electrical energy.

Control Systems for Multiple Rooms

W.E. Clark; U.S. Patent 4,014,381; March 29, 1977; assigned to Carrier Corporation describes an air conditioning system for conditioning air in a plurality of enclosed areas in a building. Conditioned air at a temperature level which may be varied is supplied to first terminal means for discharge into an area. Conditioned air at a relatively constant temperature level is delivered to second terminal means for discharge into the area. The quantity of constant temperature air discharged into the area is regulated in accordance with the temperature level therein.

When the temperature of the variable temperature air supply is at a relatively warm level, the quantity thereof discharged into the area is regulated in accordance with the quantity of constant temperature air discharged into the area. As the quantity of constant temperature air discharged into the room is increased, the quantity of warm air discharged into the room is decreased; and as the quantity of constant temperature air discharged into the room is decreased, the quantity of warm air discharged into the room is increased.

Referring to Figure 5.5a, air conditioning system **10**, which may be described
as a central station type, includes an air conditioning equipment section generally
designated by the numeral **12**, and a conduit system **14, 15,** for conducting conditioned primary and secondary air respectively, to each of the areas or rooms
provided within a common enclosure and served by the system. Equipment section **12** may be located in a basement or on the roof of a building.

For the purpose of this process, primary air may comprise fresh air or ventilating
air drawn from the outdoors, or a mixture of outdoor air and return air treated
in section **12**, while secondary air may comprise return air from the areas being
conditioned and treated in section **12**. The apparatus for conditioning the primary air preferably includes a filter **24** to remove foreign matter entrained in
the air, heating or reheating coil **26** to elevate the temperature of the air flowing
in the primary air system or circuit and a cooling or dehumidifying coil **25** to
remove excess moisture and to cool the supply air as required, arranged in series
flow relationship and encased within a suitable housing **28**. The passage of primary air over coils **25** and **26** is regulated by dampers **59** and **60**.

The portion of the central station equipment regulating the secondary air preferably includes a suitable filter **30** to remove foreign matter entrained in the air
and a dehumidifier or cooling coil **31** to remove the excess moisture and/or cool
the supply air, arranged in series flow relationship and encased within a suitable
housing **32**. Chilled water is supplied to coils **25** and **31** via suitable means.

Figure 5.5: Air Conditioning System

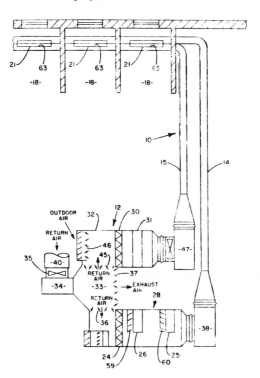

a.

(continued)

Figure 5.5: (continued)

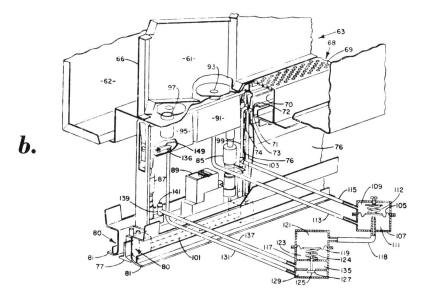

b.

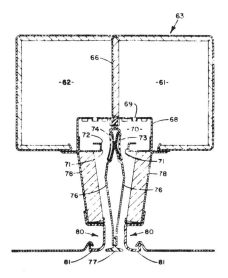

c.

(a)   Schematic of air conditioning system
(b)   Perspective view of an air conditioning terminal with a control
          therefor, illustrated partially in section and partially in sche-
          matic
(c)   Sectional view of the air conditioning terminal illustrated in
          Figure 5.5b

Source: U.S. Patent 4,014,381

Housings **28** and **32** are connected by duct **33** with return air exhaust fan **34**. The inlet of return air exhaust fan **34** is connected with return air plenum **40** which is connected by suitable means with the areas or rooms being served by the air conditioning system. Preferably, inlet air control vanes **35** are provided to vary the flow of air through fan **34**. Adjustable members **36** are provided to vary the flow of return air to the primary air conditioning apparatus. The exhaust dampers **37** connect exhaust duct **33** with the outdoors. Dampers **37** control the volume of return air discharged to the atmosphere. Housing **28** connects with primary air fan **38**.

Conduit means **14** conveys primary air from fan **38** to the areas or rooms being conditioned. Housing **32** is connected to the outlet of return air fan **34**. Preferably, adjustable dampers **45** are provided to vary the flow of supply air to the secondary air conditioning apparatus. Adjustable dampers **46** are provided to regulate the flow of outdoor air to the secondary air conditioning apparatus. Conduit means **15** conveys air from fan **47** to the area or rooms being conditioned. Conduit means **14** and **15** provide the primary air and secondary air respectively to each air terminal **21** disposed in each of the respective areas or rooms **18**.

Referring to Figures 5.5b and 5.5c, there is shown a preferred form of terminal and a control. In a typical system of the type to which the process relates, a separate terminal will be disposed in each individual space or room being conditioned. Conduit means **14** and **15** terminate in a plenum section **63** of each terminal. The plenum section is ordinarily lined with a sound absorbing material, such as a glass fiber blanket. A baffle or partition member **66** divides the plenum into first and second sections or portions **61** and **62** for respective connection to conduit means **14** and **15**. The baffle thereby maintains the primary air separate from the secondary air so that there is no intermixing.

An air supply distribution plate **68** having a plurality of openings **69** is provided to evenly distribute the supply air from plenum **63** to distribution chamber **70** which is defined by the top and side walls of distribution plate **68**. A portion of the distribution plate is disposed on either side of baffle **66** so primary air moves into a first portion of distribution chamber **70** and secondary air moves into a second portion of the chamber.

The bottom of distribution chamber **70** includes aligned cutoff plates **72** which are provided with a curved surface **71** for engagement by bladders or bellows **73** and **74** to form a damper. By varying the inflation of the bladders, the area of the opening between each of the bladders and cutoff plates may be varied to thereby regulate the quantity of conditioned air discharged into the area or space being conditioned.

The bladders are adhesively mounted on a central partition assembly comprised of opposed generally convex plates **76** and a diffuser triangle **77**. The plates have a V-shaped recessed area so the bladders are completely recessed within the plates when deflated. This provides a large area between the active walls of the bladders and the cutoff plates for maximum air flow therebetween. Further, the recessed bladder provides a smooth surface along plate **76** to minimize air turbulence.

The damper mechanism is disposed a substantial distance upstream from the discharge openings in the terminal to provide sufficient space therebetween to absorb any noise generated by the damper mechanism. For maximum sound absorption, downwardly extending walls **78** which form air passages in conjunction with plates **76** are lined with a suitable sound absorbing material, such as a glass fiber blanket. Outlet members **80** having outwardly flared portions **81** are affixed, as by welding, to walls **78**. For a more detailed explanation of the air terminal, reference may be made to U.S. Patent 3,867,980.

The terminal further includes a control section comprising a first regulating device **85**, a second regulating device **87**, and a thermostat **89**. Preferably, regulators **85** and **87** are of the type described in U.S. Patent 3,434,409 and thermostat **89** is of the type in U.S. Patent 3,595,475.

Regulator **85** is responsive to the pressure of the secondary air supplied via conduit means **15** to plenum portion **62**. Thermostat **89** is suitably operably connected to regulator **85** for a reason to be more fully described hereinafter. A filter **91** is provided to filter the secondary air passing from plenum section **62** to the regulator via opening **93**.

Similarly, a filter **95** is provided to filter the primary air passing to regulator **87** from plenum section **61** via opening or orifice **97**. Openings **93** and **97** are provided on opposite sides of baffle plate **66**. Regulator **85** is suitably joined via line **99** to the bladder regulating the discharge of secondary air from plenum section **62** through the terminal. Similarly, regulator **87** is joined via lines **101** and **103** to the bladder regulating the discharge of primary air from plenum section **61**. Regulators **85** and **87** are provided to generate a control signal indicative of the pressures of the secondary and primary air in the respective plenum sections.

The regulators increase a control signal supplied to the bladders to thereby increase the inflation thereof as the air pressure in the plenums increase and operate to decrease the magnitude of the control signal supplied to the bladders as the pressure of the air in the plenum sections decrease. By varying the inflation of the bladders or bellows in accordance with changes in supply air pressure, a relatively constant quantity of air may be discharged from the unit or terminal irrespective of variations in supply air pressure.

As noted before, thermostat **89** is associated with regulator **85**. Thermostat **89** is preferably a bleed type thermostat which operates to reduce the pressure signal supplied from regulator **85** to the bladder as the temperature of the space increases above the design level to thereby decrease the inflation of the bladder and increase the quantity of conditioned air supplied into the space. If the temperature of the space falls below the set point temperature, the thermostat bleed closes to increase the magnitude of the control signal supplied to the bladder so that it approaches its maximum value as determined by the supply air pressure. The resultant increase in the magnitude of the signal will cause the bladder to inflate to decrease the quantity of secondary air supplied to the space.

As noted before, the supply of primary air into each space, as controlled by regulator **87** is normally substantially constant. The primary air temperature is varied in accordance with ambient temperature to offset transmission gains or losses. Accordingly, a thermostat is generally omitted from the control section regulating the supply of primary air into the space.

There are times during the winter season when solar radiation negates transmission losses and the supply of warm primary air is not actually required. Previously, the supply of warm air in systems of the type described could not be regulated in accordance with actual temperature conditions in the space. That is, unless a separate thermostat were provided, only operable during the heating season, the supply of warm air was maintained at a constant level irrespective of actual requirements in the separate areas.

To overcome the foregoing problem, the process includes a pneumatic valve 107. Valve 107 includes a bellows or diaphragm 105 separating the valve into an upper section 109 and a lower section 111. A line 113 communicates lower valve section 111 with plenum section 62 so that the lower surface of diaphragm 105 is responsive to the pressure of the air in the plenum section. A conduit 115 communicates upper section 109 with a portion of the system that is at the pressure of the bladder or bellows controlling the discharge of secondary air, for example, into line 99. Accordingly, the upper surface of diaphragm 105 is subjected to bladder pressure.

Control valve 107 further includes an adjustable spring 112 or similar means to provide an additional force on the upper surface of diaphragm 105. A line 118 communicates lower valve section 111 with a second valve 117. Valve 117 has a diaphragm 119 separating the valve into upper and lower sections 121 and 123 respectively. Adjustable spring 124 provides a second force on the lower surface of diaphragm 119. A valve 125 is provided to control the flow of air through orifice 127. Section 129 of valve 117 communicates with a line 131 having air at primary air supply pressure flowing through. Line 131 terminates at one end in orifice 136 provided in filter 95.

Opening of valve 125 permits the primary air to flow from section 129 to section 135 and then via line 137 to a connection point 139 in valve 87 located between bleed opening 141 and the connection point for the air supplied to the inflatable bellows via line 101. A valve member 149 is provided to selectively open line 131 for the passage of air therethrough. Preferably, valve 149 is responsive to the temperature of the supply air and opens when the supply air temperature is at a relatively warm level.

In related work, *G.E. Cobb; U.S. Patent 4,019,566; April 26, 1977; assigned to Carrier Corporation* describes an air conditioning system for conditioning air in a plurality of enclosed areas in a building. Conditioned air at a relatively cold temperature level is supplied to a terminal for discharge into an area. The quantity of relatively cold temperature air discharged into the area is regulated in accordance with the temperature level therein. A control signal is generated, the magnitude thereof being indicative of the quantity of conditioned air discharged into the area. The magnitude of the control signal is monitored.

Warm air supply means is activated when the magnitude of the control signal indicates that the quantity of the relatively cold temperature air being discharged into the area has decreased below a predetermined level.

Multiroom Installations

A process described by *F. Westergren; U.S. Patent 3,935,898; February 3, 1976; assigned to Aktiebolaget Svenska Flaktfabriken, Sweden* relates to an air conditioning system which controls the heating effect of light fixtures by cooling the fixtures with discharge air from the room.

The process is characterized primarily in that the air flow through a room is adjusted between a highest value adapted for the maximum cooling demand and a minimum value adapted for the necessary ventilation demand, in such a manner, that the discharge air is divided into a constant portion constituting the minimum flow for adequate ventilation and a second portion varying with the cooling demand, which is passed through or around the fluorescent tube fixtures.

The variable portion of the discharge air is adjusted by a thermostat-controlled adjusting device between highest value and zero, in which latter position the electric effect supplied to the fixtures is utilized for room heating whereby the range of adjustability is increased, and that the supply air flow is varied by a thermostat-actuated adjusting device so as to be the total of the constant and the variable discharge air flow. An advantageous example of the device is characterized in that the constant portion of the discharge air flow is passed through a bypass duct.

Thus, in the process the lighting and ventilation have been integrated in one unit which results in a simpler and less expensive installation system and thereby renders it possible that this technically valuable ventilation system can be applied more widely.

Zone Temperature Responsive Load Transmitters

M.E. Demaray and R.G. Attridge, Jr.; U.S. Patent 4,042,013; August 16, 1977; assigned to Ranco Incorporated describe an air conditioning system which is controlled for automatically heating and/or cooling a multiple zone building. In the system, each zone includes a zone temperature responsive load transmitter. Each load transmitter produces an output command signal for controlling operation of damper units and heating and cooling equipment. The command signal configuration is modified to avoid operation of heating and cooling equipment throughout a relatively wide "no load" band of zone temperatures centered around the zone set point temperature. The command signal is highly responsive to sensed zone temperature changes beyond the no load band so that the zone temperature tends to be maintained in the no load band.

An air conditioning system **10** constructed according to a preferred example of the process is schematically illustrated in Figure 5.6a. The system **10** provides conditioned air to three separate zones of a multiple zone building, which itself is not illustrated. The zones are referred to as zone **1**, zone **2**, and zone **3**, and only zone **3** is illustrated schematically. The system **10** includes an air circulating duct network **12**, a blower **14** for providing a forced flow of air through the duct network, air heating equipment indicated by the reference character **16**, an air cooling equipment **18**, both of which have portions disposed within the duct network for heating or cooling air flowing through the network and a control system generally indicated by the reference character **20** which governs operation of the system **10**.

The system **10** is, for the most part, schematically illustrated and is described only briefly. Many components of the system **10** are shown and described in greater detail in the literature.

Figure 5.6: Air Conditioning System

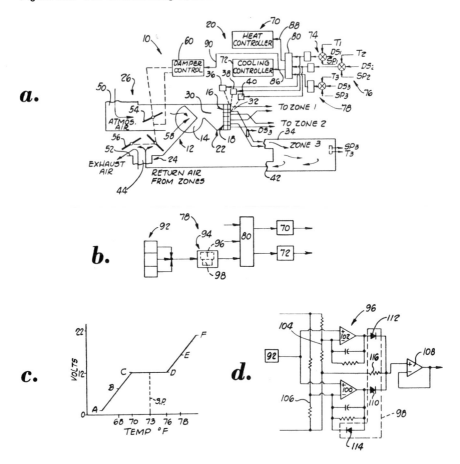

(a) Schematic diagram of an air conditioning system
(b) Schematic diagram of part of an air conditioning control system
(c) Graphic representation of a configuration of a command signal voltage versus sensed zone temperature produced in a system
(d) Schematic illustration of a command function generator for producing the command signal of Figure 5.6c

Source: U.S. Patent 4,042,013

The heating equipment can be of any suitable or conventional construction, but for the purpose of this description is considered to be constructed from a plurality of electrical resistance heaters which are operated in stages to govern the amount of heat transferred to the air in the system.

The air in the system is preferably mechanically chilled by compressor-condenser-evaporator refrigeration equipment, not shown. The cooling equipment is operable in stages to govern the amount of heat absorbed from the air in the system. One or more evaporators of the refrigeration equipment are disposed in the duct for cooling the system air.

The air duct network 12 comprises an air delivery duct system 22 for directing air from the blower 14 to the respective zones through the heating and cooling units 16, 18, respectively, a return air duct system 24, only partly shown, for receiving air exhausted from the zones, and a ventilation system 26 by which atmospheric air is admitted to the system 10 while a corresponding amount of air from the return air duct system 24 is exhausted from the system. The system 10 is of a type known as a constant volume system in that a constant flow rate of air continuously circulates in the system and each zone is continuously provided with an unvarying flow rate of air.

The delivery duct system 22 comprises a blower plenum section 30 in which the heating and cooling units 16, 18, are disposed so that air moving through the plenum 30 towards the zone passes across either the heating unit or the cooling unit, a zone damper section 32 at the discharge side of the heating and cooling units, and three discharge ducts 34 (only one of which is partially shown) for directing air from the damper section 32 to each associated respective zone.

The damper section 32 includes three actuatable damper pairs, one pair for each zone. The damper pairs are actuated by respective zone damper control units 36, 38, 40 in accordance with temperature requirements of associated zones. The damper pairs for each zone enable complementary dampering of air flowing to that zone from the heating unit 16 and the cooling unit 18. The damper pair for each zone has one limit position in which all of the air flowing to the zone passes across the heating unit, a second limit position in which all of the air flowing to the zone passes across the cooling unit, and intermediate positions in which the flow of air to the zone consists of a mixture of air which has passed across the heating unit and the cooling unit, with the proportions of the mixture being determined by the position of the damper pair.

The return duct system 24 comprises zone exhaust branches 42 (only one of which illustrated in connection with zone 3) communicating each zone to main return duct 44 which directs the combined zone exhaust air flows to the ventilating system 26.

The ventilating system 26 comprises an atmospheric air intake duct 50 through which atmospheric air is introduced into the system 10, an exhaust duct 52 through which air from the return duct 44 is exhausted to atmosphere from the system, and a dampering arrangement for controlling the flow of air through the intake and exhaust ducts 50, 52.

The control system 20 governs operation of the heating and cooling units 16, 18, the zone damper control units 36, 38 and 40, and the damper control unit 60 in response to sensed conditions of air circulating in the system 10 as well as sensed atmospheric air conditions. The control system is preferably an electrical system and includes an air heating controller 70 for governing the heat transfer to system air from the heating equipment 16, an air cooling controller 72 for governing heat transfer from the system air to the cooling equipment 18, individual zone load transmitter systems **74, 76, 78** for sensing the respective zone loads and producing temperature related command signals for governing operation of the controllers **70, 72,** and a logic unit **80** interposed between the load transmitter systems and the controllers.

In the preferred case, the command signals are low amperage dc analog signals, and the heating and cooling controllers are constructed to operate the heating and cooling equipment in stages in response to appropriate changes in command signal values transmitted to them. The load transmitter systems are such that as the zone air temperature rises, the magnitude of the command signal value tends to increase positively with respect to a reference value. As the zone temperature is reduced, the command signal value likewise tends to be reduced.

The logic unit **80** enables the control system **20** to satisfy the heating requirements of the coolest zone and the cooling requirements of the warmest zone while the requirements of the intermediate zone are satisfied by operation of its associated zone damper control unit alone. The command signal from the warmest zone has the most positive voltage level, the command signal from the coolest zone has the least positive voltage level, and the command signal from the zone of intermediate temperature has an intermediate voltage level.

The logic unit **80** is connected to the outputs of each zone load transmitter and functions to transmit the command signal from the warmest zone to the cooling controller **72** via an output conductor **86** and to transmit the command signal from the coolest zone to the heating controller **70** via a conductor **88**. The command signal from the remaining intermediate zone (or zones, if more than three zones are present in the building) is blocked by the logic unit, but remains effective to govern the positioning of the zone damper unit for that zone.

Atmospheric air is introduced in quantity to the system to effect "free" cooling of the zones by the outside air. The introduction of atmospheric air to the air conditioning system is variably controllable by the command signal from the warmest zone. For this purpose a conductor **90** interconnects the ventilating damper control unit **60** to the logic unit output conductor **86**.

The load transmitter systems **74, 76, 78** are all identical and therefore only the system **78** associated with zone **3** is described. Referring to Figure 5.6b, the load transmitter system **78** is illustrated as including a zone temperature condition sensor **92**, and a command function generator **94** which coact to produce a command signal configuration as illustrated by Figure 5.6c.

The zone temperature condition sensor **92** produces a signal which is the algebraic summation of a zone air temperature signal T, a discharge duct air temperature signal DS, and a zone **3** set point temperature signal SP. The zone air temperature signal T is preferably provided by a sensing circuit including a thermistor, or equivalent element, which is disposed in the zone. The discharge duct air temperature signal is preferably provided by a circuit including a thermistor or other equivalent sensing element situated in the discharge duct **34** adjacent the zone damper section **32**. The zone set point temperature signal is preferably provided by a circuit including a manually adjustable potentiometer which is controlled, within limits by an occupant of the zone. The zone and duct sensing circuits are schematically illustrated in Figures 5.6a and 5.6b and may be of any suitable or conventional construction.

The command function generator **94** includes a command signal generator **96** which responds to the input zone condition signals and a command signal modifier **98** which substantially varies the command signal under certain zone temperature conditions. A typical command signal produced by the command function generator **94** is illustrated by Figure 5.6c. The command signal of Figure 5.6c is illustrated in terms of voltage and sensed zone temperature and, in the illustrated case, varies between limits of 2 and 22 V dc.

The operation of the heating and cooling equipment and damper units as controlled by zone **3** is graphically demonstrable from Figure 5.6c. The heating and cooling equipment operate, respectively, in first and second command signal voltage value ranges approximately indicated by the line segments **A-B** (heating) and **E-F** (cooling). The damper units operate in a third voltage value range approximately indicated between the line segment ends **B-E**. A no load band region of the command signal is approximately indicated by the line segment **C-D**.

Assuming zone **3** is at a lower sensed temperature than either of the other zones, the zone **3** command signal value varies between, 2 and 10 V; the heating equipment is modulated or operated in stages by the heating controller to respond to the zone **3** command signal in the first value range.

When the zone **3** sensed temperature conditions rise sufficiently (e.g., to a level between 68° and 70°F), the heating equipment is no longer operated and the zone damper units are operated to progressively reduce the proportion of the air entering zone **3** which has passed across air heaters. When the command signal reaches the location indicated at **C** (12 V) the zone damper unit is conditioned to introduce equal proportions of air which has passed the air heaters and air coolers.

At this juncture the zone temperature is about 70°F and the command signal enters the no load band region (line segment **C-D**) during which the sensed zone temperature may vary widely without materially changing the command signal value. This region extends equally from each side of the zone set point temperature (73°F in the illustration) and enables the zone temperature to float through a 6°F range without any heating or cooling equipment being operated.

Further sensed zone temperature increases result in the command signal increasing along the line segment **D-E** during which the zone damper unit is operated to progressively increase the proportion of zone air which has passed the air cooling equipment. At the same time, assuming that zone **3** is the warmest zone, the atmospheric air damper unit begins to be progressively opened toward its maximum open position. This permits zone **3** to be cooled by atmospheric air.

If the atmospheric air temperature increases to a predetermined level the atmospheric air damper unit is abruptly operated to its minimum open position.

The cooling equipment is operated in the command signal value range indicated by the line segment **E-F**. Preferably multiple air coolers are operated in stages and/or modulated to maintain the zone temperature within limits.

A command function generator circuit capable of producing the command signal of Figure 5.6c is illustrated by Figure 5.6d. The command signal generator includes a pair of signal amplifiers **100, 102** having their inverting inputs connected to respective reference voltage sources **104, 106,** their noninverting inputs connected to the zone sensor **92** and their outputs connected to a high input impedance buffer amplifier **108** via the command signal modifier **98.**

The reference sources **104, 106** provide different levels to the amplifiers so that when an input signal having a given level is present on the input line, the amplifier **100** tends to produce a higher output voltage level than the amplifier **102.**

The command signal modifier circuitry is effective to control the input to the buffer amplifier **108** from the amplifiers **100, 102** to form the no load band region (line segment **C-D** of Figure 5.6c) of the command signal. The modifier circuitry includes diodes **110, 112, 114** and a resistor **116.** The diodes **110, 112** have their anodes connected to the respective output terminals of the amplifiers **100, 102** and their cathodes connected to the noninverting input of the buffer amplifier and to a reference voltage via the resistor **116.**

When the amplifier **100** is conducting in response to low-level input signals the diode **110** is rendered conductive via a circuit to the reference voltage (preferably zero volts) through the resistor **116.** The voltage drop across the resistor **116** provides an effective input signal to the buffer amplifier and back biases the diode **112.**

When the output level from the amplifier **100** slightly exceeds the reference level supplied to its inverting input by the circuit **106,** the diode **114** becomes conductive and clamps the amplifier output level against further rise even though the sensed zone temperature conditions cause increasing input signal levels. Accordingly, the buffer amplifier output level remains substantially constant through a range of zone temperatures (the line segment **C-D** of Figure 5.6c).

Open-Cycle Air Conditioning Process

A process described by *W.F. Rush, J. Wurm and R.J. Dufour; U.S. Patent 4,014,380; March 29, 1977; assigned to Gas Developments Corporation* is directed to an air conditioning system which can be operated both in a cooling mode and in a heating mode. An air conditioning apparatus for use in the process comprises an enclosure which defines an incoming air passageway for air to be treated and a separate regenerative air passageway, means for passing an air stream through each of these passageways, a sensible heat exchanger means within the enclosure and adapted for transfer of thermal energy from one passageway to the other passageway, and desiccant means for transfer of moisture from the air treatment passageway to the regenerative passageway spaced toward the exhaust to the ambient atmosphere of the regenerative air stream.

A first evaporative cooling means is provided near the exit port of the conditioned air passageway to the room and a second evaporative cooling means is provided near the entrance port of the regenerative air passageway for use in the cooling mode.

The improvement of this process comprises a low-temperature heater means situated in the regenerative air passageway between the sensible heat exchanger means and the desiccant means and capable of supplying heat to an air stream flowing within the regenerating air passageway in an amount frequently sufficient to vaporize a major amount of the moisture carried by the desiccant means into the regenerating air passageway. The heat source for the low-temperature heater means can be solar, waste heat recovery, etc. Any heat source above the temperature of the air being conditioned after exiting the heat exchange wheel is useful. In the heating mode, any heat source above the temperature of the room is useful. Additionally, a high-temperature heater means may be situated in the regenerating air passageway between the low-temperature heater means and the desiccant means which is capable of raising a portion of the heated air stream to a final regeneration temperature for the desiccant. In the cooling mode of the open-cycle air conditioner, four basic steps are involved: (a) adiabatic drying of a moist air stream to be conditioned by a desiccant means, (b) removal from the conditioning air stream of sensible heat resulting from the adiabatic drying step, (c) adiabatic saturation of the conditioning air stream with water to provide conditioned air having the desired temperature and humidity, and (d) regeneration of the sensible heat exchanger means and desiccant means.

The process may be operated in the full recirculating mode wherein air from a conditioned room passes through the air treatment passageway and returns to the air conditioned room while ambient air from the atmosphere is passed through the regenerative passageway and exhausted to the ambient atmosphere. The process may also be operated in the full ventilation mode wherein ambient air from the atmosphere is conditioned for introduction to the room to be conditioned and exhaust air from the conditioned room is utilized as the regenerative air and exhausted to ambient atmosphere. Combinations of the recirculating and ventilation mode may also be utilized.

For operation of the apparatus and process in the cooling mode, it is preferred to utilize the recirculating mode of operation while for operation in the heating mode it is preferred to use the ventilating mode of operation. The cooling mode can also be operated under the full ventilating mode or a combination of ventilating and recirculating modes, while the heating mode can be operated at less efficiency under the full recirculating or combination of ventilating and recirculating modes.

For use as a cooling air conditioner, Figure 5.7a schematically shows an apparatus for use in the process in the full recirculating mode. The regenerative air stream is ambient atmospheric air below 120°F dry bulb and below 95°F wet bulb. The regenerative air stream enters the apparatus through a humidifier adding moisture to the stream up to the saturation point for the purpose of cooling the incoming air. The moisture saturated regenerative stream is then passed through a heat exchange wheel and heated to about 130° to 200°F average, under general cooling air conditioning conditions.

The heat exchange wheel rotates in the direction shown at from 2 to 12 rpm, about 5 to 7 rpm being preferred under general cooling conditions. The full regenerative air stream from the heat exchange wheel is passed through the first heat source which may advantageously be at any temperature above the temperature of the regenerative air stream leaving the heat exchange wheel. Therefore, many economical sources are suitable for the first heat source, including heat from solar sources, boiler heat, waste process heat, and the like. After passage through the first heat source, the regenerative air stream is split into two portions, the first portion passing directly through the drying wheel and the second portion passing through a second heat source prior to passage through the drying wheel. The second heat source is at a higher temperature than the first heat source and provides sufficient heat to the second portion of the regenerate stream to regenerate the drying wheel. The final regeneration temperature of the drying wheel is that sufficient to drive off essentially all of the adsorbed water of the desiccant on the drying wheel, usually greater than 212°F and preferably on the order of 212° to 400°F.

The drying wheel rotates in the direction shown at about $\frac{1}{10}$ to $\frac{1}{2}$ rpm, preferably about $\frac{1}{6}$ to $\frac{1}{4}$ rpm. The first portion of the regenerative air stream passing through the drying wheel raises the temperature of the desiccant and depending upon its temperature, may drive off adsorbed water, while the second regenerative air stream is of sufficient volume, governed by the angle of exposure of the drying wheel, to regenerate the desiccant at a satisfactory temperature. The regenerative air, after passing through the drying wheel, is exhausted to the ambient atmosphere.

The air stream to be treated, as shown in Figure 5.7a, is room air, which is adiabatically dried by the drying wheel, passes through the heat exchange wheel where a major portion of the sensible heat is removed, followed by adiabatic saturation with water by a humidifier to provide conditioned air having desired temperature and humidity.

Figure 5.7b shows an apparatus for use in the process in the full ventilation mode. The apparatus of Figure 5.7b is the same as that of Figure 5.7a. In the process shown in Figure 5.7b, the regenerative cycle input is conditioned room air which, after serving as the regenerative stream, is exhausted to the ambient atmosphere and the input for the air stream to be conditioned is from the ambient atmosphere. Otherwise, the operating conditions are similar to the operation of the apparatus of Figure 5.7a with differing temperature and humidity conditions.

By way of a specific example of use of the cooling apparatus in a full recirculating mode, as shown in Figure 5.7a, an air stream from the conditioned room enters the apparatus at 80°F dry bulb and 67°F wet bulb (standard American Refrigeration Institute conditions), passes through the drying wheel where it is dried to less than about 0.003 lb of water per lb of air, raising it in temperature to about 148°F. The treatment stream then passes through the heat exchange wheel wherein it is cooled to about 80°F and is further cooled by evaporative cooling by passing through a humidifying device and exits from the apparatus to the room at about 56.5°F dry bulb and about 53°F wet bulb. Countercurrently with the treatment stream passage, regeneration takes place in the regenerative air passageway where a countercurrent stream is taken from the ambient atmosphere, as shown in the regeneration cycle shown in Figure 5.7c. All of the air flows are at the rate of 56.1 lb/min and the Coefficient of Performance is calculated to be 0.73.

Figure 5.7c shows that for the open-cycle air conditioners operated in the cooling mode prior to this process, under the above conditions, a gas input of 100 cfh was required. Utilizing a first heat source of 230°F, the gas input requirement for the second heat source is reduced to a total gas input (estimated) of 21 cfh, almost one-fifth of the former gas input requirement. The first heat source raises the temperature of the preheated sector of the drying wheel in Figure 5.7c to a temperature sufficiently high to accomplish, frequently a major portion of the regeneration. This reduces the sector of the drying wheel through which the air heated by the second heat source passes. Utilization of the apparatus for cooling under full ventilation mode, as shown in Figure 5.7b, reduces the Coefficient of Performance somewhat, dependent upon ambient temperature conditions and room exhaust temperature conditions. For these reasons, it may be preferred to use the recirculating mode as diagrammatically shown in Figure 5.7a when the apparatus is used to cool air, but combination of the recirculating and ventilation modes may be desirable to provide fresh air to the conditioned room.

With regard to the construction of the various elements of the air conditioning apparatus, the drying wheel is preferably a rotating wheel made of a corrugated asbestos sheet and impregnated with a hygroscopic substance such a lithium chloride, silica gel, crystalline aluminosilicates (molecular sieves), and similar substances. The corrugated sheet is then wrapped around a mandrel and wound into wheel shape to form a disc comprising a plurality of parallel channels which permit rapid drying of an air stream flowing therethrough and regeneration. Typical lithium chloride desiccant wheels are disclosed in U.S. Patent 2,700,537.

Figure 5.7: Air Conditioning Process

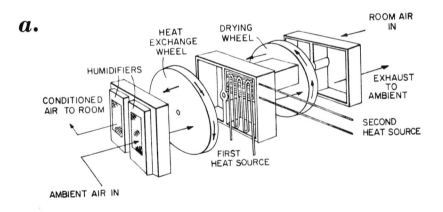

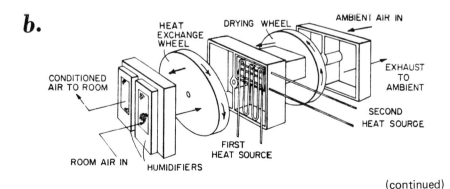

(continued)

Figure 5.7: (continued)

c.

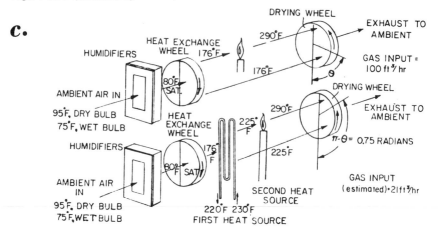

(a)   Recirculating mode
(b)   Full ventilation mode
(c)   Nonsolar versus solar and gas input for regeneration cycle

Source: U.S. Patent 4,014,380

The heat exchange wheel is preferably a wheel of substantially the same diameter as the desiccant wheel. The heat exchange wheel can be made of expanded aluminum honeycomb, aluminum foil, fibrous materials such as asbestos coated with a hydrophobic material, or the like, which permits the passage of an air stream with very little pressure drop there-across yet which presents a large heat exchange area.

The evaporative cooling means may be pads made from cooling tower fill such as corrugated paper sheets impregnated with a suitable bactericide, e.g., phenolic resin. The pads are positioned vertically and water from a reservoir pan situated below the pads is recirculated. As the water runs down the pad by gravity, an air stream passes in a tortuous path through the pad and becomes humidified.

The first heater or low-temperature heater means may be a heat-exchange coil through which a suitable heat transfer fluid is circulated. The external heat source for raising the heat transfer fluid to a desired temperature can be solar heat, waste heat from another unrelated operation, heat generated by a nuclear reactor, electrically generated heat, or the like.

The second heater or high-temperature heater means conveniently can be an open flame burner or it can be another heat exchange coil within which a heat transfer fluid is circulated at a relatively high temperature so that the portion of air stream passing over this heat exchange coil can be heated to the necessary final regeneration temperature.

The process is very well suited for use in combination with a solar heat source which provides heat input to the low-temperature heater means for both cooling and heating.

S.A. Weil; U.S. Patent 3,880,224; April 29, 1975; assigned to Gas Developments Corporation describes an improved adiabatic saturation cooling machine of the open-cycle type and method of operation in which the capacity of the machine is increased.

Selective Recycle of Room Air to Air Conditioning Unit

A process described by *F. Horowitz; U.S. Patent 4,062,400; December 13, 1977; assigned to The Port Authority of N.Y. & N.J.* relates to an air conditioning method and system, and in particular to an air conditioning system which utilizes the principles of natural heat convection in order to increase the efficiency of operation of the system.

The principle of natural convection of air currents is well known. According to that principle warm air tends to rise so that in an enclosed space such as a room the temperature of air near the ceiling is greater than the temperature of the air near the floor of the room. This principle is advantageously used in accordance with the present process to conserve energy during both heating and cooling of an enclosed space by an air conditioning system. The present process is particularly adapted for use in air conditioning systems where outside air is continuously supplied to the system to replenish and supplement air in the building.

In accordance with one example of the process, during the cooling season when cooled or chilled air is to be supplied to the room or rooms of a building, return air for the air conditioning unit is withdrawn from the rooms near the floor so that the coolest air from the room is returned to the air conditioning unit. This return air is then chilled with the outside air supplied to the air conditioning unit and returned to the room or enclosure. At the same time, air must be removed from the room and discharged to the atmosphere in an amount substantially equal to the amount of outside air being continuously supplied to the room. This spill air is exhausted from near the top or ceiling of the room.

In this manner the warmest air from the room is eliminated, while the cooler air in the room is returned to the air conditioning unit. Thus, the amount of energy required to cool the air returned to the air conditioning unit is reduced as compared to commonly proposed systems, with a resulting reduction in energy consumption.

On the other hand, during the heating mode of operation of the system, the warm air from the ceiling area is returned to the air conditioning unit and heated there with outside air before being returned to the room. In this mode, the spill air is removed from the region of the room floor and discharged to the atmosphere. Thus, valuable heat contained in the returned air is conserved so that less heat is required to be produced in the air conditioning unit to heat the return air and outside air to the desired level. On the other hand, the cooler air in the room which will require greater heating if returned to the air conditioning unit, is removed from near the floor of the room and discharged to the atmosphere to compensate for the outside air being continuously supplied to the room.

Air Distribution Ceiling and Plenum Chambers

According to a process described by *B. Sacks; U.S. Patent 3,980,127; Sept. 14, 1976; assigned to Patco Inc.* rooms of a building are provided with an air distributing ceiling and plenum chambers. Treated air is pumped to the plenum chambers where it passes through the ventilating ceiling to heat or cool an enclosed area below. A recirculation system permits withdrawal of all of the treated air from the enclosed area. The withdrawn air is recirculated after mixing with freshly treated air. The air is treated by passing a portion thereof over a heating or cooling coil while the remainder bypasses the coil. The mixture of treated and bypass air is pumped at a high rate into the plenum chambers where a static pressure such as one inch of water exists.

Control Device Utilizing Motion Feedback System

E.N. Caldwell; U.S. Patent 4,007,775; February 15, 1977; assigned to Robertshaw Controls Company describes a heat exchange system having a source of heat exchange output fluid for effecting a heat exchange function and having a source of return fluid resulting from the output fluid providing its heat exchange function. A thermally operated element controls the amount of flow of the output fluid from the source in relation to the temperature of the thermally operated element. A sensing device senses the temperature effect of the heat exchange function in relation to a predetermined temperature that the heat exchange system is to provide, the sensing device directing one of the output fluid and the return fluid to the thermally operated element to cause the element to change the amount of flow of the output fluid when the temperature effect deviates from the predetermined temperature a certain amount whereby the thermally responsive element is subject to a relatively wide swing in temperature for large control movement in relation to a relatively narrow swing in temperature at the sensing device.

The system includes feedback means controlled by the thermally operated means and being operatively associated with the control means of the sensing means to provide for modular operation of the thermally operated means whereby the thermally operated means can cycle over a relatively short stroke of operation rather than either a fully "on" or a fully "off" operation as in common control systems.

In particular, one example of the process provides a feedback arrangement wherein a balance beam is provided for controlling a leak port with such balance beam being operatively interconnected to the thermally operated means by spring means so that the thermally operated means will provide motion feedback to such balance beam that is being pivoted by changes in the sensed temperature condition.

Humidification Function

E.R. McFarlan; U.S. Patent 3,891,027; June 24, 1975 assigned to Alden I. McFarlan describes an air conditioning system which utilizes excess heat which would otherwise be discharged from the system to vaporize water and humidify the air. The system utilizes heat produced internally of the system to make up for heat losses so that heat is discharged from the system only when there is an excess of heat in the system, and auxiliary heat is added only when there is a deficiency of heat in the system. When there is an excess of heat in the system water is vaporized into the air, reducing the load on the refrigeration equipment. The heat of vaporization remains in the system.

Reduction of Frost Buildup

According to a process described by *K. Bergdahl; U.S. Patent 3,980,129; Sept. 14, 1976* recuperative heat exchangers are disposed in the supply and exhaust ducts of a ventilating system. To reduce frost buildup at the exit end of the exhaust-duct heat exchanger, the heat exchanger is disposed so that the exhaust air flows through it alternately from one end to the other, and vice versa, when the temperature is below freezing. The reversal of flow of the exhaust air is accompanied by a reversal of the flow of heat exchange medium so that the efficiency of heat exchange is not adversely affected.

Referring to Figure 5.8a, a liquid/air heat exchanger **1** is disposed in a heat-insulated housing **2**, which is provided with an air inlet **3** to be connected to the spent or exhaust air duct. The housing also is provided with air outlets **4** and **5**, respectively, to be connected also to the spent air duct. Within the housing are provided two dampers **6** and **7**, and at the air outlets **4** and **5** are provided dampers **8** and **9**.

Figure 5.8: Heat Exchange in Ventilation System

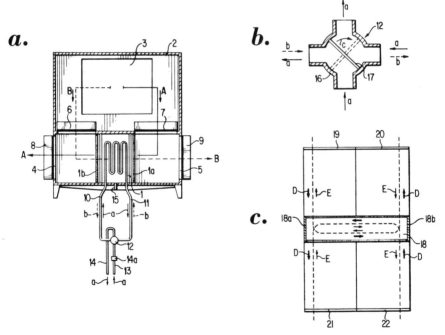

(a) Section through a liquid/air heat exchanger in the spent or exhaust air duct
(b) Detail of a liquid valve
(c) Section through an air/air heat exchanger in the spent or exhaust air duct

Source: U.S. Patent 3,980,129

The heat exchanger **1** with heat-transfer liquid lines **10** and **11** is connected via the liquid valve **12** to the liquid lines **13** and **14**, which in turn are connected to a heat exchanger in the supply air duct. The numeral **14a** relates to a transmitter inserted in the liquid line **13** for sensing the temperature of the liquid flow. The heat recovery installation operates as follows. The spent air, in the first operation case, flows in through the air inlet, passes through the heat exchanger **1** from one end **1a** (the entrance end) to the other end **1b** (the exit end) and flows out through the air outlet **4** as schematically indicated by the arrow **A**. The dampers **7** and **8** are then open, while the dampers **6** and **9** are closed.

In this case of operation, heat-transfer liquid flows from the supply air heat exchanger through the liquid line **13** via the valve **12** and line **10** into the heat exchanger **1**. From the heat exchanger liquid flows through the line **11** and via the valve **12** and line **14** back to the heat exchanger in the supply air duct. The flow direction of the liquid is indicated by the fully drawn arrows **a**. In particular the heat exchange medium flows from the end **1b** of the exchanger to the end **1a**. The heat exchanger, thus, operates in a countercurrent manner.

When the outside temperature is low, the water vapor condensed from the spent air on the heat exchanger gradually develops into ice. This ice formation takes place in the coldest zone of the heat exchanger which is the exit end **1b** of the heat exchanger where the spent or exhaust air leaves and the heat exchange liquid flow enters the heat exchanger.

In order to melt the ice formed, according to the process, the direction of the two media through the heat exchanger is periodically reversed, so that the counterflow is maintained. As shown, this is effected thereby that the dampers **7** and **8** are closed, and the dampers **6** and **9** are opened at the same time as the direction of the liquid through the heat exchanger is reversed by adjusting the valve **12**. The spent air now passes through the heat exchangers in the direction indicated by the arrow **B** from the new entrance end at **1b** to the new exit end at **1a**, and the liquid flows to the heat exchanger through the line **11** and therefrom through the line **10** from the new exit end at **1a** to the new entrance end at **1b** as indicated by the arrows **b**.

The direction of the liquid through the lines **13** and **14**, however, remains unchanged. Owing to the countercurrent relationship being maintained in the heat exchanger, but now with the directions of the two media reversed, that zone in the heat exchanger which in the preceding case of operation was hottest now is the cold zone, while the zone, which previously was the coldest one, with resulting ice formation, now will be the hot zone. Hereby, the ice formed melts, and the melt water can flow out through the drain hole **15**. At the same time as the ice melts, new ice, of course, will form, but this takes place in the zone now being the cold zone.

After the ice is melted, the flow direction of the media again is reversed, and a new ice melting and ice formation phase, respectively, commences. Owing to the mode of operation the heat recovery can be maintained without interruption and with maintained high efficiency degree. The intervals between the two phases suitably can be controlled automatically by means of program devices according to known art. The temperature conditions at which ice formation can take place are sensed by means of transmitters. Such a transmitter **14a** is shown in the liquid line **13**.

Figure 5.8b shows in detail an example of the liquid valve **12**, which comprises a housing **16** provided with four connections for liquid lines. Within the housing, a valve disc **17** is provided, which is reversible to two positions. The valve disc being in its first position, the liquid flows through as indicated by the arrows **a**. Upon reversing the valve disc as indicated by the arrow **c**, the direction of the liquid is reversed as shown by the arrows **b**. The reversal of the valve disc is effected by means of an adjusting device, which may be of pneumatic, hydraulic or electromechanic type.

In Figure 5.8c, the process is shown applied to a direct-acting air/air heat exchanger **18**. The spent or exhaust air duct is connected to the connecting flanges **19** and **20**, while the supply air duct is connected to the flanges **21** and **22**. As shown, the air media flow through the heat exchanger in a countercurrent flow, but it should be mentioned that air/air heat exchangers with cross-flow operation can also be applied advantageously.

In the first operation case, the supply air and the spent air flow in the direction indicated by the arrows **D**. The exhaust air flows through the exchanger from left to right from the entrance end **18a** to the exit end **18b**, and the supply air flows from right to left from the exit end **18b** to the entrance end **18a**. Thereby freezing commences in that zone of the heat exchanger **18b** where the spent air leaves and the supply air enters the heat exchanger. By reversing the two air flows through the heat exchanger as indicated by the arrows **E**, by means of a damper arrangement not described here in greater detail, the ice formed at **18b** is melted by hot spent air being blown thereagainst. Due to short intervals between the reversals, the temperature of the supply air leaving the heat exchanger is varied only insignificantly.

According to a process described by *J. Carrasse and D. Desbrosses; U.S. Patent 3,867,979; February 25, 1975; assigned to Societe Generale de Constructions Electriques et Mecaniques (Alsthom), France* to prevent icing on the vaporization surfaces of the condenser in a heat pump air conditioning system, the temperature of the air in contact with the condenser is raised by applying additional heat.

Basically, the mean temperature of air in contact with the condenser is artificially increased over the mean temperature which would result from heat exchange by the condenser with air at outside temperature entering the space to be heated during those periods of time when the vaporization temperature of the heat exchange fluid, in heat exchange with the air leaving the space, is sufficiently low to be subjected to risk of icing of the heat exchange surface itself.

The mean temperature of the air in contact with the condenser can be raised by various means; in accordance with one example, an external, separate and independent heat source is provided, preferably an electrical resistance unit; in accordance with another case, a portion of the exterior air is bypassed and reunited with a portion of the air which passed the condenser, so that the condenser receives additional heating by the lesser amount of heat being removed, due to the bypass; or, a portion of the heated air which already has contacted the condenser is recycled and fed back to the condenser unit itself to preheat the air entering into the condenser, thus supplying additional heat to the condensing heat exchange step. The latter two examples, in which no additional heat is supplied, which is a cost factor, are applicable when the outside ambient temperature is not less than the critical temperature.

OTHER PROCESSES

Water Flow Through Light Construction

G. Duchene; U.S. Patent 3,874,441; April 1, 1975 describes a method for the thermal and acoustical protection of a light construction in which one or more compartments are defined by walls, panels, flagstones, etc. The process comprises creating from a source whose temperature is different from the natural temperature within the construction, a bidimensional flow of water through a porous layer the two surfaces of which are in contact with fluid-tight surfaces and which is disposed in one or more of the walls, panels or flagstones so as to be in a state of heat exchange with the compartments to be protected while being protected, if desired, from the exterior medium by a thermal insulator. The structure of the layer is such that the flow is of a capillary type, that is to say, the water in motion is practically without pressure.

According to this method, the idea is basically to leave outside the construction, for example, in the ground, the mass considered necessary for the thermal protection and to expose in elevation or above the ground only a small part thereof which is at each instant flowing in the porous layers.

During a warm period, whereas the thermal inertia of a heavy wall retards the change in temperature of the internal surface of a premises when there is a variation in the surrounding temperature because the heat received is diffused by conduction in the inner layers of the wall, this process resides in receiving the heat received from the interior in the water flowing by capillary action through the porous layers and diffusing it by causing it to flow in a large mass maintained for example in contact with the ground; the flowing water also receives possibly a small amount of heat which has passed through the insulator, which perfects the efficiency of the latter.

The heat-to-weight ratio of a heavy wall and even a very heavy wall may be easily equalled by the heat-to-weight ratio of the reserve of water employed to which is possibly added the heat-to-weight ratio of the ground in thermal contact with this reserve. It is indeed known that at a depth of one meter the ground has a remarkable temperature stability.

The method can be carried out in two ways, namely by recycling the water and by the use of discarded water.

In the first case, a mass of water contained in one or more cisterns or flexible vessels is buried outside or inside the building, for example, in the basement. A suitable raising apparatus, for example, a low-flow electric pump, draws off the fresh water from the bottom of the reservoir and conducts it to the roof top where it is distributed in the porous layers of the roof, the ceilings, floors, partition walls or supporting walls. At the downstream end of the layers the water is returned to the reservoir by the effect of gravity.

Calculation shows that, for temperate climates, a total mass of water sufficient to ensure the comfort in summer is sometimes a very small fraction of the mass that the building would have had if it had been constructed of masonry. For example, 80 m^3 of water is sufficient for a youth club of 150 m^2 area whose super-structure would weigh 200 MT for a traditional construction as against about 10 MT for a light construction.

The manner of carrying out the method by utilization of subsequently discharged water is somewhat similar to the foregoing case, considering that the notion of a reserve of water is applied to the upstream end of the water circuit. In this case, there is employed either the water of the water mains of the town which is supplied under pressure to the top of the roof so that there is no need to install a pump, or water from a river, pond or lake in the vicinity. The water is discharged by way of the rain water discharge circuit of the building.

Calculation shows that in respect of temperate climates, there is required only a very small amount of water to ensure thermal comfort in the summer. For example, in the case of a youth club on the ground floor having area of $150 \, m^2$, the under-roof of which alone is treated, at the hottest hour of the day ($32°C$ outside temperature) and for an occupation of the premises by fifty persons, if the mains water has a temperature of $20°C$, a temperature of $26°C$ is maintained within the premises by a flow of water not exceeding $1 \, m^3/hr$.

The method also permits affording a thermal protection in the winter in addition to, or instead of, a traditional heating system by the use of a reserve of heat-insulated hot water brought to the required temperature for example by electric heating.

Not only can water be heated in winter, but it can be cooled in summer by passage through a forced evaporation cooling means. This arrangement would be employed when there is neither sufficient space for housing a buried reservoir nor an abundant supply of water for the application of the method employing water which is subsequently discharged.

The material of the porous layers may be of very diverse types. There may be employed, a cellular material or foam having open pores or a fibrous material defining intercommunicating pore-like cavities constituting capillary network such as a felt or a nonwoven web, for example, in the form of strips having a thickness of 3 to 10 mm and a width to suit the supports, each of the surfaces of the layer being provided with a fluid-tight film, for example, a thin sheet of polyethylene, polyvinyl or aluminum. The lateral edges of the strip are closed by a welding or an adhesion of the edge portions of the fluid-tight sheets. In the case of a fibrous material, the fibers may be synthetic (for example, polyester), mineral (for example, glass or rock wool), metallic (for example, aluminum), plant or animal fibers. They must be capable of conserving their characteristics during a long period of immersion in the thermal fluid at ambient temperature.

The fibers are interconnected by their shapes (for example, wavy fibers), by mechanical treatment (for example, lashing for the nonwoven materials) or by the addition of a resin so that there is no danger that they be carried along by the current of fluid.

The nonwoven webs manufactured and sold by the Rhone-Poulenc Company under the name Bidim seem to be particularly suitable for the porous layers according to the process.

If the fluid does not "wet" the fibers, each initiation of the system operation will be preceded by the addition to the fluid of a nonfoaming wetting product (for example, Tenatex), so that the "front" of the fluid progresses by capillarity from the supply end to the fluid collector.

The fluid-tight sheets may be secured to the felt by high-frequency spot-welding (synthetic fibers) or by adhesion (mineral fibers connected to the resin) or by clips or fasteners. In the latter case, the fluid-tight sheet is reinforced by a metal netting some of the wires of which are cut and inserted into the felt. Such a reinforced sheet is manufactured by Bekaert.

The order of magnitude of the flow through the porous layers is 1.5 liters of liquid per minute for a strip having a width of one meter.

The order of magnitude of the rate of flow of liquid is 0.5 liter per minute. Thus, the order of magnitude of the thickness of liquid flowing is 3 mm and its weight is of the order of 3 kg/m².

Vacuum Technique

A process described by *T. Xenophou; U.S. Patent 3,968,831; July 13, 1976* makes use of heat gradients to control the actual temperature at any required time, this being achieved by having a variable degree of vacuum in the walls of the building or structure at appropriate localities so that the amount of heat transfer through the walls can be varied according to the extent of the vacuum existing at the time.

Thus, according to one form in an area where excessive day temperatures are involved, the vacuum can be brought nearer to normal pressures at night and, therefore, there will be a transfer of low temperature into the building at that time when ambient temperatures are relatively low, but by then increasing the vacuum the lower temperature can be held within the structure throughout the next day and thus make living conditions much more comfortable.

Similarly in the winter where a building reaches reasonable temperatures during the day it is possible at this stage to allow the inside of the building to assume this temperature by removing the vacuum from the walls and allow heat transfer, but before there is a substantial loss of temperature the vacuum can be again introduced and the building will then remain at a temperature which is consistent with the daytime temperature or at least closely related thereto.

The actual construction of such a building can, of course, be considerably varied but it is envisaged that the walls of the building can comprise aluminum or similar panels which have a space between them, but with stiffening members to allow the space to be lowered in pressure to the required vacuum value without collapsing the panels, the construction thus including infill material or a cellular nature or bats which are sufficiently porous but incompressible to allow the vacuum to be generated. A series of projections could also be used which should, however, be of a nature such that heat transfer through this infill material does not take place to any great extent under vacuum conditions.

The various hollow sections of the building so formed which can comprise the outer walls and the inner walls as required and can include double-glazed windows or the like, can be assembled and fitted together on the site, but the cavities within the walls are of a sealed nature and, therefore, by connecting these to a vacuum pump the required vacuum can be maintained at those times when this is desirable.

According to a modification, walls in which low heat transfer is required could have a preapplied vacuum and the panels could then be sealed, and in such a case, of course, it would be possible to maintain a permanent seal in the walls and thus make them of a low heat conductive nature. Heating or cooling of the building could then take place by drawing in outside air of the required temperature at the appropriate time and thus again controlling the temperature because if the rooms are raised to the required temperature when ambient temperature is of the correct value, it will be obvious that the walls containing the vacuum will maintain this temperature within reasonable limits for a considerable time.

By using aluminum or aluminum clad walls it is possible to prefabricate these sealed panels in a factory and to evacuate and seal them and to simply assemble the structure as required.

Radiant Panel Construction

S. Vinz; U.S. Patent 4,121,653; October 24, 1978 describes a method of constructing rooms each having a floor, walls, and a ceiling and being provided with a radiant heating system or a cooling system which comprises radiant or cooling panels, respectively, connected to the walls, the ceiling, or the floor of that room and forming a thermal medium cavity wherein there may circulate a thermal medium which transmits heat to the radiant panels or which receives heat from the cooling panels.

In the construction, at least the walls and the ceiling are prefabricated components capable of bearing static loads and the interiors of the walls and the ceiling, which interiors face the remainder of the room and the radiant or cooling panels, contain respective pairs of adjacent parallel troughs of curved or part-polygonal cross section, the troughs in the walls being vertical and the ceiling being applied so that they merge in streamlined manner with the troughs in the ceiling, and the panels are suspended from supports on the walls and the ceiling with interposition of a resilient packing.

The process is thus based primarily on use of two components, viz, a prefabricated load-bearing component and the radiant or cooling panels, which can also be prefabricated, and both of which can be readily assembled on site. The parallel troughs in the walls and ceiling ensure satisfactory convection, and at the same time they act as heat reflectors like concave mirrors, reflecting the heat towards the radiant panel. The resilient packing for the supports eliminates heat and sound bridges between the walls or the ceiling and the radiant or cooling panels.

The supports are preferably comprised by a support web which runs across the entire width of, and transversely to, the respective troughs and which is inserted into ribs between those troughs. Claw-like support means mounted on the radiant or cooling panels are preferably engaged over the support web, and the resilient packing is preferably provided between support means and the support web.

Thermal baffles are preferably mounted on the rear of the radiant or cooling panels, preferably extend into the thermal medium cavity, and also preferably constitute mountings for a thermal medium pipe spaced from the radiant or cooling panel. The thermal medium pipe is also attached to the walls or the ceiling by way of thermal baffles, which at least partly embrace the pipe and which pass through the radiant or cooling panel and which are bent over to bear on the front of the panel.

Recovery of Waste Heat from Furnace Exhaust Gases

A process described by *G.S. Trump; U.S. Patent 4,079,778; March 21, 1978* is directed to a heating system capable of cooperatively capturing otherwise wasted heat of exhaust gases generated by a furnace.

According to the process, hot exhaust or flue gases generated by a furnace are conveyed by a first perforated conduit towards a flue. A second fluid-carrying conduit is coiled about the first conduit in relatively close proximity thereto so as to permit gases escaping through the perforations of the first conduit to contact the outer surfaces of the second conduit. Heat is transferred from the exhaust gases through the walls of the second conduit to the fluid. This heated fluid is conveyed to a remote location whereupon it is caused to lose its acquired heat to a space sought to be heated; it is then returned in a substantially closed-loop arrangement to the coiled portion of the second conduit. Exhaust gases that have been permitted or caused to escape from the perforated first conduit are contained by a third conduit which conveys them to the flue and which houses the first conduit and the coiled portions of the second conduit.

The flow pattern of escaping exhaust or flue gases is controlled according to the process wherein a plurality of spaced sleeve members are positioned intermediate and preferably in contact with one or the other or both of the first conduit and the coiled portion of the second conduit. The presence of these sleeve members over preselected perforations in the first conduit serves to restrict and thus inhibit the flow of exhaust gases through perforations beneath the sleeve members, thereby causing relatively higher flow rates within these restricted areas. The result is a unique flow pattern of gases around the coils. The sleeve members, in addition to causing this flow pattern whereupon the rates of heat transfer as between the exhaust gases and the fluid are controlled, further function as heat "sinks" which hold heat imparted to them by these same gases and conduct them directly into the walls of the second conduit coils.

In other examples, deflecting vanes are provided to yet further alter and control the flow of gases which contact the fluid-carrying coils. In all cases, however, contamination of the fluid by the often impure flue gases is prevented via the presence of closed second conduit system.

Fireplace Systems

A process described by *G.E. Kamstra and R.J. Powers; U.S. Patent 4,143,638; March 13, 1979* relates to heating systems of the forced air type and more particularly to a fireplace type heat exchange unit adapted for connection into the ducts of a forced air heating system.

Essentially, the heat exchange unit includes a firebox positioned within a housing or enclosure so as to define a heat transfer space. The front opening of the firebox is closed by closure means to eliminate smoking and to close off the normal draft to the firebox. The enclosure seals around the firebox and the firebox is provided with a plurality of heat exchange fin means positioned thereon to channel and guide the air entering the enclosure from the forced air system over and around and under the firebox resulting in highly efficient transfer of heat from the hot combustion gases and the hot firebox to the forced air.

A combustion air control means is positioned on the front of the unit to reduce the amount of combustion air entering the firebox to a level sufficient to support combustion. Damper means are positioned in the chimney connected to the top of the firebox at the smoke dome to control and limit the rate at which the hot combustion gases escape through the chimney. A plenum means is positioned between the damper and the firebox to collect the hot combustion gases thereby further increasing the heat transfer efficiency.

A fireplace heat exchange system used in accordance with the process is capable of increasing flue stack temperatures from the typical 250°F to temperatures in the range of 600° to 700°F. The combustion air entering the firebox is reduced from the typical range of 100 to 250 cfm to the range of 10 to 25 cfm which is sufficient to support combustion of the wood or other material positioned within the firebox.

According to a process described by *H.F. Edwards; U.S. Patent 4,137,896; February 6, 1979; assigned to Sunbeam Corporation* a combination fireplace enclosure and heat-exchanger unit for providing supplemental heat to areas external to the fireplace have been developed. The unit includes an enclosure frame having double-paned glass closure doors, and a steam-heating system disposed at the upper portion of the frame.

The heating system comprises a boiler device mounted at the rear of the frame and above the fire, and a heat-exchanger device disposed at the front of the frame. The heat-exchanger device has a steam passage which is connected by a steam line to the boiler. The passage is sloped with respect to the horizontal, such that condensate from the exchanger device can flow by gravity back to the boiler device to be converted into steam. The entire system is vented to the atmosphere whereby there is no danger of pressure buildup; accordingly the need for relief valves is obviated. Means providing an inlet to the system enable measured quantities of water to be added periodically, as required. The enclosure and heat-exchanger unit are completely self-contained, and accordingly can be readily installed with a minimum of tools, and with no external plumbing connections being required.

Automatic Environmental Control in Greenhouses

W.L. Enter; U.S. Patent 3,905,153; September 16, 1975 describes a complete automatic temperature control system which is particularly adapted for installation to control temperatures in larger types of buildings, arenas and the like. The system is particularly adapted for controlling temperature in and about greenhouses, such temperature response being formulated in relation to various parameters peculiar to a horticultural operation. The automatic control system not only functions to maintain ambient daytime temperatures according to incident solar radiation, but it also serves to automatically regulate nocturnal temperatures in accordance with accumulated daily incident solar radiation and to optimally adjust temperatures in relation to time during and after insecticidal operations in the environment. The apparatus derives a plurality of electrical output indications relating to each of instantaneous temperature, solar radiation, accumulated daily radiation, executive control modification, and insecticide modification, whereupon the electrical outputs are summed and level detected to control related heating and cooling apparatus which serves the environmental area or building.

SOLAR AND
GEOTHERMAL ENERGY PROCESSES

INTRODUCTION

Solar radiation appears promising as a potential energy source, and much effort has been expended in developing various mechanisms to convert solar radiation to useable power, such as thermal or electrical power. For obvious reasons, energy storage is an important adjunct to any solar radiation power system. Such storage can occur either immediately following collection of solar radiation (thermal storage) or following conversion of solar radiation into, for example, electrical energy (electrical storage).

To be effective and practical, a storage system must be highly reversible. It should be simple, self-regulating, service-free and capable of extended cycle life, high efficiency and rapid charge-discharge cycles.

Thermal energy can be stored by the heating, melting or vaporizing of a material, and the energy becomes available as heat when the process is reversed. For example, the heat capacity of a substance (sensible heat) may be utilized; common examples of such substances include water and crushed rock. Such an arrangement, however, requires extensive insulation in order to retain heat until the period of nonillumination, when it will be needed.

Another example of a thermal energy system employs the latent heat evolved upon transition of a substance from one phase to another (e.g., solid to liquid state or liquid to vapor state) at a constant temperature. Such a phase change storage system requires, for example, low melting eutectic materials having high latent heats. However, as heat is withdrawn, the material fuses onto cold surfaces and heat must then be conducted through frozen material. Thus, in addition to the insulation requirement noted above, a material of high thermal conductivity is desirable. As a consequence of the requirements of high latent heat, high thermal conductivity and extensive insulation, economic considerations play an important role in the selection of a suitable material.

This chapter also describes a number of processes related to recovering resources from below the earth's surface, particularly heat, and particularly geothermal energy resources. It is generally recognized by geologists that within a distance of 20 miles beneath any point of the earth's surface, including the ocean floors, temperatures reach levels that would usefully power heat engines. The only previous barrier to the recovery of such energy resources was the difficulty of providing adequate heat-exchanging means at depths with substantial geothermal resources. Once that barrier is overcome at a certain point of the earth's surface, that point becomes a suitable site for geothermal energy recovery operations.

Geologists believe that the easily available geothermal resources which have been successfully recovered in the past, existed close to the earth's surface as the result of natural heat-exchanging mechanisms at greater depths. These natural heat-exchanging formations make available only a minute fraction of the potential resources lying within 20 miles of the surface. The earth's total geothermal resources greatly exceed the world's energy needs, not only for today but for the foreseeable future. These resources have been almost totally inaccessible to past methods and apparatus, not because the drilling apparatus had not reached a high state of development, but because of the inherent limitations, both technological and economic, of the past drilling methods. It is a most important challenge for the research community to make the geothermal resources of the earth generally accessible to meet the world's great need for energy, particularly for nonpolluting energy, which is an essential requirement of life and prosperity.

SOLAR ENERGY STORAGE AND TRANSFER SYSTEMS

Mercury Vaporizer

A. Jahn; U.S. Patent 3,905,352; September 16, 1975 describes a process which employs a planetary or orbital system wherein a unitary mercury boiler or vaporizer is axially disposed of one or more orbitally moving solar ray reflecting and concentrating units mounted upon a turntable, track or the like. Each of the reflector units comprises a multiplicity of individual reflective mirror structures mounted for individual adjustment in a common frame and arranged in banks which, through constant control in a suntracking relation throughout orbital travel, effectively concentrate and reflect sun rays during a variable trajectory of the sun upon and against an effective heat-receiving area of the unitary and axially mounted boiler or vaporizer.

The mercury vaporizer is preferably mounted on a fixed location axially of the orbital travel of the reflective unit or units and preferably is provided with a revolving heat-insulated shutter mechanism which travels with the platform or support for the unit or units and exposes at any time only a fraction (approximately 40%) of the sensitized surface heat-receiving area at the top portion of the boiler.

The individual reflectors of each unit are preferably of flat surface type, each being mounted on a ball and socket joint or otherwise to permit and adapt itself to adjustment or tilting on a plurality of axes. Each of the multiple reflector units is supported in an upstanding rectangular frame which itself is mounted for swinging or tilting to an elevation mechanism from approximately –10° to +40° in elevation.

The tilting of the rectangular upstanding unit frame or frames, where two or more units are employed in orbit is controlled by an available sun sensor and its control and driving mechanism which will orient the elevation mechanism for the rectangular frame within the range of minus and plus degrees to vertical defined above.

The same sun sensor used to orient the elevating mechanism may well serve to individually adjust at slightly varying angles the multiplicity of reflectors in each unit tilting the same substantially as required to keep each mirror focused.

A second sun sensor and sensing control and driving system is employed for controlling the orbital travel of the rotary platform or track-mounted annulus analogous structure for supporting one or more of the orbital reflecting units.

Suitable commercially manufactured sensor, control system and drives are available on the American market as for example, the sensors, control systems and drives, manufactured and sold by Adcole Corporation, including for special use a commercial model called Zea Model 11866 sensors. In this model system, the analog output of the system is zero when pointing directly at the sun, is positive on one side of the sun, and negative on the other side of the sun. The motor drive controller will drive the system in the appropriate direction until the sensor output is zero.

The process provides in addition to the system and systems generally described in the foregoing synopsis a full system and plant for generating at high efficiency electricity for municipal and industrial uses. The process preferably includes a storage medium for heat which comprises a mixture of comminuted salt and comminuted copper to constitute a storage cell. Through the variable admixture of these ingredients, the rate of conductivity and transfer of heat from the exterior of the cell and to the core and oppositely from the cell outwardly to a boiler or other medium supplied, can be quite accurately controlled.

The collecting and concentrating mirrors working on a circular track with 360° traverse with the sensors will keep the multiplicity of mirrors throughout the seasons of the year in highly efficient focus with the sun and with concentration of rays on the heat-absorption portion of this stationary axially disposed boiler or vaporizer.

In working out careful calculations of results obtainable, it is contemplated that production by two units of multiple mirror elements causes the production of heat to 1800°F temperature on the absorbant vital portion at the top of the special boiler. Mercury vapor will leave the unit at nearly 1500°F and arrive at the power plant or storage unit near 1400°F.

In use for municipal or large industrial purposes, it is proposed to place a large number of these stations or systems per acre of land available. This would comprise for a fairly good sized municipality the use of perhaps fifty stations per acre. The heating of three square feet to 1800°F at each evaporator head would produce for fifty stations, 150 square feet at substantially 1400°F surface. This area should produce at least 300 lb of steam per hour at 750°F and 400 lb pressure. This amount of steam is known to be productive of 300 kW per acre. 100 acres would produce at least 30,000 lb of steam or about 30,000 kW of electricity.

Tracking Solar Energy Collection System Using Metal Hydrides

A process described by *C.G. Miller and J.B. Stephens; U.S. Patent 4,065,053; December 27, 1977; assigned to U.S. National Aeronautics and Space Administration* provides a tracking solar energy collection system utilizing a fixed, linear, ground-based primary reflector and a movably supported collector.

The process also provides secondary reflectors for refocusing the solar energy reflected from a fixed concentrator into concentrated beams of solar energy.

In the process, a large fixed primary reflector is constructed at ground level by slip-forming in concrete or stabilized dirt a trough with a segmented one-dimensional circular cross-section profile. This profile is covered with an inexpensive light-reflective material. The axis of the primary reflector is optimally aligned with respect to the sun path in the area. A heat-absorbing structure is movably supported above the primary reflector. The support mechanism transversely shifts the heat-absorbing structure to track the changing position of the sun's image diurnally and seasonally, keeping the structure at the changing line focus of the primary reflector.

The heat-absorbing structure carries secondary reflectors that either direct off-angle solar energy to the structure or refocus the line focus of the primary reflector into discrete spots of intense solar energy. These secondary reflectors are constructed so as to maximize absorption and minimize heat emission from the heat-absorbing structure. Building the solar energy collection system in stages, each stage designed for optimum efficiency within a certain temperature range, provides a more efficient and cost-effective overall system.

Endothermic-exothermic reaction chambers are provided utilizing metal hydrides, particularly magnesium hydride (MgH_2).

Metal Hydrides

According to a process described by *G.G. Libowitz; U.S. Patent 4,040,410; August 9, 1977; assigned to Allied Chemical Corporation* a reversible, closed thermal energy storage system is provided for storing thermal energy produced by a source of heat and for supplying heat to a living space. The storage system employs a metal hydride that dissociates at elevated temperatures to form metal plus hydrogen gas. The hydrogen gas is stored in a separate container either as compressed gas or as a secondary, less stable storage hydride. The hydrogen is subsequently recombined with the metal to reform the metal hydride and give off heat, as needed. The storage system is particularly useful in solar radiation heating systems.

The storage system for supplying heat to the living space comprises:

(a)   means for containing a material, the containing means exposed to an external source of heat, the material comprising at least one metal hydride capable of reversible dissociation into metal plus hydrogen gas and having a heat of formation of at least –5 kcal/mol, the external source of heat adapted to heat the material to a temperature sufficient to cause the dissociation of the material;

(b)   means for storing the hydrogen gas separate from the metal and apart from the living space;

(c)   means for communicating between the containing means and the storage means; and

(d)   means for controllably recombining substantially all of the hydrogen gas with the metal in the containing means to regenerate substantially all of the metal hydride to produce heat, with additional means for transporting the produced heat to the living space.

One major advantage of using a metal hydride as a thermal storage material is the ability to control the rate of heat evolution. For most thermal storage materials, the heat is evolved spontaneously. This is true for both the sensible heat materials (e.g., water, crushed rock, etc.) and the phase change materials (e.g., low eutectic temperature salts). Consequently, a large amount of insulation is required to prevent too rapid a heat loss.

However, in the systems described herein, in which the hydrogen gas is separated from the hydriding metal, the stored heat may be recovered when required, or it may even be stored indefinitely, with no need for insulation. Thus, with efficient metal hydride systems, thermal energy may be stored over an extended period of time, possibly even storing summer heat for winter use.

Metal-hydrogen systems are also noncorrosive and they can undergo indefinite cycling with no chemical degradation, which is a problem with existing phase change materials. An additional advantage of metal hydrides resides in their high thermal conductivities, which permit efficient heat transfer to the heat transport fluid.

The metal employed may comprise any of the hydride-forming metallic elements or a hydride-forming alloy. Specific examples include $FeTiH_{1.7}$ and the $VH-VH_2$ system.

Where two metal hydrides are employed comprising a primary metal hydride for dissociation and a secondary metal hydride for storage, the primary metal hydride should have a relatively high heat of formation, while the secondary metal hydride should have a low heat of formation. When the recovery of stored heat is desired, hydrogen is bled from the secondary metal hydride and is permitted to recombine to form the primary metal hydride. In some instances, a portion of the heat produced may be used to further dissociate the secondary metal hydride to generate additional hydrogen gas.

Photochemical Fluids

D.H. Frieling, S.G. Talbert and R.A. Nathan; U.S. Patent 4,004,573; January 25, 1977; assigned to Battelle Development Corporation describe a process for the collection, retrieval, and utilization of solar energy, wherein a photochemical fluid, containing an isomerizable compound, is passed through a collector for exposure to solar radiation to transform the isomerizable compound to a higher energy level isomer. The irradiated fluid leaving the collector passes in heat exchange relationship with photochemical fluid entering the collector and then proceeds into a trigger reactor means. In the trigger reactor means, higher energy level isomer in the irradiated fluid is triggered, such as by heat and/or catalyst contact, to revert to an isomerizable composition of a lower energy level isomer with exothermic release of heat in excess of that requisite for maintaining conversion of higher energy level isomer to lower energy level isomer.

The trigger reactor means is positioned in close proximity and/or desirably contiguous to or an integral portion of another heat exchanger through which is flowed a material adapted for storage at an elevated temperature to receive that exothermic heat excess being released upon the higher to lower energy level isomer conversion. The photochemical fluid, after passage through the trigger reactor means and the other heat exchange means, proceeds to and through the first exchange means for passage to the collector and repetition of the just-described procedural sequence. The storable material exiting from the other heat exchanger is of elevated temperature with a sensible heat content thereof available for usage, after storage immediately, in manners known to the art for utilizing heated material in useful applications such as residential hot-water supplying and residential heating and cooling.

A typical useful geometrical isomerizable compound is an organic compound which is selected from the several classes of compounds of: indigo and thio-indigo derivatives; stilbene derivatives; cyanine-type dyes; and modified aromatic olefins.

Endo-5-Norbornene-2,3-Dicarboxylic Acid Anhydride

A process described by *J.C. Powell; U.S. Patent 4,100,091; July 11, 1978; assigned to Texaco Inc.* relates to the storage of thermal energy by the allotropic change of endo-5-norbornene-2,3-dicarboxylic acid anhydride.

One troublesome problem encountered with energy forms such as solar energy, which is intermittent in nature, is storing the energy until it is needed.

The process is based on the discovery that endo-5-norbornene-2,3-dicarboxylic acid anhydride (NDAA) can store thermal energy at about 93°C and release such energy at temperatures below 40°C. The crystal structure of NDAA has been extensively studied at about room temperature and the transformation of the endo to the exo isomer has also received much attention in the literature.

The endotherm onset for the compound of the process is about 70°C below its melting point (about 164° to 166°C) observed by differential scanning calorimetric (DSC) studies. After passing through the endotherm, the endotherm could not be observed if the material were cooled above about 48°C and reheated. Thermogravimetric analysis on a fresh sample showed less than 0.1 wt % loss up to 140°C, thereby ruling out a sublimation effect.

The compound can be synthesized by a Diels-Adler reaction between maleic anhydride and cyclopentadiene [L.F. Fieser, *Organic Experiments,* D.C. Heath (Boston) 1964, Chapter 15, p. 83.]

In studies leading to this process, DSC was used to measure the transition points of heat uptake and release. The endothermic onset was 95.0±0.5°C initially for fresh NDAA. On subsequent cycles of cooling and reheating, the onsets were 91±0.5°C at a heating rate of 10°C/min. Upon rapid cooling, the onsets of the exotherm were 40.0°, 36.5°, and 35.0°C (all ±0.5°C), respectively, in succeeding cycles.

The energy absorbed was measured using the melting of pure indium as a standard. The ΔH for NDAA averaged 3.3±0.3 kcal/mol (20 cal/g). This is compared to the allotropic change for orthorhombic sulfur to monoclinic sulfur at 95.6°C: 0.088 kcal/mol (2.74 cal/g). The exothermic energy appeared to be of the same order of magnitude as that absorbed.

The 13.12 mg sample of NDAA was as much as 9°C cooler than the temperature of a reference thermocouple at 97° to 100°C when heated at 10°C/min and as much as 50°C hotter than the reference at 36° to 44°C when cooled.

Infrared specroscopy indicated no chemical changes were associated with the up-take or release of heat. Spectra were recorded of a mineral oil mull of NDAA at 28°C, 82.5° to 88°C, 101° to 102°C, and again at 27°C in that order in the 2.5 to 15.0 micrometer region using an electrically heated AgCl cell. There was a slight shift in the intensities of three absorption peaks between 11.8 and 12.9 micrometers which occurred gradually during heating and which did not change upon cooling even after standing overnight. This is attributed to a physical change in the mull.

Polarized light microscopy (PLM) showed an allotropic change at the endothermic transition point. Using a petrographic microscope equipped with a hot stage, the following observations were made: At room temperature NDAA crystals are anisotropic (bright and colored) in the dark field of crossed polarized light. Upon heating, the crystals lose their double refraction (become dark) between 92° and 97°C. Small crystals, near or in a mull (including the mull previously heated in the IR experiment discussed previously) gave a sharp crystal structure change at 94.5°C at 2° to 4°C/min. Under normal light the NDAA undergoes no significant change.

Two-Phase Refrigerant Mass

J.E. Garriss and D.E. Garriss; U.S. Patent 4,128,123; December 5, 1978 describe a method for transporting heat from an available source to a point of use, as in solar heating systems to circulate a sensible heat-carrying medium such as air, water, a chemical compound, or the like, which is heated at the source, pumped to the point of use where the heat is extracted, and returned to the source to repeat the process.

The process involves a multipurpose, passive refrigerant cycle for transporting heat energy from a source location where it is available to a destination heat sink where it can be used or stored for future use. The heat energy intended for transport and the difference in temperature between the source and sink power the system cycle. The operating principle of the cycle is based upon the interactions of heat energy with the refrigerant heat vehicle. A two-phase refrigerant mass is thermodynamically interposed in the system between the heat source location and the heat sink location with the liquid phase positioned to receive heat from the source.

The vapor phase of the refrigerant mass contiguous with the liquid phase is made to extend in a continuous body of vapor from the surface of the liquid in the heat source location to the heat sink location where it is positioned to transfer heat to a cooler medium. The properties of the refrigerant at saturation are identical throughout the mass, the specific values being dependent upon the saturation temperature of the liquid phase and the type of refrigerant.

The positioning of the refrigerant mass in simultaneous heat transfer contact with both the source and the sink establishes the conditions desired to obtain a dynamic state of concurrent evaporation and condensation in different areas of the same refrigerant mass at an equilibrium saturation pressure and temperature relative to the temperature of the liquid phase. Under the steady state operating conditions of the heat transport function of the cycle, the gross total heat content of the refrigerant mass remains essentially constant at an equilibrium value while latent heat is simultaneously added to, and substracted from the gross total. The balances between the distribution of energy and matter in the refrigerant mass are maintained by migrational movement of vapor bearing latent heat from the point of evaporation at the source to the point of condensation at the sink.

From the description of the heat transport function, it will be apparent that the movement of the refrigerant mass in the vapor phase, and the subsequent change of state to the liquid phase at the heat sink location make it necessary to provide a second function for the continuous return of the condensed liquid phase refrigerant from the heat sink location to the heat source location to close the cycle of operation. The method and power for this function are also obtained from the interaction of the system refrigerant with the heat energy available from the source.

According to this process, a relatively small portion of the refrigerant mass in the liquid phase and vapor phase is confined and isolated from the transport area of the cycle, and is positioned to receive heat from the source to increase its saturation temperature and pressure to a value higher than the pressure existing in the transport areas of the cycle. The high pressure thus generated is hydraulically applied to move refrigerant liquid from the sink location to the source location against the opposing resistance of flow friction, counterpressure, and gravity head.

Individual applications of this process for use in transporting heat energy from a heat source location to a heat sink location will require obvious variations in the design of components, and in the method of automatic control due to inherent characteristics of the system. These variations are dictated by the magnitude of temperature difference available between the heat source and the heat sink in combination with the relative elevation of the heat source evaporators to the elevation of the heat sink condenser. The available temperature difference places a limit on the maximum allowable difference in height between the heat source evaporators and the heat sink condenser when the condenser is at a lower elevation than that of the evaporators.

The magnitude of required temperature difference diminishes to zero as the elevation of the heat sink condenser changes from an elevation below that of the source evaporators to elevations above the source evaporators until the static head exerted by the height of the return liquid column is sufficient to return the liquid refrigerant by gravity alone.

Thermal Siphon

N.O. Movick; U.S. Patent 3,951,204; April 20, 1976 describes methods and apparatus for thermally circulating a liquid, and more particularly a method and apparatus for thermally cycling a liquid in a direction counter to that of a normal thermal siphon without substantial limitation as to height.

An apparatus for thermally pumping heat from an elevated to a lower position is illustrated in Figure 6.1a and generally designated by numeral **10**. Externally, apparatus **10** includes an enlarged expansion chamber **12** at the upper portion thereof, having a filling and pressure-regulating valve **13** preferably located at the upper portion of expansion chamber **12** and communicating with the interior thereof. Located immediately adjacent expansion chamber **12** and in communication with an external stand pipe **15** is heat input means shown as fins **16**. Fins **16** are adapted to supply heat from a source to external stand pipe **15**. At the bottom portion of the external stand pipe there is provided a means for extracting heat shown as heat extraction fins **17**.

Figure 6.1: Thermal Liquid Circulating Process

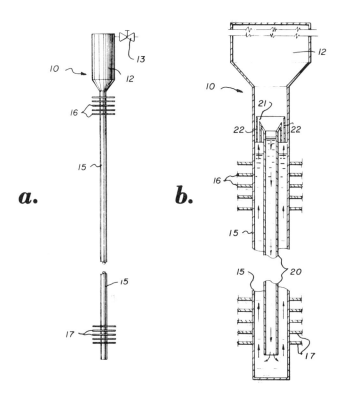

(a) Simplified view of apparatus.
(b) Sectioned, more detailed view of the apparatus of
     Figure 6.1a.

Source: U.S. Patent 3,951,204

Specific operation of the apparatus will be readily understood with reference to Figure 6.1b which illustrates the operational details of the apparatus. An internal stand pipe **20**, having an internal collector **21** with inlet pipe **22** disposed at the upper portion thereof, and at the lower portion of expansion chamber **12**, is included. It will be noted that internal stand pipe **20** and external stand pipe **15** are in communication with one another at a location below heat extraction fins **17**. Also, it will be noted that external stand pipe **15** and internal stand pipe **20** are filled with liquid to a level intermediate heat input fins **16** and internal collector **21**. Accordingly, when heat is provided through heat input fins **16**, the liquid in external stand pipe **15** is heated to a boiling condition thereby expelling liquid upward through inlet pipes **22** into expansion chamber **12**. Also, vapor is concurrently provided from external stand pipe **15** to expansion chamber **12**.

The liquid expelled from external stand pipe **15**, through inlet pipes **22** and the vapor condensed in expansion chamber **12** are intercepted by internal collector **21** and conducted into internal stand pipe **20**. As a result, the relative liquid levels between the liquid in external stand pipe **15** and internal stand pipe **20** differ as shown in an exaggerated manner in Figure 6.1b with the level in internal stand pipe **20** being higher. This, of course, induces a flow down internal stand pipe **20** and up external stand pipe **15**.

Further, as a result of the heating and condensation, the liquid in internal stand pipe **20** tends to be at a relatively elevated temperature with regard to the surroundings. As the heated liquid travels down internal stand pipe **20**, it tends to come to a temperature equilibrium with the liquid traveling upward in external stand pipe **15**. However, it is to be understood that the entire system is at a temperature sufficiently high to permit extraction of heat at heat extraction fins **17**. At external surfaces elsewhere, apparatus **10** is preferably insulated.

While the liquid in either internal stand pipe **20** or external stand pipe **15** tends to be at a relatively higher temperature in the vicinity of heat input fins **16**, and while the liquid in external stand pipe **15** tends to be at a relatively lower temperature adjacent to and somewhat above heat extraction fins **17**, for the most part, the liquid in internal stand pipe **20** and external stand pipe **15** tends to be at an intermediate and substantially equal temperature for the main portion of the concurrent length of these members. Accordingly, there is no substantial difference in density between the liquid in internal stand pipe **20** and external stand pipe **15**.

For this reason, heat can be pumped from the elevated location of heat input fins **16**, to heat extraction fins **17** located a substantial distance lower. With only localized differences in temperatures, and with substantial equality of temperature over the greater portion of the length of internal stand pipe **20** and external stand pipe **15**, little resistance to flow because of density difference exists.

The liquid utilized as the working fluid may be any boilable liquid such as water, alcohol, glycols, mixtures thereof, and like compounds such as Freon, etc. While fins **16** and **17** are used symbolically to indicate the position and general nature of the heat input and heat extraction, it will, of course, be understood that heat exchanger means of any nature including liquid to liquid and liquid to gas interfaces may be utilized.

The process permits the input of heat to an elevated location in a system. The heat unbalances the liquid level between two stand pipes by displacing liquid from one stand pipe to the other stand pipe and, thus, induces a flow between the stand pipes. This enables the system to pump heat from an elevated to a lower position. Further, the stand pipes are in thermal contact with one another, and preferably concentric, to permit thermal equilibrium between the liquids in the two pipes. Thus, differentials in pressure as a result of densities of liquid are avoided. Even though the two stand pipes circulate liquid at an essentially common temperature over most of the length of the stand pipes, the liquid is at a temperature sufficiently elevated to permit extracting of heat at the lower location. Finally, since the liquid level does not reach to the communication between the stand pipes at the upper level, reversal of the temperature relationship between the upper location and lower location will not induce a reverse flow. This is useful, for instance, in the instance of solar heating to preclude reverse pumping of heat from a heat sink to a solar collector which functions as a radiation cooler at night.

In related work, *N.O. Movick; U.S. Patent 4,116,379; September 26, 1978* describes a domestic heating system that utilizes heat collectors such as solar collectors or fireplace heat collectors, and effectively transfers the heat from these collectors for storage or usage without additional energy input. Thermal pump apparatus is utilized for transferring the collected heat to a lower level of the building where the heat can be more easily distributed or stored. A heat exchanger is utilized to extract heat from the thermal pump apparatus at the lower level and transfer the heat to the supply piping or ducting of a conventional forced air or hot water heating system. A heat storage tank can be included in the system for storing the collected heat that cannot be used immediately.

Sulfur Dioxide-Containing Unit

E.J. O'Hanlon; U.S. Patent 4,089,366; May 16, 1978 describes a method for transferring heat downwardly to an out of sight area. This is particularly applicable to the reception, transport, and storage of solar heat in a manner that prevents its prompt loss back to the sky if the sky clouds over.

Likewise when incorporated in the roof of any shelter structure, it can provide internal solar warmth to the structure and at a minimum construction and maintainence cost.

Figure 6.2 shows one of the units for receiving and sending heat downwardly. In Figure 6.2, numeral **1** is an upper sheet metallic container, **2** is a lower sheet metal container positioned under it, **3** is a metallic passageway joining container **1** and **2** together and reaching almost to the top of container **1** and almost to the inner bottom surface of container **2**. All tubular and container contacts and connections are permanently sealed and made leak-proof so no liquid sulfur dioxide or sulfur dioxide fumes can possibly escape.

Numeral **4** represents a sheet metal surface fastened by welding or brazing to the bottom of container **2**, numeral **5** being a similar sheet of metal fastened by welding or brazing to the top of container **1**. Around the outer surfaces of containers **1** and **2** and tubular passageway **3** is packed heat insulation material **6**, in this case rigid polyurethane foam.

Within the upper container **1** there is shown a supply of liquid sulfur dioxide **7**. It need not fill the entire container but could occupy at least the bottom one-third.

As previously noted, the upper outlet of tubular passage 3 approaches near, but does not touch the inner upper top surface of container 1, while the bottom opening of tubular passage 3 approaches near but does not touch the inner bottom of container 2. The upper surface of sheet metal plate 5 is blackened to better receive and absorb heat from the rays of the sun.

When container 1 is partially filled with sulfur dioxide the apparatus is ready to function to send heat downwardly during the daytime. If it is solar heat, then as the blackened surface metal sheet 5 gives its heat to container 1 this causes the sulfur dioxide in container 1 to vaporize sending the solar heated vapor downward through the metal passageway 3 into container 2. Since container 2 is cooler than container 1, the vapor of liquid 7 condenses and becomes condensate 70. Then at nighttime, container 1 cools drawing the liquid 70 back up into container 1.

Figure 6.2: Solar Heat Transfer Unit

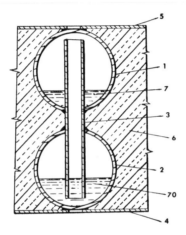

Source: U.S. Patent 4,089,366

Particulate Phase Change Material and Polymeric Resin

E.S. Dizon; U.S. Patent 4,111,189; September 5, 1978; assigned to Cities Service Company describes a device for the collection of solar radiation and the storage of thermal energy including a housing reservoir and a phase change matrix disposed in the housing reservoir. The phase change matrix comprises a polymeric material and a particulate phase change material. The particulate phase change material, having a melting point below about 100°C, is discretely dispersed in the polymeric material. A heat exchange means is positioned within the matrix for moving a heat absorbing medium and is in communication with the outside of the reservoir.

Figure 6.3a illustrates the prior art of a system for solar radiation collection and thermal energy storage, in which a solar radiation collector, generally illustrated as 11, comprises a black body material, such as a surface coated with black paint or a surface made up of anodized metal.

Figure 6.3: Solar Radiation Collector and Thermal Energy Storage Device

a.

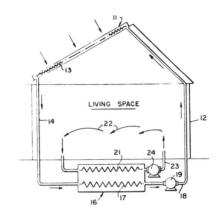

b.

c.

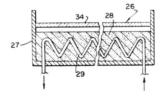

(a) Schematic illustration showing a prior art solar radiation collection and thermal energy storage system.

(b) Schematic illustration showing this process, a combined solar radiation collection and thermal energy storage device.

(c) Side elevation, in section, showing the solar radiation collection and thermal energy storage device.

Source: U.S. Patent 4,111,189

The collector **11** is mounted upon a house, generally illustrated by **12**, absorbs solar radiation, and converts the radiation into thermal energy. A primary heat absorbing medium is circulated through a first heat exchanger means **13** within the collector **11** to absorb the converted thermal energy. The primary heat absorbing medium is then transported by conduit means **14** to a thermal energy storage bin, generally illustrated by **16**, wherein the thermal energy is subsequently conducted through a second heat exchanger means **17** to a heat storage material such as water, rock, or phase change material situated within storage bin **16**.

Water or rocks, as a heat storage material, utilize the sensible heat of storage of the water or rocks. The phase change storage material utilizes the latent heat of fusion of the phase change material for heat storage. After the heat is transferred to the storage bin, the cooled primary heat absorbing medium is transported from the thermal energy storage bin via conduit means **18** to the collector by a primary pump or blower **19**.

As needed within the house, a secondary heat absorbing medium is circulated through a third heat exchange means **21** positioned within the thermal energy storage bin to absorb the stored thermal energy. The secondary heat absorbing medium transports the thermal energy from the storage bin to a heat distribution means **22** via conduit means **23** by a secondary pump or blower **24**. The heat distribution means is positioned within the house so as to optimize the heating thereof.

The primary and secondary heat absorbing mediums may be water, air, or other fluids. The type of fluid determines the type of pump or blower **19** and **24**, the type of heat exchange means **13, 17,** and **21**, and the type of heat distribution means within the house.

Referring in particular to Figures 6.3b and 6.3c which illustrate this process, a solar energy device, generally illustrated as **26**, is mounted upon a house so as to optimize the collection of solar radiation. The solar energy device comprises a housing reservoir **27**, and a phase change matrix **28**. The matrix **28** includes a polymeric material and a particulate phase change material. The particulate phase change material is discretely dispersed in the polymeric material. The phase change material is any solid material which melts endothermally to produce a liquid and reversibly and exothermally recrystallizes, thereby releasing the previously absorbed thermal energy. The solar radiation strikes the surface of the solar energy device **26** and is converted to thermal energy.

The converted solar radiation, in the form of thermal energy, is stored in matrix **28** by the melting of the phase change material. A heat exchange means **29** is positioned within the phase change matrix **28**. The thermal energy, when needed within the household, is removed by a heat absorbing medium circulated through the heat exchange means **29** situated within the solar energy device **26** and transported through a conduit means **31** by a pump or blower **32** to a heat distribution means **33** positioned in the house. Thus, the thermal energy storage bin, heat exchange means **17** and **21**, conduit means **18** and **28**, the primary heat absorbing medium, and the blower or pump in Figure 6.3a, the prior art, have been eliminated.

The phase change material of matrix **28** should have a melting point of under 100°C. Preferably the melting temperature should be from about 50° to about 80°C. The material should preferably have a latent heat of fusion above about 50 cal/g so as to limit the amount of the phase change material needed for adequate storage. The material must be immiscible in polymeric materials. Preferably, the phase change material is selected from a group consisting of polyethylene glycol, tritriacontane, pentacosane, camphene, myristic acid, methyl oxalate, stearic acid, and tristearin. More preferably, the phase change material should be polyethylene glycol having a molecular weight of from about 4,500 to about 20,000. The most preferred phase change material is polyethylene glycol having a molecular weight of about 6,000.

The polymeric material should have good heat transfer characteristics. It should not be miscible in the phase change material, but should be compatible with the phase change material to the extent that a fine particulate phase change material may be enclosed in small and discrete portions throughout the polymeric material. The polymeric material should preferably be resistant to degradation by solar radiation and have a melting or softening point greater than the operating temperature of the device. The preferred polymeric material is a curable polymeric liquid. More preferably, the polymeric material should be selected from a group consisting of a polymeric polysulfide, the polymeric polysulfide having a formula $RS_2-(RS_2)_n-SH$, polymeric silicone rubber having a formula:

$$CH_3-\underset{\underset{CH_3}{|}}{\overset{\overset{CH_3}{|}}{Si}}-O\left[\underset{\underset{CH_3}{|}}{\overset{\overset{CH_3}{|}}{Si}}-O\right]_n\underset{\underset{CH_3}{|}}{\overset{\overset{CH_3}{|}}{Si}}-CH_3$$

wherein n is from about 500 to about 5,000, polymeric butyl rubber and other polymeric sealant rubbers.

More preferably, the phase change matrix is mixed with carbon black. The carbon black should be a high structure and high surface area carbon black, such as paint grade black or conductive grade black.

The most preferred composition is about 45 wt % of polymeric polysulfide, about 6 wt % of carbon black and about 45 wt % of polyethylene glycol.

The device additionally includes one cover plate **34**. More preferably, the cover plate is of a transparent material and is spacedly situated to form an air space between the cover plate and the phase change matrix.

Example: Polyethylene glycol with a molecular weight of 6,000 was pulverized into a finely divided state of about 20 mesh and mixed with the following: 54 wt % of curable liquid polymeric polysulfide, with a molecular weight of 4,000, about 3 wt % powdered lead peroxide dispersed in oil, a curing agent, and about 10 wt % of paint grade carbon black. The temperature of mixing was maintained at 40°C and completed in 15 minutes at which time the liquid matrix was poured into a circular metallic reservoir measuring 7.5 cm in diameter and 6 mm in depth. The matrix was cured at 38°C, the curing requiring about 4 hours.

The matrix was tested by placing the housing reservoir in a suitably insulated glass cover box and exposing the matrix to solar radiation where it obtained a maximum temperature of about 78°C after 200 minutes of exposure. The reservoir was then removed from the solar radiation and from the insulated box, and allowed to radiate the absorbed heat freely for 210 minutes until it reached a temperature of about 36°C. The results showing the absorption of the heat and the reversible desorption of the absorbed heat are given below.

Time	Temperature, °C
0	24
20	35
35	44
50	49
65	52
80	54
95	55
110	63
125	68
170	76
200	77
230	75
245	66
260	57
275	51
290	46
305	46
320	46
335	46
350	44
365	43
380	41
395	38
410	36

Particulate Bed of Magnesium Hydroxide

G. Ervin, Jr.; U.S. Patent 3,973,552; August 10, 1976; assigned to Rockwell International Corporation describes a cyclic method of storing and recovering thermal energy utilizing a particulate bed of a decomposable heat storage material selected from the group consisting of the hydroxides of magnesium, calcium, and barium.

The bed of heat storage material is confined within a container adjacent a water-permeable wall of the container. Thermal energy of chemical decomposition is stored by heating the bed of selected hydroxide to a temperature within the range of from 300° to 900°C and above the decomposition temperature of the selected hydroxide for a time sufficient to decompose at least a part of the selected hydroxide to form the corresponding oxide and water vapor. The water vapor is withdrawn by passing a carrier gas into contact with the water-permeable wall of the container to absorb the water vapor permeating therethrough. The stored thermal energy is recovered by passing a water-laden carrier gas into contact with the water-permeable wall, whereby the water vapor permeates through the wall into contact with the oxide to reform the selected hydroxide and generate heat of reaction which is removed by the carrier gas.

Sorption System Using Molecular Sieve Material

A process described by *D.I. Tchernev; U.S. Patent 4,034,569; July 12, 1977* relates to a system for the utilization of low-grade heat such as solar energy or the waste heat of a power generating plant by utilizing the large variation of the sorption capacity of molecular sieve zeolite, and other sorption materials, such as activated carbon and silica gel, with variations of temperature. In particular, the system relates to a system which converts small variations in absolute temperature to relatively large variations of gas pressure which is utilized to produce mechanical or electrical energy or cooling in refrigeration.

This is accomplished due to the extremely strong temperature dependence (exponential up to the fifth power of the temperature of gas sorption and desorption on certain materials such as exist in the molecular sieve zeolite family. The large pressure differential is used in the construction of a solar energy cooling system utilizing such materials. Two different approaches are described, one utilizing constant temperature across the molecular sieve and the other using a temperature gradient which is developed.

Due to the extremely strong temperature dependence, a change in temperature from 25° to 100°C can desorb better than 99.9% of the gas at constant pressure. Alternatively, at a constant volume, the same change in temperature causes an increase of pressure as high as four orders of magnitude. However, although the preferred material is a molecular sieve zeolite, the process can also use other solid sorbents such as activated carbon or silica gel. In such materials, the sorption capacity for gases is a strong function of temperature and accordingly to this extent they can be utilized in substantially the same fashion as the zeolites.

Two approaches to the use of solar energy are described, the first being to construct the roof of a building with panels made of absorbent material and to saturate them at ambient temperatures with the working gas. When the panels are heated by solar heat, they desorb the gas, the pressure increases and the subsequent gas expansion produces the desired cooling effect. The gas is then collected in a separate container which preferably is also provided with a sorbing material and during nighttime when the roof panels cool by radiation, they may be recharged to saturation again by the working gas and ready for a new cycle during the following day.

The sorption capacity of commercial zeolites is on the order of about 20 to 40 pounds of gas for each 100 pounds of such material. Using existing values of activation energies of between 4 and 10 kilocalories per mol, the theoretical cooling capacity for each 100 pounds of sorbent material is between 10,000 and 20,000 Btu. Thus, it will be appreciated that the existing roof area of a typical house is sufficient for a reasonably efficient cooling system.

The roof panels may be made by pressing and sintering the molecular sieve materials into the proper shape and sealing them in a container capable of withstanding pressure. Two types of containers are described. One with a glass cover in which the solar energy is absorbed directly by the molecular sieve panel which has preferably been darkened on one surface with, for example, carbon black to increase the absorption of solar energy; the other container is constructed completely of a darkened metal and absorbed energy is conducted to the absorbent material on the interior by a structure similar to the familiar honeycomb structure which surrounds the molecular sieve on all sides. Although this latter structure uses indirect heating of the molecular sieve material, it is capable of higher working pressures and, therefore, of higher operating efficiency.

Liquid Aquifer Energy Storage System

A process described by *W.B. Harris and R.R. Davison; U.S. Patent 3,931,851; January 13, 1976* generally provides steps for collecting and storing hot and cold water in underground aquifers during summer and winter months, respectively, and making the same available for heating during winter months and cooling during summer months, respectively. The method includes heating or cooling water, passing the hot or cold water to an aquifer for storage, removing the hot or cold water from the aquifer as required, removing heat or cold from the water, and returning the warm or cool water to an aquifer.

The term aquifer is used here in its commonly accepted sense, to wit, a water-bearing bed or stratum of permeable rock, sand, or gravel capable of yielding considerable quantities of water to wells or springs.

Thus, in the process, hot water produced by a solar heater or cold water from a cooling pond is pumped into an underground porous formation. The hot or cold water which is pumped into the formation displaces any water which already exists in the formation until a large hot or cold zone is created. The first time this is done, a significant quantity of heat or cold will be used to change the temperature of the rock formation. In subsequent cycles, the amount of heat or cold lost to the heating or cooling of the surrounding core rock will be reduced significantly. In most instances, the operation of the system will require two zones in the water-bearing formations for both the heating and cooling portions of the system.

GENERAL SOLAR HEATING PROCESSES

Water Ponds and Thermal Siphon for Enclosures

According to a process described by *H.R. Hay; U.S. Patent 3,903,958; Sept. 9, 1975* enclosure temperatures are modulated by water heated by solar energy and cooled evaporatively to ambient air. Control means include moving exterior insulation, enclosing or exposing the water, using forced air, and providing special means for heat storage and transfer. Water ponds horizontally disposed atop the enclosure, or in floor plenums and frequently in direct thermal exchange with underlying space, or water circulating in walls by thermosiphon action may be used separately or in combinations with the control means.

Figure 6.4 shows the effect of a structure with a roof pond on natural wind patterns as determined by smoke tests or wind velocity measurements. Low velocities of 1 to 10 miles per hour cause air flow essentially parallel to the ground until the air impinges an obstructing surface such as on the outside walls **104** of an enclosure. Then, air passes partially around the obstruction and partially across its top. Air flow over the structure with a flat roof **102** and roof pond **154** produces an air space **115** of little or no air movement. This condition adversely affects water evaporation from ponds **154** and retards cooling the enclosure with which the pond water is capable of thermal exchange.

Wind scoops on ships, and on housetops in Hyderabad, Pakistan, direct air flow to interiors of enclosures for ventilation and for evaporation of water directly from the skin of people.

Figure 6.4: Design of Solar Energy Temperature-Modulated Enclosure

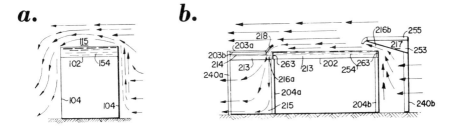

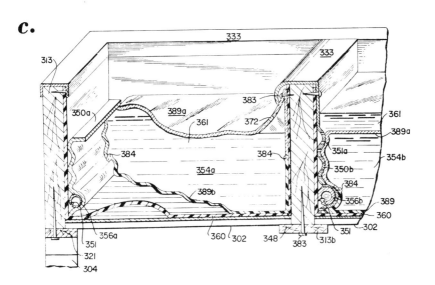

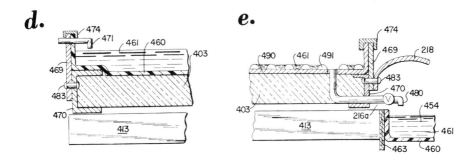

(continued)

Figure 6.4: (continued)

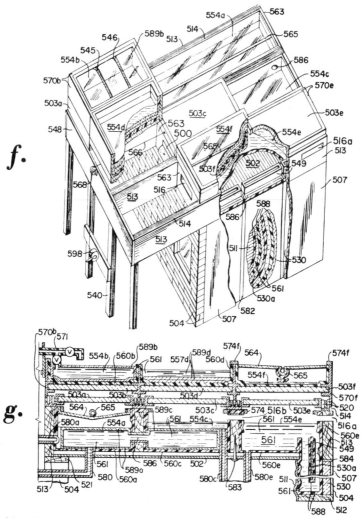

(a) Diagrammatic sectional view of an enclosure obstructing air flow which assumes the general pattern indicated by arrows.

(b) Diagrammatic cross section showing means for causing air movement to approximately parallel the upper surface of an enclosure.

(c) Fragmentary diagrammatic and partially cutaway view of means for retarding or increasing evaporation of liquid confined above an enclosure.

(d)(e) Fragmentary sectional representations of two means for confining liquid to be cooled by evaporation on top of movable insulation.

(f) Diagrammatic perspective and partially cutaway view of an enclosure showing details of roof ponds and wall construction.

(g) Diagrammatic sectional view of details of the roof construction and fragmentary details of the wall construction of Figure 6.4f.

Source: U.S. Patent 3,903,958

Prior art has also used wind scoops to direct dry air against damp sufaces, such as earthenware jugs to cool water. Instead of such separate, unsightly and costly scoops, this process uses the wall of the structure, and the ground, as major portions of the scoop and adds a deflector near the top of the wall at approximately roof-pond level. A portion of the deflector may protrude from the wall to trap wind impinging thereon. For this purpose, one may combine the function of the deflector with one or more useful devices, such as a water heater, a solar still, a rain collector, a shading projection, or the movable thermal barrier often a part of the means of this process for controlling temperature.

Figure 6.4b shows an example in which wind flow, indicated as arrow lines, impinging upon wall 204b is trapped under member 217 and is caused to flow through a passage 216b in a direction approximately parallel to and in close proximity with the water surface of roof pond 254. With properly shaped deflectors, such as those of 217 and 218 of Figure 6.4b, air velocities at least as high as those prevailing before impingement on wall 204b obtain near the surface of pond 254. The significance of this is evident from the fact that when 44 Btu of heat is lost from a square foot of pond surface by evaporation into still air, 75.5 Btu are lost with the air moving only 2 miles per hour. This difference can readily lower pond temperatures and convert a condition of thermal discomfort within an enclosure cooled by the pond to one of comfort.

The example of Figure 6.4b comprises space shown partially enclosed by walls 204a and 204b made of concrete or any material capable of supporting wood or metal roof beams 213, shown extended beyond wall 204a to post 240a, supporting roof pond 254 confined above metal roof sheets 202, fastened under beams 213 by screws not shown. Roof pond 254 is confined between two beams 213, crossing the enclosure, and end closures 263 between the beams and shown here as above walls 204a and 204b. The ponds are lined with a suitable material such as black, flexible polyethylene film shown better in Figure 6.4g, crossing over beams 213 and end closures 263 to provide a weather-tight roof above the enclosure.

Atop the beams 213 are aluminum trackways 214 of types standard for overhead closet doors in which insulation panels 203a and 203b, having suitable metal framing with attached wheels, can be moved with a drawcord passing over pulleys to a winch all as better shown in Figures 6.4f and 6.4g. The insulating panels 203a and 203b, preferably of rigid polyurethane foam about 2 inches thick and framed in extruded aluminum channels and painted or clad to prevent deterioration by solar irradiation, may be stacked, as better shown in Figure 6.4f, over the extended beams when roof pond 254 is to be exposed for evaporation cooling and they may completely cover pond 254 when desired.

To the right in Figure 6.4b, wind-trapping member 217 is supported by posts 240b and fastened to beams 213 or to wall 204b in a manner leaving preferably an elongated opening 216b, usually not less than 1 inch nor more than 1 foot in height, between pond closures 263 and the under side of wind-trapping member 217 which may have its extremity near the pond, turned down slightly to further cause air flow through passage 216b to parallel the water surface of pond 254 and to have high velocity.

Wind-trapping member 217, of Figure 6.4b, is shown comprising in combination element 255 which may be an elongated water heater or solar still of types to be later described, and below this an elongated water tank shown as triangular shape **253**. The upper surface of member **217** may be provided for rain catchment and the member may shade a walkway between wall **204b** and posts **240b**, though, in other instances, no such multiple purpose need be served by member **217**.

At the left of Figure 6.4b, deflector **218** is a piece of elongated metal, plastic, wood, or other suitable material fastened by screws or other means to the framing of top movable insulation panel **203a** as better shown in Figure 6.4e. The space **216a** between two beams **213** forms a passage for upward-flowing air impinging on wall **204a** when prevailing winds are from the direction opposite that shown by the arrows. Deflector **218** causes air flow through passage **216a** to be directed parallel to the surface of the pond after passing over end closure **263**.

In operation, cooling of the enclosure is obtained by exposing pond **254** at night by moving insulating panels **203a** and **203b** from an extended position overlying pond **254** to a position so far to the left of wall **204a** as to leave an open passage **216a** under deflector **218**. If prevailing winds are as shown by arrow lines of Figure 6.4b, air will be deflected by member **217** toward pond surface **254** and travel substantially parallel thereto until at least partially trapped by deflector **218** which causes the air to flow toward the ground and to resume the course it had prior to impinging on wall **204b**. Air deflected around the enclosure sides, rather than over it, tends to converge again at the left side of Figure 6.4b, adding to the creation of a partial vacuum near wall **204a** at ground level which also helps draw air through passage **216b** and thereby parallel to pond **254** in a manner conducive to create a higher air velocity.

When winds reverse direction from that shown by Figure 6.4b, arrows, deflector **218** and member **217** still act to increase evaporation and cooling of exposed water in pond **254**. When the pond is adequately cooled, or about one hour after sunrise, the movable insulation **203a** and **203b** is positioned over pond **254** to prevent absorption of solar energy or ambient heat which would warm the pond. Pond water, often cooled several degrees below minimum morning air temperature by nighttime evaporation, can then keep space within a suitably designed enclosure at comfortable temperatures throughout days with 110°F ambient air.

The ponds may absorb heat from this space through a metal ceiling **202** which is the pond support, or pond water may be transferred by pump, thermosiphon, or other suitable means to a remote device, such as a fan coil or radiant panel for heat exchange with the space. Such a thermosiphon system is shown in Figure 6.4f and further detailed in Figure 6.4g.

Wind deflectors of the type of **218** may be mounted directly to roof beams **213**, or to the fascia of buildings and extend outward from the walls to scoop and direct wind on all sides of a roof to benefit from winds originating from any direction. Moreover, the enclosure walls may be extended beyond the enclosed area to largely eliminate wind deflection around the sides of the building and to increase air flow across the roof pond.

If the system for deflecting natural wind across roof ponds does not provide adequate evaporation cooling, forced air may be employed. An electric fan, or blower unit of the standard type, mounted to discharge air across uncovered ponds at night is an effective means for cooling the water. With much greater advantage, it has been found that a fan or blower can be mounted to force air over the pond while the insulation is positioned above the pond. In Figure 6.4f, an opening **516** through end closure **563** and the pond liner (not shown), and above the surface of pond **554e**, can admit unsaturated air which undergoes and causes cooling while passing across the pond surface and under insulation panels **503f** and **503e** to an outlet **516a** to the exterior of the building or while passing to outlet **516b** leading to a second plenum. A fan or blower shown diagrammatically as **500** mounted on end closure **563** of pond **554c**, may be used to force the air through an associated opening similar to **516**, or natural air currents may be trapped under a suitable deflector (not shown but of the type of **218** of Figure 6.4b) mounted above opening **516** to direct an increased flow of air through opening **516**.

Although these examples function effectively with the fixed insulation over ponds to form plenums with zones for air flow, it is preferred to use movable insulation in order to collect solar energy for heating an enclosure on winter days. Insulation which is movable may be either positioned to expose the pond, if that is desired, or left over the ponds to form the plenum, or plenums, if that position is indicated.

The air directed through the plenum formed by the surface of pond **554e**, insulating panels **503f** and **503e**, and beams **513** of Figure 6.4f is cooled and humidified. During very dry days conventional means may discharge this conditioned air into the enclosure underlying pond **554e**, or some other enclosure.

Air forced by fan blower **500** of Figure 6.4f, cooled in passage over ponds **554c** and **554e**, may be caused to flow out openings **516** and **516a** to form a cool air curtain near walls **504** and windows or doors therein which are usually sources of high heat infiltration.

Evaporation of water from roof ponds can be increased by a wicking device whether the ponds are covered or uncovered; and forced air may be introduced under the surface of the liquid if desired. In another example, a multistoried building may be cooled without involving insulation movable or fixed. The floor design may include a plenum lined with plastic, metal, or other water-tight material to confine water to be evaporated by air forced through a zone in the plenum; the cooling effect can then be transferred by conduction or by circulating the water through a fan coil to modulate temperatures of space both above and below the plenum floor.

For example, a second-story floor of concrete with downward extending beams closes three sides of a plenum; an acoustical ceiling of metal may be fastened to the underside of the beam to complete the plenum and support an enclosed tray of water about one inch deep. A float valve may be used to maintain water level in the tray and an overflow provided for occasional flushing or to drain excess water. Air blown into the plenum may be exhausted from its zone of movement through ducts for any purpose earlier mentioned. Both the air and the water will be cooled by water evaporation and will act while within the plenum to cool the room below and the room above.

Figure 6.4f shows an example of the process in which a thermosiphon means within a wall effects heat transfer between space within an enclosure and a roof or plenum pond. This is of particular value to cool the space when there is inadequate heat exchange through structural elements of the pond; or, it can be used to cool enclosures not underlying the pond.

As shown in Figures 6.4f and 6.4g, walls **504** and wall plates **521** support beams **513** on the underside of which are fastened by nails **583**, screws, or other means, corrugated metal sheets **502** forming the ceiling of the enclosure and the bottom of roof pond **554e** having a black plastic liner **560e** over which water **561** is preferably maintained about 6 to 7 inches deep when the pond is to be used for winter heating. For clarity, it should be mentioned here that ponds **554a, 554b, 554c, 554d,** and **554f** do not relate to the thermosiphon action in pond **554e**, but involve other parts of the process.

Pond **554e** is shown covered by the insulation panels **503e** and **503f** in their extended position in Figure 6.4f; the panels are shown in Figure 6.4g as if stacked over pond **554e**. The insulation of panels **503e** and **503f** is best shown in Figure 6.4g as being framed in an aluminum extrusion such as **570f** to which wheels **520** are fastened to move the panels in trackway **514** atop beams **513**. Movement of the panels may be made independent of each other or they may be made to move in an interlocking manner so that a single device such as the winch **598** and drawcord **566** move all insulation panels above the ponds, and to position them over a carport or other suitable areas best shown in Figure 6.4f, where support is provided by extensions of beams **513** and by posts **540**.

At least a portion of one wall of the enclosure is formed into a thermosiphon element which contains water **561** in common with roof pond **554e**. The thermosiphon wall may be variously constructed; it is shown in Figures 6.4f and 6.4g as having an inside surface of metal sheets **511** with corrugations horizontally disposed and fastened by lag screws to wall studs **582** shown only in Figure 6.4f. The outer surface of the wall may consist of ¾ inch thick exterior grade plywood sheathing **507**, likewise fastened by lag screws to studs **582**. Other parts will be described in relation to the thermosiphon action.

As best shown in Figure 6.4g, thermosiphon action is established through one or more U-shaped lengths of lay-flat plastic tubing **588** sealed to the roof pond liner **560e** at seals **584** with one of the legs doubled and extending upward to form a chimney **549** rising somewhat above middepth of pond **554e**. Chimneys **549** are shown best in Figure 6.4f on both sides of a stud **582** terminating at the level of the bottom of the pond, where the upper ends of the studs are covered by the liner **560e** and a channel **586** is formed between the chimneys. These channels permit cold water at the bottom of the pond **554e** to pass around the chimneys **549** and, owing to the higher density of the cold water, to flow down the short leg of the U-shaped siphon tube **588** shown toward the outside of wall **504**.

So that cold water in the short outer leg absorbs less heat through exterior sheathing **507**, a 1.5 inch thick sheet of rigid polyurethane insulation **530** is between them. Also, a 0.5 inch thick sheet of the insulation **530a** separates the short and long legs of the thermosiphon tube to retard heat transfer and maintain the temperature differential causing thermosiphon action. Insulation **530a** may extend above the bottom of pond **554e** to help form the chimney **549** and stops short of the bottom of the wall cavity where the U-shaped thermosiphon rests between the studs on floor plate **512** or on cross-bracing, or other suitable support between the studs, at a higher level.

The long leg of the thermosiphon, as shown in Figure 6.4f and 6.4g, starts at the floor plate **512** and ends at the top of chimney **549**. Within the wall cavity, it directly contacts corrugated metal sheets **511** forming the inside surface of the wall. Heat from the room readily passes through metal sheets **511** and the thin, 20 mil, flexible plastic and is absorbed by water **561** in the long leg of the thermosiphon where, by virtue of the lower density of warm water, it rises while the colder water from the short leg replaces it in thermosiphon action. It has been found that as little as 0.5°F differential in the short and long legs is sufficient to initiate thermosiphon action. The warmed water in the long leg flows out chimney **549** and stratifies adequately in the upper portion of pond **554e** to permit the more dense colder water at the bottom of the pond to pass down the short leg of the thermosiphon.

The thermosiphon device of Figures 6.4f and 6.4g is substantially limited to cooling an enclosure having a roof or plenum pond or other high-level reservoir of cold water. Thermosiphon cooling depends upon cooling pond **554e** at night by moving the insulation panels **503e** and **503f** from the daytime position over the ponds to a nighttime position not over the ponds. Used in conjunction with a plenum and forced air evaporation of liquid, however, thermosiphon action can be maintained at a more constant temperature level than the cyclic diurnal cooling of an open pond permits. Thermosiphon cooling is inexpensive since corrugated sheets of various thickness to withstand normal head pressures from the water are standard; seamless polyethylene lay-flat tubing can be used for the U-shaped thermosiphon; and rigid polyurethane is thermally efficient in the thin sheets allowing practical construction.

Office Building

A process described by *D.W. Pulver; U.S. Patent 3,935,897; February 3, 1976* involves solar heating, particularly methods for collecting solar energy for heating an office building during the winter and cooling during the summer. Such buildings have been designed to maximize shielding of vision areas from the sun's rays, while utilizing a series of collecting panels, exposed to the sun's rays to transfer solar energy to heating and cooling systems.

According to this process, heat loss through the exterior walls is minimized by using maximum insulation in the opaque area and double glazing vision areas. Radiant energy is collected by means of a series of heat collectors prominently displayed on the east, west and southern facades of the building. Chilling loss during the summer months is minimized by orienting the vision glasses on the north side facade, northeasterly on the east facade and northwesterly on the west facade. On the south facade the vision glasses may be shaded by superposed heat collecting panels.

A heat responsive fluid is circulated in radiant contact with the heat collector panels and then to the desired heating and cooling systems. The circulating fluid is blocked sequentially in those areas where the adjacent panels are not exposed to the sun. The heat may be used to operate a low-temperature absorption or other conventional types of refrigeration systems, as well as a building heater. Forced air may also be preheated by means of the heated circulating fluid which may be stored within the building basement.

Thus, in a typical 35-story office building, a controlled interior climate may be achieved by the following: (1) Minimization of heating and air conditioning requirements by respectively minimizing heat loss in winter through the exterior wall by means of maximum insulation in opaque areas, double glazing in vision areas, and by minimizing solar heat gain in the summer by orienting vision glass to the north on the north facade, north-northwest on the west side, and completely shading vision glass on the south side by means of angled solar collectors.

(2) Collection of all solar energy striking the building facade and roof by means of double or triple-glazed clear glass covered collectors made of coated copper or aluminum plates, with integral or attached fluid carrying channels, connected to pipes, containing a liquid (e.g., water or water and ethylene glycol) which is heated to optimum temperature, returned to a central location at the base of the building, then redistributed through the building where needed for heating in the winter, and used to operate an absorption cycle or other such known system operable under heat exchange with a heated fluid, amply illustrated by the prior art for a refrigeration equipment in the summer. An insulated storage tank in the basement stores excess hot liquid for nighttime cooling or heating; any excess hot liquid can be sold to neighboring buildings.

(3) The integration of the above two principles results in a special design which, with the exception of energy for lighting, can be a net exporter of energy, meaning that the excess of solar energy collected on sunny and mildly overcast days vs. that needed for heating and cooling the subject building can be exported in a quantity greater than that needed to be imported (gas, oil, steam or electric backup) on a day or days when solar energy is not available in adequate quantities to heat or cool subject building.

Combination Solar Water Heater and Chiller

A process described by *R.J. Rowekamp; U.S. Patent 3,886,998; June 3, 1975* relates to a method and an apparatus which can be used either to heat or cool water through solar technology merely by the addition or removal of a few materials from a basic structure, thus making it possible to expose water to sunlight in winter so as to provide hot water for heating buildings, or to chill water by exposing it to cold night air during summer so as to cool the same buildings.

The basic object is to reduce the cost of hot and cold water thus produced by using what is essentially a water chiller as a basic structure, and then adding to it during winter a glass panel at the top and an insulating panel at the bottom so the device can be converted into a solar water heater. Also provided are ways for heating or cooling several small houses or one very large building through the use of automatic controls, large storage tanks, and an enclosed collector area located in the backyard and in the midst of several houses so that it is not necessary to mount the devices on the roof of the buildings themselves, as has been the practice in most solar energy projects.

Solar-Augmented Heat Pump System

A process described by *D.N. Shaw; U.S. Patent 4,148,436; April 10, 1979; assigned to Dunham-Bush, Inc.* relates to solar-augmented air source heat pump systems employing a multicylinder reciprocating compressor.

The process involves an air source heat pump system of the type including a first heat exchanger forming an indoor coil, a second heat exchanger forming an outdoor coil, and a multicylinder reciprocating compressor. Conduit means carrying refrigerant includes a reversing valve which connects the first and second heat exchangers and the compressor in a closed series primary refrigeration loop to permit the outdoor and indoor coils to operate alternatively as the evaporator and condenser for the system, depending upon heating or cooling mode.

The improvement comprises a third heat exchanger with the conduit means connecting the third heat exchanger across the outdoor coil. Selectively operable valve means within the conduit means causes refrigerant to flow through the third heat exchanger while isolating the outdoor coil from the closed primary loop. A storage tank containing a mass of heat sink fluid is connected in a secondary closed loop including the third heat exchanger, and a solar collector is operatively connected to the storage tank for normally supplying heat to the heat sink fluid. Means are provided for sensing the temperature of the ambient air passing over the outdoor coil and the temperature of the stored heat sink fluid, and means are provided for comparing the temperatures and for operating the selectively operable control valve means.

The heat sink fluid of the storage tank may comprise glycol or other fluids, and the system may be provided with a pump and solenoid valve means within the closed loop connecting the storage tank to the solar assist evaporator coil for controlling the circulation of the glycol therebetween.

Preferably, the reciprocating compressor comprises a plurality of cylinders and the system further comprises means for automatically controlling primary loop refrigerant circulation to and from the compressor for operating the compressor in single stage with all cylinders in parallel or for placing, in response to ambient temperature drop below a predetermined value under system heating mode, at least one cylinder under high side multistage compressor operation. The system may further include means for jointly or alternatively operating the outdoor coil and the solar evaporator coil as evaporators for the heat pump system under heating mode. The system preferably includes a subcooler for subcooling condensed refrigerant within the primary loop under at least system heating mode and means for selectively returning vaporized refrigerant to the low or high stage cylinders of the compressor.

Desalination Apparatus Using Metal Balls

A process described by *K.F. Ziehm, Jr.; U.S. Patent 4,077,849; March 7, 1978* pertains to improvements in the desalination of water, particularly seawater or other salt-containing water which is nonpotable. Such types of waters are readily available on the shores of seas or oceans, gulfs and backwaters thereof and to a lesser extent in arid areas of the world in inland lakes, seas and the like.

The process utilizes a special approach by employment of solar heat-absorbing metal member(s) such as balls, rollers or a metal endless belt to initially absorb solar heat in a heating chamber. The metal member(s) pass beneath a transparent cover and become heated through absorption of the solar radiation passing through the cover. These metal member(s) with their latent heat then pass directly into a vaporizing chamber where they are sprayed with seawater or other salt or saline water.

The latent heat therein causes essentially salt-free water to vaporize. The vapors are condensed by a condenser system such as condenser tubes through which pass cooler raw seawater or other readily available fresh and/or saline water. The condensed vapors are collected, thereby providing substantially pure water.

After the metal member(s) pass through the vaporizing chamber they are encrusted with the salts of the seawater or other saline water or coated with relatively high concentrations of aqueous solutions of such salts sprayed thereon and may exit from the vaporizing chamber either wet or dry.

Accordingly, the metal member(s) are immediately rinsed in a rinse tank which preferably is supplied with warm rinse water coming from the condenser tubing. After such rinsing, the rinsed portions of the metal member(s) return to the heating chamber for solar reheating and a repetition of the vaporizing and rinsing functions.

In the case of balls or rollers, the latter are lifted by a lift conveyer out of the warm rinse water and pass by tank to the vicinity of the head or entry of the heating chamber. At this point they are again lifted and fed into the solar heating chamber.

If desired, the vaporizing chamber is operated under vacuum. For example, at 15" Hg absolute, the equilibrium vapor pressure of water at sea level is 176°F, vs its normal boiling point of 212°F.

Referring to Figures 6.5a and 6.5b, the solar desalinator **10** comprises a solar heating chamber **11** and a vaporizing chamber **12**. The heating chamber is made up of a rectangular glass cover **13** in its upper side. The function of the glass cover is to transmit most efficiently the solar rays to the metal balls or rollers beneath the glass and to retain in the balls or rollers the absorbed solar heat.

Optionally, the glass cover may have around its periphery opaque or transparent sides, e.g., the inclined sides **14**, whereby the sides and glass cover form a receptacle for collecting rain water. This rain water is of a substantially pure and/or potable nature and serves as a supplement to the output of desalinated water of the solar desalinator.

The metal balls **16** (or cylindrical rollers) are made of a metal having high heat conductivity. The most preferred metal from an economic viewpoint is aluminum. These aluminum balls (or rollers) are anodized black or other dark colors to enhance their absorption of the solar radiation.

The metal balls (or rollers) roll along a metal sheet **17** beneath the glass cover. The sheet **17** preferably has a slight, longitudinal downward pitch or slope in the direction of the movement of the metal balls to enhance their slow travel through the heating chamber and vaporizing chamber. Preferably the upper surface of the sheet has longitudinal grooves or tracks **18** to provide longitudinally parallel paths along the sheet for the metal balls.

The metal balls (or rollers) are fed to the heating chamber at the entrant end **19**, move slowly beneath the glass cover and become heated by absorption of the solar radiation. The heated balls (or rollers) then pass directly into the vaporizing chamber.

Figure 6.5:   Desalination Apparatus

a.

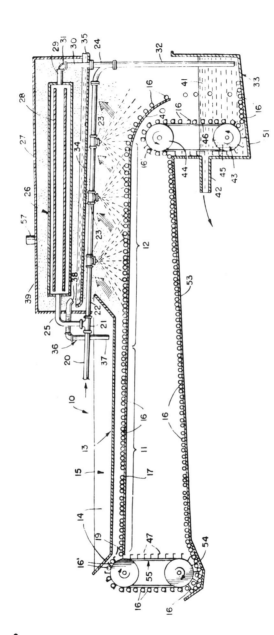

(continued)

Figure 6.5: (continued)

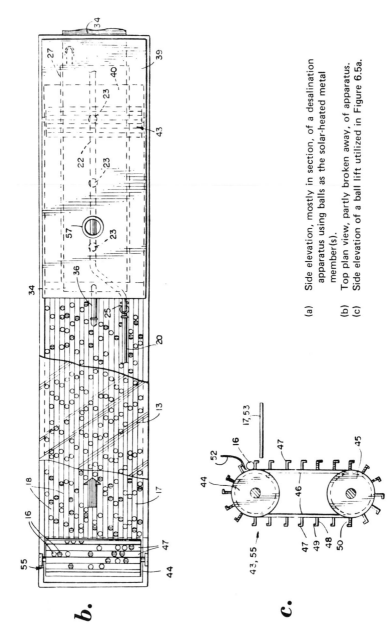

(a)   Side elevation, mostly in section, of a desalination
      apparatus using balls as the solar-heated metal
      member(s).
(b)   Top plan view, partly broken away, of apparatus.
(c)   Side elevation of a ball lift utilized in Figure 6.5a.

Source: U.S. Patent 4,077,849

This chamber has a pipe **20** for feeding seawater or other saline water to the vaporizing chamber **12**. At the T-coupling **21**, the feedwater branches into pipes **22** and **25**. The pipe **22** has a series of spray nozzles **23** and is coupled at its downstream end to a Y-coupling **24**. The pipe **25** is connected to a jacket **26** for condensing vapors, the feedwater keeping the jacket cool for condensation thereon. The jacket may comprise a shell **27** containing internal baffling or tubing **28** over or through which the cooling water flows inside the jacket. Alternatively, the condenser may simply be a series of convoluted or parallel tubes without an external jacket.

The downstream end of the jacket is connected by pipes **29** and **30** and their intermediary elbow **31** with the Y-coupling. In passing through the jacket and the pipe, the seawater or other salty or saline water is warmed by heat exchange occurring within the vaporizing chamber. This warmed water is fed by pipe **32** to a rinse tank **33**.

When the seawater or the like is sprayed from nozzles **23** onto the heated balls (or rollers) passing through the vaporizing chamber, all or a substantial portion of the sprayed water is vaporized. The vapors rise as indicated by the arrows in Figure 6.5a into the upper portion of the vaporizing chamber. Here the vapors are condensed on the cooler shell and the condensate drops into a condensate-collecting trough **34** beneath the jacket. The condensate trough has a slight pitch so that the substantially pure condensate water **35** flows by gravity out of the vaporizing chamber.

Where the heating chamber **11** embodies a rainwater collecting receptacle **15**, a tap pipe **36** may be used to convey the collected rainwater from the pipe's lower end **37** adjacent the glass cover **13** to its discharge end **38** at the head end of the condensate-collecting trough **34**. A mechanical or water pump may be provided on or in the tap pipe.

The upper portion of the vaporizing chamber is an enclosed housing **39** which is sealed against loss of vapor contained within the housing. The lower end of the housing may be open, as illustrated, or it may have depending walls, gaskets, etc., to prevent escape of vapors. The condensate-collecting trough has a width less than the width of the housing so that vapors can pass upwardly around the trough into the upper portion of the housing **39**.

The discharge end **40** of the ball-conveying plate **17** has a downward pitch or curvature above the rinse tank. The balls (or rollers) fall into the warm sea, salt or saline water **41** in the rinse tank, the water level being maintained constant at the level of the overflow pipe **42**, from which the discharge water is returned to the sea, ocean, lake, etc.

The balls **16** (or rollers) have, after passing through the vaporizing chamber, solid salt deposits or concentrated salt solutions on their surfaces. The salt deposits and/or concentrated salt solutions are rinsed off in the warm water **41** in the tank. The balls (or rollers) are then conveyed by a lift **43** from the bottom of the tank for return to the heating chamber. The lift **43** and another lift **55**, the function of which is later described, have the construction illustrated in Figure 6.5c. These lifts comprise an upper roller or pulley **44** and a lower roller or pulley **45**, one of which is driven.

An endless belt or series of side-by-side belts **46** is positioned about and driven by the rollers or pulleys **44, 45,** which rotate in the direction of the arrows shown in Figure 6.5a. A series of longitudinally elongated lifts **47** are mounted on the belt(s) **46**. Each lift comprises an elongated bar **48** connected at one edge to the belt(s) **46** and bearing an elongated lip **49** at the other edge. Each bar and lift forms an elongated seat **50** for conveyance of the balls **16** (or rollers).

As shown in Figure 6.5a, the rinse tank **33** has a bottom wall **51** which is downwardly pitched in the direction toward the lift **43** extending along one edge of the rinse tank. As the ball lift operates within the tank, the rinsed balls (or rollers) are picked up at the bottom of the tank and lifted upwardly in the seats **50**.

At the upper side of the roller or pulley **44**, the balls (or rollers) fall out of the seats, e.g., the balls **16'**. A resilient blade **52** (Figure 6.5c) projects into the path of the lifts **47**. The balls **16** are pushed by the free edge of the blade **52** off the back side of the lifts onto the sheet **17** or **53** when the following lift **47** strikes the blade **52**.

The ball return from the lift **43** of the rinse tank to the head or entrant end **19** of the solar heating chamber **11** comprises a sloping or pitched, ball **16** (or roller) return sheet **53**. The sheet **53** slopes downwardly from the rinse tank and its lift to a collector well **54**. The balls **16** (or rollers) accumulated in the well are picked up by the lifts **47** of the lift **55**, the structure of which is illustrated in Figure 6.5c. The balls or rollers are then lifted and discharged onto the sheet **17** at its entrant end **19** for another cycle.

If desired, the vaporizing chamber **12** may be operated under partial vacuum. For this purpose, a vacuum tap **57** is provided in the top wall of the housing **39**. A partial vacuum has the advantage of lowering the vaporizing temperature of the water sprayed onto the heated balls (or rollers) in the vaporizing chamber.

GEOTHERMAL ENERGY

Multiple-Completion Geothermal Mining System

A.T. Van Huisen; U.S. Patent 3,957,108; May 18, 1976 describes a system for the mining of geothermal energy in which a plurality of geothermal wells radiate from a single surface site into a subsurface geothermal reservoir. The wells can be drilled by conventional slant drilling techniques and each may contain a closed end heat exchanger which receives water and generates steam. Some of the heat exchangers are disposed vertically and others are implanted horizontally. By alternating production of the wells in a programmed cyclical manner, convective movement of the hydrothermal fluid will occur within the geothermal zone. The generated steam is collected in a reservoir at the surface site and utilized to generate electricity. The condensate from the turbine can be recycled to the wells.

Referring to Figures 6.6a and 6.6b, the system includes a plurality of geothermal wells **10** each having a first, open, upper end **12** converging toward and meeting within a surface point bounded by a closed reservoir **14**. The lower end **16** of each well terminates within a geothermal zone **18**. The ends are positioned in a predetermined pattern such as those of wells **20** and **22** which are spaced vertically from each and the ends of wells **20** and **24** which are spaced horizontally from each other.

Figure 6.6: Geothermal Mining System

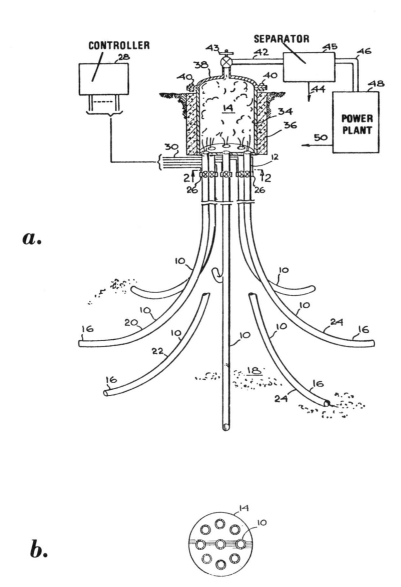

(a)   Perspective view of a multiple-completion geothermal mining system.
(b)   Sectional view taken along line 2–2 of Figure 6.6a.

Source: U.S. Patent 3,957,108

Each well contains a valve **26**, suitably a servo-controlled valve such that production from each well can be individually controlled, suitably by a time sequencer controller **28** which activates or deactivates each valve **26** according to a predetermined program by sending signals through lines **30**. The sequential production of wells **10**, especially by rotating production in a circular pattern, will promote convective movement of the hydrothermal fluid within the geothermal zone **18**, and thus, increase heat transfer and recovery and decrease the possibility of scale forming on the external surface **24** of the portion of the well casings within the zone **18**.

The geothermal, heated fluid recovered from the well collects within the reservoir **14**. The reservoir may be located on the surface but is preferably recessed below the surface to take advantage of the insulating and warming effects of the subsurface strata. The reservoir is preferably a steel vessel **34** bounded by a concrete layer **36** and includes a removable lid **38** secured by bolts **40**.

The collected steam is transferred through conduit **42** containing a pressure regulating valve **43** to a separator **45**. Condensate and solids are removed through line **44** and the steam is delivered through line **46** to the power plant **48**. The power plant can be a direct prime mover engine or a turbine generator. Condensate is removed through line **50** and may be recovered or recycled to the wells.

Geothermal fields are classified according to their production of hot water, hot water and steam, or dry steam. The system of Figure 6.6a can be used for the recovery of geothermal heat values from all of these types of fields. Hot water fields typically produce temperatures between 60° and 100°C, with gradients of 30° to 70°C/km. Because of the low enthalpy, hot water fields are not being used to generate electricity. They are being used instead for space heating and air conditioning. For electrical production most geothermal fields produce both water and steam at temperatures greater than 100°C. The highest temperature field in use is at Cerro Prieto, Mexico, at which temperatures have been measured up to 380°C. Similarly, the dry steam fields in commercial use have temperatures at 210° (The Geysers, U.S.) to 260°C (Larderello, Italy).

The geothermal fields in production have the following basic geological characteristics:

(1) A Source of Heat — In general, magmatic intrusions at shallow depths of 7 to 15 km provide a heat source at a temperature above 100°C, typically from 200° to 400°C.

(2) A Source of Water — Commercial wells produce more than 20 tph of water and steam. The best well, located in the Cerro Prieto field, produces 350 tph. The water is believed to come from surface sources rather than being magmatic water. Therefore, it is probably replenishable at a rate determined by pressure, permeability, source availability and other factors. Reinjection may be utilized to replenish the source and dispose of unwanted surface water such as condensate from the turbine.

(3) A Permeable Rock Aquifer — Almost any permeable rock can serve as an aquifer such as deltaic sand, volcanic turf, base salt flows, ignimbrite, greywacke, carbonate volcanics and limestone. Convection currents through the rock are believed to be the primary heat transfer mechanism between the magma

and upper levels of the aquifer; these can be reached by drilling. Pressures in the aquifer may be as high as 2,000 psi or more. Both thermal conductivity and permeability are critical parameters which limit the energy production of the well.

(4) A Cap Rock — A rock layer of low permeability is required above the aquifer to limit the heat transfer by convection. The heat loss transfer through the cap rock to the surface primarily by convection is very low; this allows the system to remain hot. Many systems are believed to be self-sealing due to mineral, primarily silica, depositions of the hot water flashed and the cold temperature near the surface.

All of these geothermal aquifers can be mined by direct thermal mining methods in this process which will provide the same advantages of collection of the thermal fluid at a single point, thus reducing capital investment and well installation costs and the surface area needed for converting the fluid to mechanical or electrical energy. However, the wet geothermal areas can more efficiently be mined by the downhole heat exchanger relation of the multiple-completion system. Dry geothermal areas may also be mined by either method. A man-made aquifer may be developed by explosion-stimulated methods. If the hole is hot and dry and not fractured, a large aquifer could be developed by hydrofracturing alone or in combination with explosive-induced means. The downhole heat exchanger offers the opportunity to recover heat from the dry and hot geothermal area without the need to inject water to the zone to create a hydrothermal fluid.

If the system is hot, dry and fractured, water can be introduced from the surface internal of the well casing or external of the well casing and the resultant steam collected through the annulus of the well removed and harnessed to produce energy. If the system is hot, dry and unfractured, it may be utilized as such with the downhole heat exchanger to produce steam by indirect heat exchange methods or hydrofracturing and enhancement can be practiced through additional thermal stress fracturing or by the use of high explosive or nuclear devices to fracture large quantities of hot rock.

Referring again to Figures 6.6a and 6.6b, in a system for mining wet geothermal energy, the separator **45** would be a flash unit and the wet steam is delivered to the power plant **48** while the separated salts and condensate are removed through line **44**. The salts may be separated into commercial salts for sale, may be concentrated and disposed of or may be reinjected into the zone. In a system in which fairly dry steam is directly recovered, separation of solids may be required in a cyclone separator to minimize cavitation of the turbine blades.

Hydrothermal injection systems for mining a dry field are also illustrated.

Thermit Reactions to Form Metal Conductors

According to a process described by *L.S. Bouck; U.S. Patent 4,030,549; June 21, 1977; assigned to Cities Service Company* a structure and system for the transfer of energy between a locus in a borehole and a subterranean formation are formed by (a) drilling a borehole from the surface into the formation; (b) fracturing and propping the formation by injecting a fracturing and reactive slurry into the formation, the slurry comprising a finely divided aluminum and a reactive metal oxide in a fluid carrier; (c) igniting the reactive slurry within the formation so that the aluminum and metal oxide components react with a thermit reaction to form a liquid metal within the fracture system formed in the formation by the fracturing and propping; and (d) allowing the liquid metal in the formation to cool and solidify within the fractured system.

Thus, in one aspect, a geothermally heated formation is penetrated by a bore. The formation is fractured and propped with a slurry capable of maintaining a thermit reaction. The slurry is ignited. Upon cooling, a solid iron conductor fin network from the bore into the formation in the fracture system is formed. Thermal energy is conducted by the network from the formation to a locus in the bore where it is recovered to the surface by a liquid or liquid-vapor recovery system.

According to another aspect of the process, a hydrocarbon-bearing formation is penetrated by a bore. The formation is fractured and propped with a slurry containing particles of iron oxide and particles of aluminum oxide. The metallic components are ignited in the fracture system. After a thermit reaction has occurred and cooling has occurred, a solid iron conductor fin network from the bore into the fracture system in the formation is formed. Thermal energy is conducted by the network into the formation from a surface source by a liquid or liquid-vapor heat transfer system.

Heat-Drill and Drilling Mud

According to a process described by *R.G. Clay; U.S. Patent 3,991,817; Nov. 16, 1976* a drill body having a particular shape with an attached heating element of a particular configuration is used to drill into the earth and form two shafts at the same time in the earth. The two shafts are in fluid communication through the body of the drill and are used to circulate a drilling mud through the drill body to carry off excess rock. The heating element operates at a temperature well above the melting point of the rock, melts through rock ahead of the drill body, and raises the rock through which it passes to well above the rock's melting point, raising the average temperature of the rock through which the drill body passes, however, to a selected lower degree above the rock's melting point.

The heating element passes through a plurality of rock portions spaced throughout the region to be melted, and sweeps through only a fraction of the spatial volume swept out by the drill body. The molten rock takes one of two alternate paths: it either flows into the interior of the drill body and then into the drilling mud, circulating through the drill body and then to the surface; or it flows around the exterior of the drill body to the top of the drill body, which makes two shafts in the molten rock. Means are provided for making the shafts the desired shape, for the gradual cooling of the shaft walls, and for the maintenance of the molten rock in the desired shape until the rock solidifies, leaving two permanent shafts. Means are also provided for causing the shafts formed to spiral around one another in a controlled manner.

The walls of the downflow shaft are formed with one or more grooves extending throughout the length of the shaft, shaped like a V cut into the wall. The grooves are made to facilitate later fracturing of the surrounding rock, and a concentration of thermal stress at the apex of the V cut also facilitates later fracturing and may even cause initial fractures to form at the apex by thermal stress.

The drilling mud circulating through the drill body and the shafts absorbs heat in its passage and that heat may be utilized while drilling is taking place; in particular a well-known effect due to absorbed heat is the thermosyphonic effect, which creates a driving force acting on the fluid in its direction of motion and which may in some cases be the only pumping force needed to circulate the fluid. A heat exchanger at the surface removes heat from the drilling mud for any desired use, particularly to help provide energy to the drilling operation.

Once the drill has reached the desired depth, fractures extending into the surrounding rock are introduced through the grooves hydraulically, propped open by well-known means, and partitioned by material forced horizontally into the fractures at selected depths to form heat-collecting cells. If the shafts spiral around one another, the fracture surfaces radiating from the downflow shaft will spiral in the same direction and rate as the shafts, and will resemble helical surfaces. The drilling mud, or a fluid which replaces the drilling mud, will collect heat from the surrounding rock as the fluid circulates through the shafts, the fracture cells, and the drill body.

As the drilling progresses, the thermosyphonic effect provides a pumping action to the circulating fluid which will increase to the point that a turbine may need to be placed in the flow stream to limit the rate of flow. The turbine may be placed in either the downflow shaft or the upflow stream, but the downflow does not contain the rock being removed and may be preferred. The turbine may also be used to provide power. The drill may be restarted at a later date to proceed to a lower level without any need to seal off the fractures. Fractures extending from one shaft do not form part of connecting passages to the other shaft, so all fluid circulating from one shaft to the other passes through the drill body.

Deviated Wells

A process developed by *J.L. Fitch; U.S. Patent 3,863,709; February 4, 1975; assigned to Mobil Oil Corporation* is directed to recovering geothermal energy from a low permeability subterranean geothermal formation having a preferred vertical fracture orientation. A first and a second well are provided to extend from the surface of the earth and penetrate the geothermal formation at an angle of at least 10° measured from the vertical and extend into the formation in a direction transversely of the preferred vertical fracture orientation.

A plurality of vertical fractures are hydraulically formed that are spaced laterally one from the other a predetermined distance measured in a direction approximately normal to the preferred vertical fracture orientation, which fractures intersect the wells. A fluid is injected via one well and into the vertical fractures and flowed through the fractures to the other well to absorb geothermal energy from the formation and heat the fluid. The heated fluid is produced to the earth's surface via the other well.

Another example is directed to recovering geothermal energy from a permeable geothermal formation. In accordance with this example, adjacent vertical fractures communicate with only one of the first and second wells and one vertical fracture communicates with the first well and the adjacent fracture communicates with the second well. A fluid is injected via the first well, into a vertical fracture and flowed through the formation to absorb geothermal energy and heat the fluid and the heated fluid is flowed into an adjacent fracture and produced to the earth's surface via the second well.

Geothermal Energy in an Active Submarine Volcano

A process described by *H.C. Georgii; U.S. Patent 3,967,675; July 6, 1976; assigned to AB Hydro Betong, Sweden* involves the exploitation of the geothermal energy in an active submarine volcano. In the process, an elongated, substantially vertical, columnar concrete body is arranged above the orifice of the volcano so as to extend from the water surface vertically downwards through the water and into the magma in the orifice of the volcano so that the lower portion of the concrete body is submerged in the magma. The concrete body has such a displacement and such a weight that it floats in a balanced vertical position in the water and the magma.

A coolant, preferably water, is circulated through internal cooling ducts or pipes in the concrete body from the upper end of the body downwards into the lower portion of the concrete body, which is submerged in the magma and where the coolant is heated by heat transfer from the surrounding magma. The heated coolant is then returned through internal cooling ducts or pipes in the concrete body to the upper end, where the heat content in the heated coolant is utilized.

INDUSTRIAL APPLICATIONS

AIR SEPARATION PLANTS

Regenerator

A process described by *R.M. Thorogood; U.S. Patent 4,131,155; December 26, 1978; assigned to Air Products and Chemicals, Inc.* pertains to reversible heat exchanger or regenerator systems used in an air separation plant for the production of, inter alia, pure nitrogen by the fractional distillation of liquefied air.

In a typical air separation plant using the low pressure distillation cycle it is standard practice to remove water and carbon dioxide from the incoming air by condensation on the surfaces of a reversing heat exchanger or regenerator system such as shown in Figure 7.1a.

Deposited impurities are subsequently removed by flowing gas 10 at a lower pressure in the reverse direction through the heat exchanger or regenerator 12, thereby evaporating the impurities and discharging them from the plant. The heat exchanger 12 has a flow path 31 for a nonreversing stream and two parallel flow paths 32 the ends of each of which divide into two branches.

At the warm end 6 of the heat exchanger 12, the flow paths 32 each divide into a low pressure gas outlet branch 34 and an air inlet branch 36 and, at the cold end 8 of the heat exchanger 12, each flow path 32 divides into a low pressure gas inlet branch 46 and an air outlet branch 48.

Interchange of the air flow 14 and low pressure gas flow 10 in the heat exchanger 12 is effected by a system of switch valves 16, 18, 20 and 22 at the warm end of the exchanger, and check valves 24, 26, 28 and 30 at the exchanger.

Generally, the low pressure gas 10 is a waste product from the plant, and thus minor leakage of air (typically at 90 psia) across a closed switch or check valve into the low pressure gas (typically at 18 psia) is not of importance since the resulting contamination of the low pressure gas is of no consequence. However, in some instances it is important that the low pressure gas **10** should not be contaminated with air **14** although contamination with water and/or carbon dioxide is acceptable. In particular, a gas comprising mainly nitrogen with carbon dioxide and water can be used for the deoxygenation of seawater.

In this application it is important that the nitrogen contains carbon dioxide but has a low oxygen content. The retention of carbon dioxide in the nitrogen, and thus also in the seawater, inhibits calcium carbonate deposition (scaling) by inhibiting a shift from bicarbonate in the following equilibrium.

$$Ca(HCO_3)_2 \rightleftharpoons CaCO_3 + CO_2 + H_2O$$

According to this process, a reversible heat exchanger or regenerator system comprises a heat exchanger or regenerator having at least one flow path for the reversing heat exchange fluids, which flow path has at each end, an inlet branch and an outlet branch, wherein each branch is provided with two valves arranged in series, a vent pipe in communication with the branch between the two valves and means for controlling the flow of fluid through the vent pipe. The means may comprise, for example, remotely actuable switch valves or orifices.

Figure 7.1: Air Separation Process

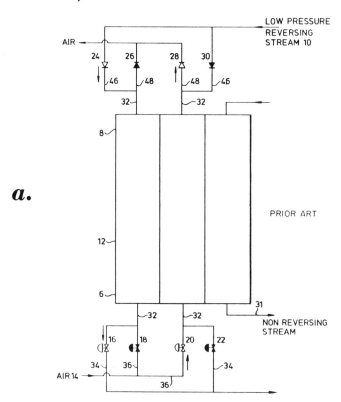

(continued)

Figure 7.1: (continued)

b.

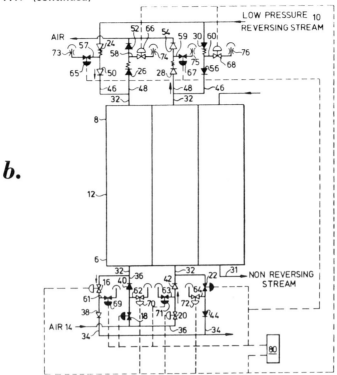

(a) Schematic drawing of the heat exchanger section of
   a typical air plant.

(b) Schematic drawing of the reversible heat exchanger
   system according to this process as it would be used
   in a conventional air separation plant.

Source: U.S. Patent 4,131,155

In use, the inlet branches are connected to sources of different fluids and, in the
preferred case the inlet branch at the warm end is connected to a supply of air
while the inlet branch at the cold end is connected to a high purity, low pressure,
nitrogen stream. By venting any leakage air before it can reach the low pressure
nitrogen stream, contamination of the low pressure nitrogen stream with leakage
air is minimized.

At the warm end of the heat exchanger the valves arranged in series preferably
comprise a remotely actuable switch valve and a check valve downstream thereof.
At the cold end of the heat exchanger both of the valves arranged in series in
the branches are preferably check valves.

In the preferred case the valves at the cold end of the heat exchanger are check
valves and each vent pipe includes an adjustable orifice fitted downstream or upstream of a remotely actuable valve. Because the valves in the branches at the
cold end are check valves, there will be a continuous flow of the low pressure

gas to vent along with any leakage gas from the closed valve. The adjustable orifices allow control of this flow to suit the leakage rate of the closed valve. The leakage of low pressure gas may also be reduced by biasing appropriate check valves closed.

The method of operating the system comprises the steps of passing a first gas (e.g., air) through the heat exchanger or regenerator via the inlet branch at one end of the heat exchanger or regenerator and the outlet branch at the other end thereof while venting leakage gas through the vent pipes associated with the outlet branch at the one end of the heat exchanger or regenerator and the inlet branch at the other end thereof; and subsequently passing a second gas (e.g., pure or substantially pure nitrogen), at a lower pressure than the first gas, through the heat exchanger via the inlet branch at the other end of the heat exchanger and the outlet branch at the one end of the heat exchanger while venting leakage gas through the vent pipes associated with the inlet branch at the one end of the heat exchanger or regenerator and the outlet branch at the other end.

For a better understanding of the process reference is made by way of example to Figure 7.1b which shows a reversible heat exchanger system in accordance with the process forming part of an air separation plant.

The conventional reversible heat exchanger system of Figure 7.1a can be converted into a system according to the process by placing a check valve in series with and downstream of each of the switch valves **16, 18, 20** and **22**, and check valves **24, 26, 28** and **30** normally used. In particular, check valves **50, 52, 54** and **56** are arranged downstream of check valves **24, 26, 28** and **30** respectively. At the warm end **6** of the reversible heat exchanger **12**, the check valves **38, 40, 42** and **44** are arranged downstream of switch valves **16, 18, 20** and **22** respectively.

Vent pipes **57** through **64** are arranged between valves **24** and **50**; **26** and **52**; **28** and **54**; **30** and **56**; **16** and **38**; **18** and **40**; **20** and **42**; and **22** and **44** respectively. The vent pipes **57** to **60** vent to atmosphere. Alternatively they may be connected to a waste pipe entering the cold end **8** of the heat exchanger **12**. The vent pipes **61** to **64** from the warm end **6** of the heat exchanger **12** vent directly to atmosphere or to a suitable warm waste pipe. The vent pipes **57** through **64** are each provided with a secondary remotely operatable valve **65** to **72** and vent pipes **57** to **60** are each provided with a variable orifice **73** to **76** respectively.

In use, each secondary valve **69** to **72** is opened when the valves in its associated branch are closed. Thus, secondary valves **70** and **72** are opened when valves **18, 40**; and **22, 44** are closed. Similarly, secondary valves **69** and **71** are opened when valves **16** and **38**; and valves **20** and **42** are closed.

At the cold end **8** of the heat exchanger **12**, secondary valves **66** and **68** are opened and closed with valves **16, 70, 20** and **72**. Secondary valves **65** and **67** are opened and closed with valves **69, 18, 71** and **22**. The valves **16, 18, 20, 22, 65, 66, 67, 68, 69, 70, 71** and **72** are all controlled by a valve timer **80**.

At the cold end **8** of the heat exchanger **12** there will be a continuous flow of the low pressure nitrogen to vent along with any leakage air from the closed valve. The flow of vent gas may be controlled by the adjustable orifices **73** to **76** to suit the leakage rate of the closed valves. The check valves **24, 26, 28** and

30 are provided with light springs so that they will remain closed against a small pressure differential between the low pressure nitrogen and the vents. In this connection it should be appreciated that with the valve in, for example, vent pipe **58** open and the orifice **74** correctly adjusted, the pressure in the vent pipe **58** will normally be only slightly less than the pressure of the low pressure nitrogen. Thus, a restricted flow path is provided which inhibits back diffusion of air into the low pressure nitrogen.

Screen Heat Exchanger

A process described by *V.G. Pronko, E.V. Onosovsky, A.V. Chuvpilo, I.N. Zhuravleva, V.A. Korneev, D.A. Klimenkov, G.M. Smirnova, V.V. Usanov, J.I. Ivanov, B.A. Chernyshev, V.D. Nikitkin, A.F. Nikolaev, M.S. Trizno, V.G. Karkozov, T.J. Verkhoglyadova, E.V. Moskalev and L.I. Yakovleva; U.S. Patent 4,147,210; April 3, 1979* provide a screen heat exchanger whose screens ensure a highly intensive heat transfer at a comparatively low hydrodynamic resistance of the heat exchanger and a high compactness of the heat-exchanging surface.

This is accomplished by providing a heat exchanger comprising spacers with holes which form passages for the fluid medium, screens made of a sheet material whose heat conduction is considerably higher than that of the spacers, the spacers and screens alternating with each other and being rigidly connected into a bank, the screen elements forming meshes, and at least two headers communicating with the passages of the bank and connected rigidly to its counteropposed sides for distributing the fluid medium among the bank passages wherein, according to the process, the mesh-forming elements are set on an angle to the plane of the screen. It is practicable for the angle of the mesh-forming elements to the screen plane to be from 45° to 90°.

Such a design of a screen heat exchanger will ensure a more favorable profile of the screen elements from the hydrodynamic point of view, the profile being constituted by a system of short parallel plates washed longitudinally or at a small angle by the flow of fluid medium. This brings about a more favorable relationship between the intensity of heat transfer and the hydrodynamic resistance.

Such screens made of a sheet material guarantee a high compactness of the heat-exchanging surface since in this case the increase in compactness is not limited by the strength of the source material. Additionally, owing to the inclination of the mesh-forming elements, the screens in the bank can fit tightly against one another or even enter one another partly, which likewise adds to greater compactness of the heat-exchanging surface.

The screen heat exchanger according to the process comprises alternating spacers **2** (Figure 7.2a) and screens **3** rigidly interconnected into a bank **1**. The spacers **2** have holes which form passages **4** in the bank **1** for the flow of the fluid medium. The passages **4** may vary in cross section and be, for example, of a slotted shape (Figures 7.2a and 7.2b).

Such passages may have different widths of the slot for the passage of the forward flow moving in the direction of arrow **C** and the reverse flow moving along arrow **D**, the fluid in this case being helium. The slotted cross sections of the passages make it possible to change within wide limits (from 1 to 5 and over) the relationship between the cross sections of the forward and reverse flow passages

4 depending on their pressure and flow rates. In its turn, this gives a possibility of producing a screen heat exchanger for the given operating conditions with the optimum weight, size and hydrodynamic resistance.

Figure 7.2: Screen Heat Exchanger

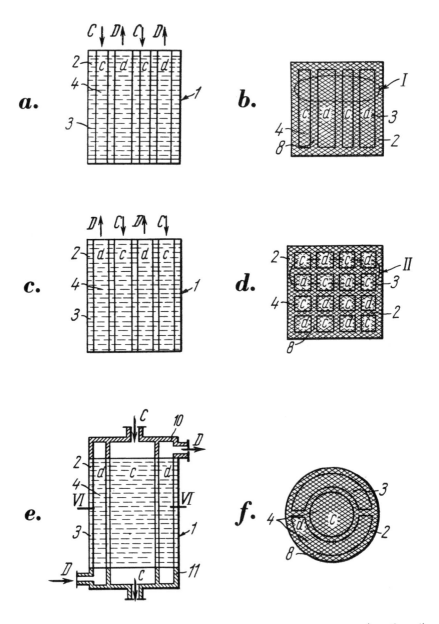

(continued)

Figure 7.2: (continued)

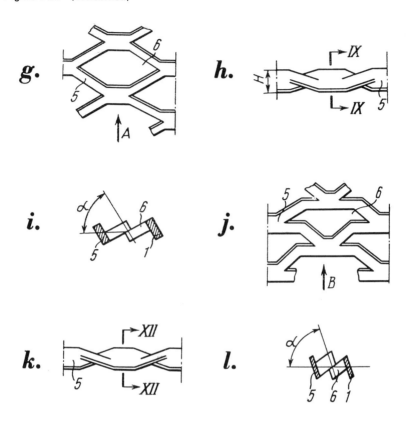

(a) Schematic longitudinal section through the screen heat exchanger with slotted passages, the headers being omitted for convenience.

(b) Heat exchanger in plan view.

(c) Schematic longitudinal section through the screen heat exchanger with square passages, the headers being omitted for convenience.

(d) Heat exchanger of Figure 7.2c in plan view.

(e) Schematic longitudinal section through the screen heat exchanger with circular and round passages.

(f) Sectional view taken on line VI-VI in Figure 7.2e.

(g) Plan view of a hexahedral screen mesh, on enlarged scale.

(h) View along arrow A in Figure 7.2g.

(i) Section taken along line IX-IX in Figure 7.2h.

(j) Plan view of an octahedral screen mesh, on enlarged scale.

(k) View along arrow B in Figure 7.2j.

(l) Section taken along line XII-XII in Figure 7.2k.

Source: U.S. Patent 4,147,210

The passages may also be of a square cross section as shown in Figures 7.2c and 7.2d. In this case the relationship between the cross sections of the passages for the forward flow moving along arrow **C** and those of the reverse flow moving along arrow **D** is equal to unity. These passages are practicable when the volumetric flow rates of the forward and reverse flows are nearly the same. In some cases the passages may be of a circular or round cross section as shown in Figures 7.2e and 7.2f.

The spacers **2** for the screen heat exchangers used in cryogenic engineering are made from thermoreactive epoxynovolak compound film characterized by good adhesion to the majority of metals. This film is made by extrusion at a high speed, approaching 60 m/hr. The material of the spacers is cheap and displays high strength at cryogenic temperatures.

In other cases of employment of the heat exchangers the material of the spacers **2** as well as that of the screens **3** is selected to suit the operating conditions of the exchanger. For example, when the heat exchanger operates under the conditions when the influence of the longitudinal heat condition is not essential, the spacers may be made from sheet flux or from a material with an applied flux.

The screens **3** are made from a sheet material whose heat conduction is considerably higher than that of the material of the spacers **2** which ensures anisotropic heat conduction of the screen heat exchanger and particularly a maximum possible lateral heat conduction which intensifies heat transfer and a low longitudinal heat conduction which reduces the transfer of heat from the hot to the cold end of the heat exchanger along its walls.

Each screen **3** is made so that its elements **5** (Figures 7.2g, 7.2h, 7.2j and 7.2k) forming the meshes **6** are inclined at an angle α (Figures 7.2i and 7.2l) to the plane of the screen **3**.

It is practicable that the angle α should be from 45° to 90°. Inclination of the elements **5** of the meshes **6** is produced when the screen is made from a sheet material on an automatic device and is determined, on the one hand, by the plasticity of the material and, on the other, by the optimum relationship for each particular application of the apparatus between its thermal and hydrodynamic properties.

The meshes **6** may have various polygonal shapes, mostly extending in one direction and depending on the shape of the cutting tool of the automatic device.

The source material for making the screens in this version of the process is constituted by a coiled sheet 0.05 to 0.5 mm thick of various metals loaded into the automatic device. Such a screen whose compactness is 2,000 to 20,000 m² of the surface per cubic meter of free volume is characterized by a sufficiently high accuracy of the basic dimensions, reaching 3 to 5%. The screen can be made of any sheet material whose elongation is not under 15%.

With respect to cryogenic engineering, the most promising tendency is the employment of electrolytic sheet copper for the recuperative heat exchangers of helium liquefiers and refrigerators and of sheet lead for the regenerators of gas-fired refrigerating machines.

Thus, due to the solution suggested in the process it is possible to provide an efficient and highly-compact screen heat exchanger with anisotropic heat conduction of its structure, consisting of standardized units and parts whose manufacture can easily be mechanized and automated.

The screen heat exchanger used as a recuperative heat exchanger in cryogenic helium installations operates as follows.

The fluid medium, in this case warm gaseous helium of a forward flow whose direction is shown by arrows **C** moves from the compressor into the header **10** wherein it is distributed uniformly among the passages **4** of the bank **1** marked with letter **c** in Figures 7.2a through 7.2f. Moving through these passages, helium flows around the screens **3** and transfers heat to them. Then the heat received by the screens **3** is transferred due to their heat conduction into the passages marked **d**.

Then the helium cooled in the passages **4c** enters the header **11** where part of the helium flows into a gas expansion machine expands thereat and is additionally cooled. The other part of the forward flow of helium is throttled in valves and returns into the header **11** in the form of a cold reverse flow combined with the flow of helium leaving the gas expansion machine; in the header **11** this helium is distributed among the passages **4** marked with letter **d**. The reverse flow of helium moves through these passages in the direction shown by arrow **D** and, flowing around the screens **3**, is heated, accumulates in the header **10** and leaves the heat exchanger.

An experimental specimen of the screen heat exchanger has undergone laboratory tests which have shown that the screen heat exchanger is characterized by a good relationship between heat transfer and hydrodynamic resistance along with comparatively small size and weight. Compactness of the heat-exchanging surface formed by the screens of various dimensions ranges from 2,000 to 20,000 m^2 of surface per cubic meter of free volume as specified above.

Temperature Balancing for Reversing Heat Exchangers

According to a process described by *A. Gotoh, T. Mizokawa, Y. Sawada and T. Katsuki; U.S. Patent 4,034,420; July 12, 1977; assigned to Kobe Steel Ltd., Japan* there is provided a method of balancing the temperature of a plurality of reversing heat exchangers each containing at least two heat exchanger cores connected in parallel relationship, each heat exchanger having a feed gas stream, a return gas stream and a product gas stream of which the paths of the feed gas stream and return gas stream are alternately changed over.

In this temperature balancing method, the temperature of a particular stream is measured each time the feed gas stream and the return gas stream are changed over in each heat exchanger. The temperature of each heat exchanger is measured over certain change-over period, and these values are used for calculating the mean temperature of each heat exchanger, this being a parameter which is controlled.

Further, the mean temperature of all heat exchangers is obtained from the mean temperatures of each heat exchanger and used as the set value with which comparison is made for balancing purpose. The parameter to be controlled and the set value are compared, and at least a certain flow is regulated in dependence on the system deviation between the controlled parameter and the set value.

According to a further feature of the process, when the degree of valve opening of any one of the temperature control valves of the reversing heat exchangers goes outside a predetermined control range, it is controlled by comparing the degree of opening with the mean degree of opening of all of the temperature control valves. This mean degree of opening is used as a set value.

The information resulting from the comparison is utilized to actuate another automatic valve other than the temperature control valves, in such a manner that the mean degree of valve openings of all temperature control valves are held within a control range. Thus, continuous control of the temperature balance is achieved by maintaining the degree of valve openings of the temperature control valves within the control range.

The change-over period of the reversing heat exchangers is automatically varied in dependence on the degree of temperature balance of the reversing heat exchangers while continuing the controls of the features as described above.

A heat exchanger system employing a number of reversing heat exchangers to which a temperature balancing method of the process is applied is utilized particularly in air separation.

FLUE GAS

Modulatable Heat Exchanger

G.R. Grimes; U.S. Patent 3,863,708; February 4, 1975; assigned to Amax Inc. describes a modulatable heat exchanger system for cooling dust-laden flue gas, such as flue gas from a molybdenum sulfide roaster having an elevated dew point.

In the modulatable heat exchanger system the conditions are automatically controlled during the heat exchange cycle to avoid condensation of fluids on heat exchanger surfaces.

In carrying out the process, the flow of hot fluid (flue gas) inside a heat exchange tube is preferred for the reason that all heat exchange from the flue gas is under control and the tubes are more accessible for cleaning when necessary.

The flow of colder fluid (air) in the annular space surrounding the heat exchange tube is also preferred in that it assures flow of the air parallel to the flow of the flue gas and also assures a substantially equal mass flow of air for each heat exchange tube. The temperature change for both hot and cold fluids is a function of the relation of the mass flow of each to the combined mass flow of both fluids.

Concurrent flow of the two fluids is desirable as otherwise longer heat exchange tubes would be required for countercurrent flow of the two fluids.

Referring to Figures 7.3a and 7.3b, a pair of heat exchanger units **10, 10A** is shown; the **A** designation indicating the same corresponding elements in the corresponding unit. The two units comprise vertically disposed shells **11, 11A** having confined therein a bank of heat exchange tubes **12, 12A** which are solidly connected to and between tube sheets or headers **13, 13A**, the bank of tubes communicating with manifolds **14, 14A** located adjacent the tube sheets or headers of heat exchanger units **10, 10A**.

Figure 7.3: Modulatable Heat Exchanger

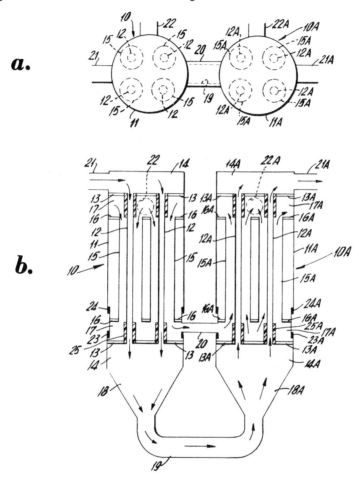

Plan and elevation views of a pair of
series connected heat exchangers.

Source: U.S. Patent 3,863,708

Surrounding each tube are annular tubes **15, 15A** which are solidly attached to and between tube sheets or headers **16, 16A**, the annular tubes communicating with manifolds **17, 17A** adjacent headers **16, 16A**.

The two units have a converging bottom **18, 18A** in the form of an inverted cone connected together by a U-shaped conduit **19**. The air manifolds **17, 17A** at the lower portion of the units are coupled together via conduit **20**. Thus, as the hot flue gas passes through the two series-connected heat exchange units by way of conduit **19**, the air passes through annular tubes **15** as well as through annular tubes **15A** by way of connecting conduit **20**.

The hot flue gas is drawn into unit **10** through inlet **21** by means of a blower, not shown, the flue gas entering manifold **14** and, hence, heat exchange tubes **12**. Concurrently with the flow of the flue gas, air is similarly drawn into unit **10** via inlet **22** into manifold **17** and down through the annular spaces surrounding tubes **12**.

The hot flue gas flows through the tubes into lower manifold **14** and via U-shaped conduit **19** into the bottom of heat exchange unit **10A** through lower manifold **14A**, tubes **12A**, upper manifold **14** and then through flue gas exit **21A** downstream of the gas flow.

Meanwhile, the cooling fluid (heated or unheated air) flows concurrently and in parallel with the flue gas through annular tubes **15** into lower manifold **17**, via conduit **20** into lower manifold **17A**, up through annular tubes **15A** into upper manifold **17A** and finally through exit **22A**.

By controlling the flow and temperature of the air independently of the flow of the flue gas, the temperature of the heat exchanger surface can be maintained at above the dew point of the flue gas being cooled and condensation of corrosive fluids within heat exchanger tubes **12** and **12A** substantially inhibited, if not avoided.

As will be noted further from Figure 7.3b, packed joints **23**, **23A** are provided to allow for differential expansion of heat exchange tubes **12**, **12A** and annular tubes **15**, **15A**. Similarly, expansion joints **24**, **24A** are provided to allow for differential expansion between shell **10**, **10A** and annular tubes **15**, **15A**.

It is preferred that the extremities of heat exchange tubes **12**, **12A** be insulated, e.g., with insulation **25**, **25A**. The insulation around the heat exchange tubes within the colder fluid manifolds **17**, **17A** is provided to eliminate problems associated with heat exchange in cross flow. With a multiple number of heat exchange tubes, the colder fluid is heated by contact with the first heat exchange tubes in the flow path.

Theoretically, at least, the hotter fluid in these first heat exchange tubes enters the parallel flow section at lower temperature than the hotter fluid in subsequent heat exchange tubes. Conversely, the colder fluid entering the first annular spaces in the flow path enters the parallel flow section at lower temperature than subsequent annular spaces. It is desirable to have uniform temperature distribution for all heat exchange tubes and substantially equal temperature differentials between the hot and cold fluids for each tube.

The shell **11** (and **11A**) serves as a base insulation on each unit, and to provide an insulating blanket of air around the annular tubes **15**, **15A**; otherwise temperature loss through the annular tube wall will result in an uneven temperature distribution between the colder fluid in the various annular spaces.

It is desirable that all components of the air circulating system be well insulated. Heat loss from the air circulating system determines the maximum available temperature of the air entering the heat exchanger, when the possibility of adding extraneous heat is neglected. The maximum available air temperature determines the maximum allowable air supply for any stated minimum hot fluid flow. In turn, the maximum available air flow determines the maximum heat exchange capacity of the heat exchange unit.

It is preferred that the heat exchange units be erected vertically. This avoids the problem of dust settling out of heat transfer surfaces when the velocity of the hot fluid is low. The heat exchanger may be constructed as a single unit or as a group of units connected in series. The heat exchanger flow and temperature control system are described in detail in the patent.

Corrosion Resistant Design

A process described by *A.K. Persson; U.S. Patent 4,147,209; April 3, 1979; assigned to SKF Industrial Trading and Development Company BV, Netherlands* relates generally to heat exchangers, and specifically to a metallic heat exchanger for transferring heat from a heat-emitting medium to a heat-absorbing medium through the walls of a number of flow tubes through which the one medium flows, these tubes being surrounded by the other medium. The heat exchanger is especially intended for use when preheating the combustion air for an oil-fired industrial furnace with its own flue gases.

Referring to Figure 7.4a, the heat exchanger comprises an outer jacket **1**, which encloses a space **2** containing a number of flow tubes **3**, **4**. The flow tubes are connected at one end to the central inlet **5** for the one medium, which in the following is assumed to be the warm medium in the form of flue gases, while at the other end they are connected to the outlet **6** for this medium. During operation the indicated medium is distributed over the flows through the tubes between the inlet and outlet. The flow is indicated by arrows.

The space **2** in the outer jacket is divided into three sections, which are separated from one another by an inner jacket **7** placed parallel to the outer jacket, and by a wall **8** connecting the inner and outer jackets. The inner jacket **7** separates the tubes **3** from the remaining tubes **4** in space **2**, and the wall **8** is provided with apertures through which the tubes **4** extend.

Figure 7.4: Corrosion Resistant Heat Exchanger

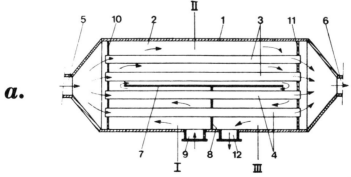

(continued)

Figure 7.4: (continued)

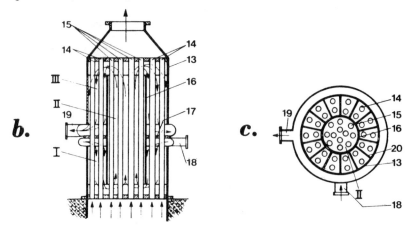

    (a)   Elevation view in section, illustrating the operation
            of the process.
    (b)(c) Longitudinal and transverse cross section views re-
            spectively through another example of the heat
            exchanger.

Source: U.S. Patent 4,147,209

Mainly to prevent high-temperature corrosion, the medium surrounding the flow
tubes (assumed in the following to be the cold medium) is introduced into the
space **2** through an inlet **9** which is located in the outer jacket close to the wall
8 at an axial distance from the end walls **10, 11** which secure the tubes **3, 4**,
said end walls defining the active heat-transferring section of the heat exchanger.

The cold medium initially enters the first section **I**, in which the counterflow
principle or mode of heat transfer is employed in relation to the medium in the
tubes **4**, and flows in the direction of the end section of the heat exchanger
where it flows around the edge of the inner jacket **7** and then between this
jacket and the end wall **10** into the second section **II**.

In this section the parallel flow or mode of heat transfer principle is employed
in relation to the medium in the tubes **3**, and the medium surrounding the tubes
flows axially to the other end wall **11** of the heat exchanger where it flows
around the opposite edge of the inner jacket **7** and then between this jacket and
the end wall **11** into the third section **III**. There the counterflow principle is
again employed in relation to the medium in the tubes **4** and the medium surrounding the tubes finally flows through the outlet **12**, which is located close to
the wall **3** on the side opposite to inlet **9**.

As illustrated in Figure 7.4a, tubes extending through region (or section) **I**
have start and finish flow portions as regards the flow of the first medium
therethrough, with start-flow portions near inlet end wall **10** and finish-flow
portions near transverse wall **8**. For producing the counterflow heat transfer
in this region, the inlet means **9** to region **I** for the flow of the second medium

is near the finish-flow portions of the tubes in this region while the outlet means for region I is near the start-flow portions of the tubes, and a similar structure prevails in region III. In region II the tubes have start and finish flow portions also and the inlet and outlet means to this region for the flow of the second medium are near the start and finish parts, respectively, for providing the parallel flow heat transfer.

In this way, the entire quantity of the colder medium flows only around a portion of the flow tubes for the warm medium. This means that when using the heat exchanger to heat combustion air for an industrial furnace with the aid of its flue gases, the air in section I cools the flue gases flowing through the tubes located in this section and in section III down to a temperature which allows heating of the air in section III close to the limit for high-temperature corrosion.

Since the entire quantity of air in sections I, II and III flows only around some of the flow tubes for the flue gas, the temperature in section I is increased by an amount considerably smaller than that by which the temperature of the flue gases is lowered. Accordingly, on exiting from section I, the entire quantity of air, at a sufficiently low temperature, encounters that portion of the flue gases which is introduced into the flow tubes of sections I and II at the highest temperature.

In section II, the heat is utilized according to the parallel flow principle, as a result of which the temperature of the tubes can be held almost constant at a value approximately equal to that at the inlet to this section. The flue gases enter section III at a temperature lower than the initial temperature and are now cooled even further. By properly distributing the number of tubes over sections I, II and III, as well as by properly locating the wall 8 between the end walls 10 and 11, such conditions can be created that high-temperature corrosion can definitely be prevented even when the quantities of gases fluctuate within broad limits.

If mainly low-temperature corrosion in the heat exchanger is to be prevented, the cold medium is simply fed in the opposite direction, i.e., first through section III using the parallel flow principle, then through section II using the counterflow principle and finally through section I using the counterflow principle.

Figures 7.4b and 7.4c show longitudinal and transverse cross sections respectively through an advantageous example of a heat exchanger in accordance with the process. The outer jacket 13 consists of a cylindrical pipe which encloses the flow tubes 14, 15 and the inner jacket 16 consists of a pipe concentric with the outer jacket and encloses the flow tubes indicated with 15. The inner and outer jackets are connected to one another by a wall 17 which divides the annular space between them, the wall being provided with apertures through which the flow tubes 14 extend.

The inner jacket 16 and wall 17 divide the space within the outer jacket 13 into three sections, which are indicated with I, II and III, wherein section II is enclosed by inner jacket 16. The inlet and outlet for the medium surrounding the flow tubes are designated with 18 and 19 respectively.

As in Figure 7.4a, the flow of the various media is indicated by arrows. The principle of operation of the heat exchanger is likewise analogous to that described in Figure 7.4a. To provide better directional control of the flow of the medium surrounding the tubes **14**, the outside of the inner jacket or, if necessary, the inside of the outer jacket can be provided with flow-directing vanes running longitudinally. This assures that the medium flowing around the flow tubes is always distributed over the cross section of the heat exchanger when passing through the three sections, thus improving the heat transfer.

If the heat exchanger is to be connected to a furnace whose flue gases have a very high temperature before entry into the heat exchanger, or, if for the purpose of energy utilization, the gases are to be cooled to a lower temperature than that attainable by using the combustion air, the inner jacket and, if necessary, the flow-directing vanes can be covered completely or partially with channels for the purpose of water cooling. The cooling water can, in turn, be used for heating rooms, for example.

If the following data were assumed, the following example results in the appropriate dimensions of a heat exchanger in accordance with the process. An oil-fired furnace at a temperature of 1000°C supplies 1,600 stm³/hr (standard cubic meters per hour) of flue gas. The heat content of the flue gases is intended to be utilized to heat 1,400 stm³/hr of combustion air from 0° to 500°C. It is assumed that the surface temperature of the surfaces contacted by the flue gases is at most the average temperature between the flue gases and air concerned.

For the heat exchanger in accordance with Figure 7.4b with flow conditions in accordance with the arrows shown in the figure, and with gas and air velocities selected such that a heat transfer coefficient of 35 W/m²K can be assumed, the result is that the outgoing gases have a temperature of 625°C, that the heat transfer surfaces are approximately 4 m² in section I, approximately 8.5 m² in section II, and approximately 4.5 m² in section III, if the lowest temperature on the flow tubes is to be more than approximately 150°C and the highest temperature on these tubes is to be less than approximately 600°C.

Preheating of Combustion Gas

E.J. Nobles; U.S. Patent 4,044,820; August 30, 1977; assigned to Econo-Therm Energy Systems Corporation has found that in order to obtain a more effective heat transfer with greater efficiency it has been found that the heat can be transferred between countercurrently flowing hot flue gas and ambient inlet combustion air by arranging a multiple array of tubes passing directly between the cooled and heated zones with a heat transfer liquid flowing through the tubes. The heat absorbed by the heat transfer liquid at each level of the flue gas is then directly transferred to the combustion air at the adjacent level.

As a result, a closer approach to countercurrent heat exchange is obtained with simple, uncomplicated equipment. The temperature level of thermodynamic availability of heat is preserved with the highest level Btu from flue gas being transferred to the highest level use in the preheated air.

The amount of heat transferred in each passage or level is small leading to a nearly constant liquid temperature. This essentially constant and controllable liquid temperature permits working close to condensation temperature without danger of condensation. The apparatus can be used in many applications where it is desired to cool a gas close to its dew point while insuring that the gas temperature does not go to or below the dew point temperature causing condensation on exchange surfaces.

ELECTRIC GENERATING PLANTS

Controlling Thermal Pollution

According to a process described by *J.L. Allen; U.S. Patent 4,009,577; March 1, 1977* thermal pollution from industrial manufacturing and electric generation plants is avoided by incorporating in the water-cooling system of such plants a heat exchanger to allow the heat load, which would otherwise be injected into the source of external water supply, to be exchanged through the cooling tower to the atmosphere. In a preferred case, the system incorporates a counterflow heat exchanger so that the temperature of all water exhausted into the source of external water supply is within the few degrees of the temperature of the external water supply.

Figure 7.5: Cooling Tower Blowdown Heat Exchange System

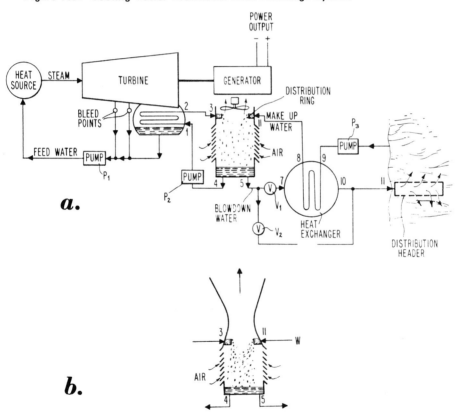

(a)   Schematic diagram of a steam-driven turbine used in conjunction with an electrical power generator.

(b)   Schematic illustration of a natural draft water cooling tower.

Source: U.S. Patent 4,009,577

A schematic illustration of the steam turbine electrical power generation system is shown in Figure 7.5a. A heat source is used to convert water into steam which then passes through a turbine providing the motive power to the generator. Some of the steam cools as it goes through the turbine and is bled off at various points and recycled as shown. A large portion of the steam passes through the turbine and is exhausted into a condenser as shown. This steam is condensed into water by the condenser and collects in the hot well of the condenser as illustrated. A pump is used to remove this water from the hot well and feed it back to the heat source for recycling.

The system just described is a closed cycle system. External water flows through the condenser entering at point **1** and exiting at point **2** as shown. The temperature of the inlet steam to the condenser and of the condensate contained in the hot well of the condenser is the same. In an operational configuration this temperature typically ranges between 78° to 85°F. The difference in energy between the inlet steam and the hot well condensate is the latent heat of vaporization which is removed by the cooling liquid entering at point **1** and exiting at point **2**. Cooling water is injected into a cooling tower at point **3** as illustrated.

The function of the cooling tower is to extract the heat load picked up by the cooling water in the condenser. Cooling water enters the cooling tower in a circular distribution header at the top of the tower and then passes through a mesh and enters the shaft of the cooling tower as small water droplets. These water droplets rain through the tower.

The water droplets which have rained down through the tower and have been cooled by evaporative loss collect as a liquid at the base of the cooling tower and are extracted at point **4** by a pump which recycles the water through the condenser.

The system just described is open-cycle and three losses occur during operation of the cooling tower. The first is the evaporative loss to the atmosphere which is necessary to achieve cooling. The second loss is termed drift water loss, which consists of water droplets that become entrained in the upward draft and are lost to the atmosphere.

The continual addition of makeup water to the cooling tower would result in the accumulation of scale and salts within the circulating water system because loss through evaporation is essentially distilled water leaving the salts behind. To avoid this undesirable buildup of scale and salts, a portion of the water is extracted. This portion, termed blowdown water, constitutes the third loss source since it is discharged to the external source of water supply.

The makeup water is thus equal in volume to the volume of water lost through evaporation and drift and the amount of water extracted as blowdown. The extraction of blowdown water is usually operationally so designed that the accumulation of salts within the tower is held to an acceptable level. This accumulation varies, dependent upon system design. Representative operational numbers would be 200 ppm of total salts contained in the makeup water compared to 600 ppm of salts contained in the water collected within the tower. Typically, the salts are constituted of calcium and magnesium sulfates.

A mechanical-draft tower is illustrated in Figure 7.5a. A fan is used to create the draft of air through the tower. Figure 7.5b illustrates an alternative configuration of a cooling tower termed a natural-draft tower.

A natural-draft tower is considerably taller than an equivalent mechanical-draft tower. The mechanical-draft tower illustrated in Figure 7.5a in an operational configuration might be 100 to 125 feet tall, where an equivalent capacity natural-draft tower would be 500 to 600 feet tall. The height of the natural-draft tower is required to obtain a differential in temperature between the air entering the bottom of the tower and exiting the top.

If the tower is 500 feet high, the differential in temperature between the bottom and the top of the ambient air surrounding the tower is about two to three degrees. The addition of hot water to the natural-draft tower will cause a further increase in temperature of the air entering the bottom of the tower and will further increase the tendency of the air to rise. The heat exchanger system which is the feature of this process, could be used with either a mechanical or natural draft air-cooling tower.

Typically, the evaporative loss of an air-cooling tower represents about 3% per any unit time of the system water circulating within the tower. Again, by matter of illustration, the blowdown is usually a third to a half of the evaporative loss.

The wet bulb temperature of the air outside the tower governs the amount of cooling that can be effected by the tower. It is not possible to effect cooling below the wet bulb temperature of the ambient air surrounding the tower. For example, assume that the wet bulb temperature of ambient air surrounding the tower is 80°F and the temperature of the external water source is 70°F. The temperature of the blowdown water exiting the tower under these conditions would be 80°F. If the temperature of this blowdown water were not decreased by passing it through the heat exchanger as illustrated, it could cause an ecological problem when exhausted in the external water supply since a differential temperature of 10°F is involved. Depending upon the climatic conditions, this differential can be as high as 30°F.

The purpose of the heat exchanger illustrated in Figure 7.5a is to cool the blowdown water by passing it through a counterflow heat exchanger such that the exit temperature of the water leaving the heat exchanger **10** is approximately the same as the inlet water to the heat exchanger from the external water supply **9**. Accordingly, the blowdown water when ultimately injected into the external water supply is at approximately the same temperature and represents no added heat load to the external water supply. The makeup water which flows through the heat exchanger to cool the blowdown water is, of course, heated in the process. This makeup water is, however, applied to the top of the cooling tower **11** and is cooled in the normal operation of the tower.

Representative figures are shown below, for a typical operational installation. It should be noted that the amount of increase in duty cycle of the cooling tower required to cool the makeup water is relatively modest compared to the overall capacity of the tower.

Unit size	10^9 watts
Circulating water system flow rate	476,600 gpm
Cooling duty of tower	7.9 x 10^9 Btu/hr entering the tower
Tower blowdown	4,500 gpm
Drift	950 gpm
Evaporative loss	12,650 gpm
Makeup water	18,100 gpm
Typical operating temperatures, $°F$	
Blowdown temperature	88.9
Wet bulb air temperature	75.0
River water temperature	75.0
Differential temperature	13.9

Using maximum makeup water yields minimum temperature rise to makeup water to tower per pound of water.

$$\frac{4,500}{18,100} \text{ x } (13.9) = 3.5 \text{ degrees of rise in makeup water temperature}$$

3.5°F is equal to 3.5 Btu/lb. The increased duty to tower resulting from the required cooling of the blowdown water is:

$$18,100 \text{ gpm x } 60 \text{ min/hr x } 8.3 \text{ lb/gal x } 3.5 \text{ Btu/lb} = 2.05 \text{ x } 10^6 \text{ Btu/hr}$$

Percentage increase in tower duty is:

$$\frac{2.05 \text{ x } 10^6}{7.9 \text{ x } 10^9} \text{ (100)} = 0.026\%$$

For the example worked out above, it should be noted that the increase in tower duty is considerably less than 0.1% even though the temperature of the blowdown water is 13.9°F higher than the river water temperature. Of course, the figures given above are by way of example only and could vary substantially depending upon the climatic conditions.

In some circumstances, it may not be necessary, or even desirable, to cool the blowdown water before discharging it into the external water source. As shown schematically in Figure 7.5a, valves **V1** and **V2** can be adjusted to direct any desired amount of the blowdown water through the heat exchanger before exhausting it through a distribution header into the external water source.

With **V1** closed and **V2** open the heat exchanger will be bypassed entirely. With **V2** closed and **V1** open, all blowdown water will pass through the heat exchanger. Valves **V1** and **V2** can, of course, be adjusted to intermediate positions to divert only a portion of the blowdown water through the heat exchanger. The blowdown water exhausted into the external water source normally will proceed under the force of gravity, so no pump is necessary on the outlet side of the heat exchanger.

The purpose of the distribution header is to distribute the blowdown water evenly into the external water supply so that no localized hot spots or concentrations of salt result. Since this process contemplates cooling the blowdown to a temperature very close to that of the external water supply, some applications may not require a distribution header. This is desirable, because the cost of such a distribution header can be substantial.

Compressible Fluid Contact Heat Exchanger

A process described by *F.R. Hull; U.S. Patent 3,915,222; October 28, 1975* is directed to the contact interchange of thermal and kinetic energy between adjacent compressible fluid streams across a virtual heat transfer surface at substantially different velocities in parallel flow. The process may find especial application as a regenerative heat exchanger in gas turbine power plants, or as the low-velocity contact-type air preheater of a steam generator or furnace.

In the process, hot low-pressure exhaust fluids and cool compressed intake fluids enter the receiver-side section of an elongate heat exchanger. Intake-fluid stream pressure energy is converted to kinetic energy within nozzle passageways of the receiver-side section. The cold high-velocity intake-fluid stream is rapidly heated in the velocity-accelerated contact interchange process by the hot low-velocity exhaust-fluid stream within the mixing section. Following the contact interchange process, the intake-fluid and exhaust-fluid streams are separated from each other by flow-dividing members and discharged from the separator-side section. Within the preheated intake-fluid stream, normal shock in supersonic flow across the inlet of the intake-fluid discharge passage is averted by the effects of variable control over characteristic length and exhaust-fluid outlet flow control.

Overall, the process provides an efficient means of effecting a substantial recovery of thermal energy present in the exhaust fluid stream of an atmospheric gas turbine power plant. Variation of the gas turbine power plant process arrangement to confine the combustion process to the turbine exhaust stream, in conjunction with use of the regenerative apparatus of this process for heating the intake fluid stream to the gas turbine, will alleviate the fouling of turbine blade surfaces by deposit of solid products resulting from the combustion process, through segregation of the combustion process from the thermodynamic expansion process within the gas turbine.

Frasch Process

According to a process described by *X.T. Stoddard and R.C. Terry; U.S. Patent 4,051,889; October 4, 1977* a Frasch process sulfur mine power plant is coupled to a steam-electric generating plant by a coupling heat exchange means. In this way, the heat liberated by condensing steam is substantially used to bring Frasch mine water up to operating temperature.

In commercial practice of the process multiplicity of heat exchangers would be employed to couple the steam-electric generating station to the Frasch process sulfur mine, but for purpose of illustration only two heat exchangers are described.

The steam-electric generating station operates by heating water to steam at, for example, 1000°F, then discharging the steam first through a high pressure turbine, second through an intermediate pressure turbine, then delivering steam at 500°F to the coupling heat exchanger. The steam-electric generating station is further equipped with an economizer so that the sensible heat from the products of combustion is transferred to inbound Frasch mine water, raising the temperature of the Frasch mine water from ambient to for example 190°F. Steam from the intermediate pressure turbine discharge is delivered to the coupling heat exchanger at, for example, 500°F with a pressure of, for example, 600 psi with a heat content, for example, of 1,200 Btu/lb. Frasch mine water at a temperature of, for example, 190°F is also delivered to the coupling heat exchanger. The steam is condensed in the coupling heat exchanger with the condensate exiting at, for example, 240°F for return to the steam-electric generating station for recycling. The heat liberated by condensing the steam is transferred to the Frasch mine water which enters the coupling heat exchanger at, for example, 190°F and exits at, for example, 330°F. Sufficient pressure is maintained on both sides of the heat exchanger to keep the condensate and the water below the bubble point.

In planning for the coupling of a steam-electric generating plant with a Frasch process sulfur mine, the primary limitation is the maximum capability of the Frasch mine to receive water. This maximum capability is then matched with the maximum expected delivery of useable heat from the steam-electric generating station.

With this match-up the coupling heat exchanger can be designed and the two facilities can be tied together. The steam-electric generating station can then be operated to match the varying power demands during each 24 hour period, and the Frasch process sulfur mine can be operated successfully within these variations. The savings in fuel costs as a result of operating the plants coupled together approximates the cost of fuel for the Frasch plant should it be operated as a separate plant.

Movable Shutters to Reduce Wind Activity for Cooling Equipment

According to a process described by *L. Heller, L. Forgo, G. Bergmann and G. Palfalvi; U.S. Patent 3,933,196; January 20, 1976; assigned to Transelektro Magyar Villamossagi, Hungary* movable shutter-type elements are provided for eliminating the diminishing effect of the wind on cooling equipment erected in open air.

The movable opening and closing elements are adjusted by a control or regulating device according to actual wind conditions, to effect a damming of pressure in the path of the streaming air in front of the heat exchanging surface, enabling the wind energy to be used to aid in increasing the cooling effect. The opening and closing of elements rotatable around a vertical axis may be used to utilize to a greater extent the wind energy in addition to their wind reflecting effects.

Referring to Figures 7.6a, 7.6b and 7.6c, the cooling equipment illustrated is of the natural draft variety and includes as its essential element a chimney 1. A heat exchanger 2 is so arranged that air, illustrated by arrows 7, streams through it horizontally. Movable opening-shutter elements 3 are positioned to rotate about a vertical axis and form a cylindrical jacket by encircling the damming-up field, illustrated herein as ring shaped.

As depicted in Figure 7.6b with the chimney 2 removed, the direction of the wind is shown by arrows 6, and as is appreciated, movable opening-shutter elements are made to rotate about a vertical axis to most favorably direct the wind into the damming-up field illustrated as 4.

Adjustment of the opening-shutter elements is effected in any well known manner such as hydraulic, pneumatic or hand operated controls or adjusting apparatus, which receive impulses from devices sensing both the speed and direction of the wind, and data which characterizes the performance of the apparatus (e.g., temperature of the cooling water) to thereby effect the most favorable cooling effect by the adjustment of the opening-shutting-up elements. During still-wind conditions, the opening-closing-up elements are fully opened.

Figure 7.6: Reduction of Wind Activity for Cooling Equipment

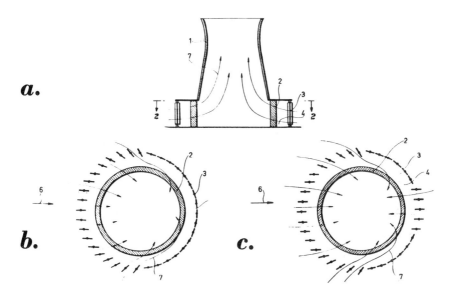

    (a)   Sectional view of cooling equipment including a
             a heat exchanger.
    (b)(c) Sectional views taken along the line **2–2** in Fig-
             ure 7.6a with the shutter elements shown in
             different adjusted positions.

Source: U.S. Patent 3,933,196

In a strong wind, the adjustment as shown in Figure 7.6b is used, the essential element of which is that the shutters on the side opposite to the wind direction are totally closed, and the remaining shutters are adjusted to angles varying along the circumference so as to direct the wind toward the inside room. Thus, as is appreciated, the directing effect of shutters is utilized as well.

As a result of shutting up the side opposite to the wind direction, the motion of the wind is transformed into a damming up pressure damming field **4**. Since the air **7** entering the cooling equipment cannot escape elsewhere, it flows through the heat exchanger and the chimney into the open air. Thus the damming-up pressure of the wind is added to the draft and ameliorates the working condition of the cooling equipment.

In weaker winds the effect of closing up the side opposite to the wind direction is poor, and in such cases the increasing pressure of the dam point which rises upon the action of the wind flowing around the cooling apparatus on the side opposite to the direction to the wind is utilized by opening some of the shutters proximate to the dam point as is depicted in Figure 7.6c.

GASIFICATION PROCESSES

High Temperature Thermal Exchange Process

H.P. Meissner and F.C. Schora; U.S. Patent 4,037,653; July 26, 1977; assigned to Institute of Gas Technology describe a process which provides direct thermal exchange between a flowing gaseous stream and a heat-carrying liquid. When used for thermal exchange, the heat-carrying liquid should have a vapor pressure as low as possible and preferably negligible over the temperature range used.

Thus, evaporation of the heat transfer liquid is kept to a minimum and heat entering or leaving the heat-carrying liquid serves to change its temperature rather than to supply the heat for its vaporization. The heat-carrying liquid is selected to be chemically nonreactive with components present in the flowing gas stream. In another use of the process, the heat-carrying liquid may be selected to undergo desired chemical reactions with specific components of the flowing gas stream to remove undesired components from the gas stream.

Generally, the process is carried out by generating droplets of heat-carrying liquid in a flow passageway, the surface of these droplets providing large areas for thermal exchange contact with the gas flowing in the passageway.

A shower of droplets can be generated by throwing the liquid through the flowing gas by impellers or rotating disks partially immersed in a liquid pool at the bottom of a chamber or a splash condenser or the like. Alternatively, such a droplet shower can be generated by a submerged gas jet. Again, these droplets might be generated from suitable sprayheads or nozzles located in the chamber, through which the heat-carrying liquid is circulated. The liquid droplets passing through the gas may exchange heat with the flowing gas in the process and are then recycled. The heat-carrying liquid can be maintained within a predetermined temperature range by circulating this liquid in heat-exchange relationship with a second gas stream for desired thermal exchange so as to regenerate it with respect to the desired thermal exchange of the first gas stream.

Depending on the composition of the gas streams being processed, suitable materials for operation at higher temperatures include metals and other inorganic and organic materials which remain molten and have low vapor pressures and low viscosities over the temperatures of interest.

Selection of a suitable heat-carrying liquid depends upon the composition of the gas stream being processed. Thus, a gas containing for example, 5% H_2O and 5% CO_2 can be cooled or heated in the range of about 327°C by use of molten lead (MP 327°C), since lead will not react with such a gas. Molten magnesium (MP 651°C) on the other hand would be suitable if removal of H_2O and CO_2 from the above gas were desired, since magnesium would react extensively with the H_2O and CO_2 present. If the desired purpose were to remove H_2O and CO_2 from this gas simultaneously with heat transfer, then magnesium would be preferable over lead.

Figure 7.7: High-Temperature Thermal Exchange Process

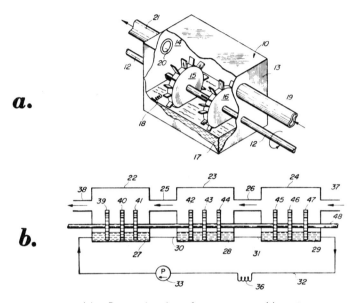

a.

b.

(a) Perspective view of an apparatus with parts broken away to show interior detail.
(b) Schematic elevational view of an apparatus using a countercurrent series of similar units.

Source: U.S. Patent 4,037,653

Referring to Figure 7.7a, enclosed chamber **10** is provided with through-shaft **12** suitably journaled in opposing walls **13** and **14**. Paddle wheels **15** and **16** are fixedly mounted on shaft **12** and rotate therewith. Pool **17** comprising a molten or liquid substance is present at the bottom of enclosure **10** to a level such that the lower portions of paddle wheels **15** and **16** are immersed therein.

Heater or cooler means **18**, also submerged in pool **17**, serves as an external heat sink. Pool **17** may be maintained at the desired temperature by heat exchange through the chamber walls or by a heat exchange surface within the pool. Hot gas to be cooled or heated is supplied to chamber **10** via conduit **19**, and exits from chamber **10** via exit port **20** and conduit **21**.

In operation, pool **17** is maintained at a desired, predetermined temperature by cooler or heater means **18** and shaft **12** is driven so as to rotate paddle wheels **15** and **16**, thus generating a spray of liquid droplets in the confined gas flow passageway defined by chamber **10**. The resulting large surface area of the drop-lets provides a very rapid and effective heat transfer with a gas stream which is passed through chamber **10**. Chamber **10** is filled with droplets of a primary liquid flung upward by rapidly rotating wheels, which are partly submerged in pool **17** of liquid filling the bottom part of the vessel. The droplets fly up through the gas, and then fall back again through the gas to pool **17** below, exchanging heat with the gas in the process. Some of the droplets strike the top of chamber **10** and drop off, falling back through the gas.

The primary liquid in the pool of this exchanger can in turn be cooled or heated with a second circulating liquid travelling through coils **18** submerged in liquid pool **17** and maintained at the desired temperature. This second liquid can be molten salt or molten metal, or hydrocarbon oil, with heat carried away as sensible heat in this liquid. Alternatively, water can be vaporized at an appropriate temperature and pressure in these coils. Alternatively, the primary heat transfer liquid itself could be withdrawn and circulated through external cooling coils.

Methods of adding instead of removing heat to the primary liquid are readily apparent. A high heat flux between the gas and primary heat transfer liquid can be attained because of high concentration of liquid droplets which can be maintained by several methods in spray chamber **10**.

The primary heat transfer liquid selected can be a molten metal or a molten salt, depending upon desired properties. When the gases involved carry entrained liquids or solids, then a portion of such liquids or solids will be picked up in the primary heat transfer liquid. Removal of such material, when insoluble in the heat transfer liquid, can be accomplished by filtration, by skimming of the solids from the top of the liquid phase, or similar methods.

The advantages of the apparatus of Figure 7.7a over the usual regenerator and recuperator thermal exchangers are many. Structural problems are minimized, since no part of this apparatus need exceed safe temperature limits. Thus, rotating wheels **15** and **16** and shaft **12** can, if necessary, be internally cooled, as can the spray chamber walls. The heat transfer itself is excellent, especially since the droplet surfaces are continually renewed, thus fouling is not a problem. Continuous operation is easily attained and very high temperature operations can be performed.

As an example, the free space in chamber **10** above the liquid is about 1,000 ft^3. Nitrogen gas is passed through the chamber, entering at about 1000°F and atmospheric pressure. The cooling liquid is molten lead maintained at about 600°F, having a density of 630 pcf and a heat capacity of 0.02 Btu/lb°F. The droplets generated by the paddle wheels have an average droplet diameter of about 0.05 in. Depending upon speeds of rotation, and hence on the number and diameters of droplets present per cubic foot, the heat transfer coefficient between the droplets and the gas ranges from 10 to 500 Btu/hr°F per ft^3 of gas space.

At a superficial entering gas velocity equivalent to about 3 ft/sec through the chamber cross-section at the hot end, the gas may be cooled about 200°F and exits from the chamber at about 800°F. While the gas flow and liquid in an apparatus as shown in Figure 7.7a may be in cocurrent or countercurrent relations, ralatively little countercurrency will be attained in a single chamber, unless it is of excessive length, due to the vigorous mixing which occurs.

While the example illustrated in Figure 7.7a is primarily a single stage operation, the process can also be operated in cocurrent or countercurrent series as shown in Figure 7.7b. Chambers **22**, **23** and **24** are connected in series and communicate via gas conduits **25** and **26**.

Similarly, liquid pools **27**, **28** and **29** are connected via pipes **30** and **31** and the liquid contained therein is recirculated via pipe **32** by means of pump **33**, through heater or cooler **36**. Paddle wheels **39** through **47**, inclusive, are mounted on shaft **48** which is journaled in the walls of chambers **22**, **23** and **24** and is driven by a suitable prime mover. With higher numbers of chambers or stages connected in series, higher degrees of countercurrency can be attained.

High Temperature Gasifier

J. Kummel, H. Dressen, W. Danguillier, P. Gernhardt, W. Grams and S. Pohl; U.S. Patent 4,098,324; July 4, 1978; assigned to Dr. C. Otto & Comp. GmbH and Saarbergwerke AG, Germany describe a system for cooling high-temperature gasifiers and method for its operation wherein cooling conduits extend vertically through the walls of the gasifier and are connected at their ends to a closed water circulation system which incorporates heat exchangers for removing heat from the system.

Boiling in the system is prevented, and good heat transfer characteristics are achieved, by maintaining a high pressure system, at least 40 bars, coupled with a flow velocity of between 5 and 7 m/sec at a maximum internal diameter of the cooling conduits of 51 mm. Means are provided for maintaining the temperature of the water exiting from the cooling conduits at least 10°C below the boiling point at the pressure and flow rate of the system.

The cooling tubes in the walls of the gasifier are covered by a ramming compound of refractory material held in place by expanded metal or mesh which curves around each tube but is spaced therefrom and welded to webs interconnecting the tubes in a gas-tight wall. The expanded metal supports the ramming compound but at the same time does not come into direct contact with the tube walls and thus eliminates points of maximum thermal loading where boiling might occur.

With reference to Figure 7.8a, the cooling walls of a slag bath generator are designated generally by the reference numeral **1**. The cooling walls surround the gasification chamber **2** of the gasifier which is supplied with coal duct, oxygen and water vapor as indicated generally by the arrows, reference numeral **3**. The particular fuel which is being gasified is immaterial. The kind of material gasified only affects the quantity of the gasification medium which is produced. The slag bath and the flame jet of the generator which is directed to the bath enable any desired temperature up to 2500°C to be obtained.

The cooling walls **1** comprise a plurality of vertical or horizontal cooling tubes which impart the longitudinal sectional shape shown diagrammatically in Figure 7.8a to the space surrounded by the cooling walls **1**, the cooling tubes merging at the top end into a ring main **4** and at the bottom end to a ring main **5**. The cross section of the tube wall is shown in Figure 7.8c. A circulating pump **6** pumps sufficient cooling water into the bottom ring main **5** to insure that cooling water at a temperature of 200°C flows from the main ring **5** at a velocity of

5 to 7 m/sec through the cooling tubes, which have a maximum internal diameter of 51 mm, into the top ring main **4**. Assuming an operating temperature of 1700° to 2500°C in the gasifier, the cooling water is heated through 25°C while passing through the tubes with a cooling water pressure of 40 bars. The cooling water pressure and the velocity prevent boiling of the cooling water as explained above.

Figure 7.8: Water-Cooled, High Temperature Gasifier

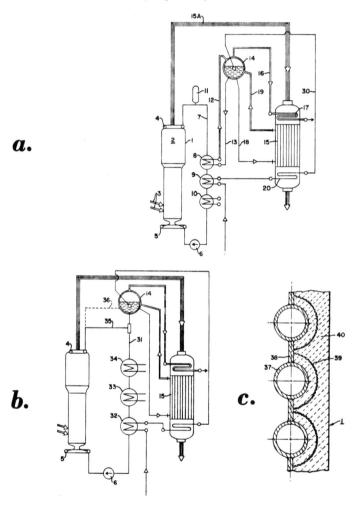

a.

b.

c.

    (a)   Schematic illustration of a slag bath generator
            cooling system.
    (b)   Schematic illustration of a modified slag bath
            generator cooling system.
    (c)   Illustrates part of the tube wall of a slag bath
            generator utilizing wire mesh to hold a ram-
            ming compound in place.

Source: U.S. Patent 4,098,324

The heated cooling water from the top ring main **4** is supplied through a circulating line **7** to several heat exchangers **8**, **9** and **10** which are connected in series. After leaving the last heat exchanger **10**, the cooling water is again returned to the cooling tubes at a temperature of 200°C. A compensating vessel **11** which is a pressure accumulator, is connected to the circulating line **7** between the top ring main **4** and the first heat exchanger **8**. The compensating vessel **11** equalizes the change of volume of water when this is heated.

The cooling tubes with the ring mains **4** and **5**, the circulating line **7**, the heat exchangers **8**, **9** and **10** and the circulating pump **6** form a closed cooling water circuit which comprises the primary circuit of a two-pressure system shown in Figure 7.8a. In the heat exchanger **8**, the cooling water circuit is connected to a second circuit which comprises the secondary circuit of the two-pressure system. The secondary circuit includes a riser **12** and a downcomer **13**. The riser **12** and the downcomer **13** connect the heating surfaces of the heat exchanger **8** to an exhaust steam drum **14**. The riser **12** extends into the drum interior above the water level of the exhaust steam drum and the downcomer **13** enters the interior of the drum below the water level.

The heat exchanger, therefore, is supplied with water from the exhaust steam drum by natural circulation (i.e., because of differences in specific gravity). The water begins to boil in the heat exchanger **8** at a pressure in this case of 25 bars, to be returned as a water-steam mixture through the riser **12** into the exhaust steam drum **14**.

The exhaust steam drum **14** is part of the waste-heat boiler **15** connected to the output of the gasifier **2**. The steam collected in the exhaust steam drum is supplied through a duct **16** to a superheater **17** situated in the waste-heat boiler **15** and then escapes. The cooling water is also supplied through a downcomer **18** to a cooling system situated downstream of the superheater **17** in the waste-heat boiler **15**, and this cooling water flowing through the downcomer **18** is returned to the exhaust steam drum through a riser **19**. The riser **19** and the downcomer **18** are both connected to the exhaust steam drum **14**, the point of entry of the downcomer **18** being substantially lower than that of the riser **19**.

The waste-heat boiler **15** is supplied with raw gas from the slag bath generator through conduit **15A** after the gas has undergone intermediate cooling to a temperature of 850° to 900°C. The superheated steam discharged from the superheater **17** can be supplied for any desired purpose. The waste-heat boiler **15** is a medium pressure boiler given a steam drum pressure of 25 bars.

The heat exchanger **9** which follows the heat exchanger **8** on the cooling water system is intended for preheating the feed water for drum **14**. It is connected in series with another feed-water preheater **20** in the waste-heat boiler, the feed water which is preheated in this manner being supplied to the exhaust steam drum through a duct **30**.

The heat exchanger **10** is optionally connected to a low-pressure boiler or comprises such a boiler. By controlling the cooling medium flowing through the heat exchanger **10**, the temperature of the water entering the ring main **5** can be controlled to insure that the water, when it exits from the ring main **4**, has a temperature below the boiling point of the water at the pressure utilized.

The slag bath generator cooling system shown in Figure 7.8b differs from that of Figure 7.8a in that the riser **12** and the downcomer **13** are replaced by a suction line **31** which is connected to the exhaust steam drum **14**, the other connections being the same. The suction line **31** contains several serially connected heat exchangers **32, 33** and **34** of which the heat exchanger designated by the numeral **32** corresponds to the heat exchanger **9**, the heat exchanger designated by the numeral **33** corresponds to the heat exchanger **10** and the heat exchanger **34** being any other heat exchanger. The heat exchangers **33** and **34** can be utilized to alter the temperature of the water entering the ring main **5** to insure that the boiling temperature is not reached at the pressure utilized.

The suction line **31** is connected to the circulating pump **6** which pumps the cooling water into the bottom ring main **5** as in the example illustrated in Figure 7.8a. The heated cooling water discharged from the top ring main **4** is again supplied directly to the suction line **31** through a duct **35**, preferably at a position between 2 and 3 m below the exhaust steam drum.

The cooling water discharged from the ring main **4** can also be supplied to the exhaust steam drum **14** directly through a duct **36** shown in broken lines. Connection of the generator cooling circuit on the water and pressure side to the exhaust steam drum in both cases insures that the generator cooling circuit and the waste-heat boiler are operated at the same pressure.

In Figure 7.8c, the individual cooling tubes **37** of the cooling walls **1** are shown in cross section. The cooling tubes **37** are welded into a gas-tight wall by means of webs **38**. These webs **38** extend perpendicularly between the cooling tubes **37** and are welded thereto. They extend along the entire lengths of the tubes **37** so as to provide the aforesaid gas-tight wall. A grid of expanded metal or mesh **39** curves over each cooling tube **37** and is tacked by spot-welding to the middle of two adjacent webs **38**.

The expanded metal **39** functions as a retaining means for the ramming compound **40** which is applied to the cooling tubes **37** on the gas chamber side (i.e., the right side shown in Figure 7.8c). In this manner, the expanded metal or mesh **39** acts to secure the ramming compound **40** in place, but at the same time, it does not make direct contact to any of the cooling tubes **37** so as to form hot spots on the tube walls where vapor can form. This is in contrast to the prior art techniques wherein fingers or rods were welded at the individual tubes **37** to act as a supporting means for the ramming compound **40**.

NUCLEAR POWER PLANTS

Thermal Power Plant with Gas-Cooled Nuclear Reactor

R. Naegelin; U.S. Patent 4,147,208; April 3, 1979; assigned to Sulzer Brothers Limited, Switzerland describes a heat exchanger which acts as a recuperator and is constructed of a plurality of identical subassemblies which can be manufactured and tested in a plant and shipped as individual units to a construction site.

Each subassembly includes a plurality of straight tubes which interconnect with spherical shells at each end and which serve to carry a flowable medium. Each subassembly also includes a guide tube about the straight tubes which is open at both ends to convey a flowable medium over the straight tubes but which has an outer flange which cooperates with similar flanges on the other subassemblies to block any flow of this medium over the outside surfaces of the guide tube from one end to the other.

Figure 7.9: Heat Exchanger-Recuperator

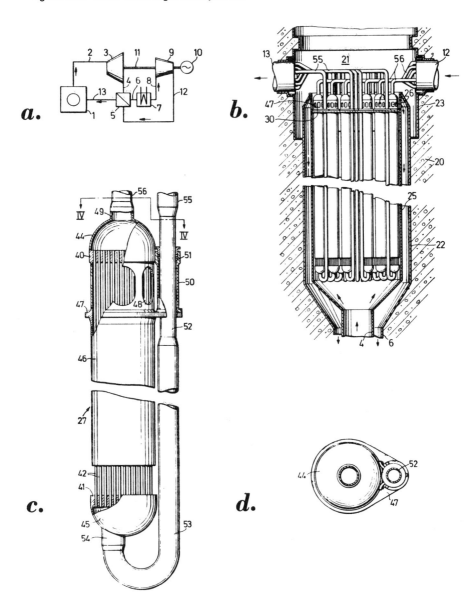

(continued)

Figure 7.9: (continued)

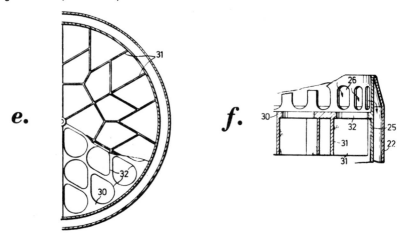

e.

f.

(a) Schematic view of a thermal power plant in which a heat exchanger is used as a recuperator.
(b) Vertical cross section through the heat exchanger.
(c) Broken view of a subassembly of the heat exchanger.
(d) Top view of the subassembly according to Figure 7.9c.
(e) Horizontal cross-sectional view of the heat exchanger of Figure 7.9b without subassemblies.
(f) Vertical cross-sectional view of a detail of Figure 7.9b without subassemblies.

Source: U.S. Patent 4,147,208

Referring to Figure 7.9a, a thermal power plant has a gas-cooled nuclear reactor **1** connected via a hot gas line **2** to a gas turbine **3**, the discharge of which is connected via a line **4** to a recuperator **5**. From the recuperator **5**, the gas flows via a line **6** through a cooler **7** and a line **8** to a compressor **9** which, together with the gas turbine **3** and an electric generator **10**, is mounted on a common shaft **11**. The gas compressed in the compressor **9** then flows via a line **12** to the secondary side of the recuperator **5** and from there back to the nuclear reactor **1** via a line **13**. The recuperator **5** is preferably housed in a concrete structure of the nuclear reactor **1**, as is shown in detail in Figure 7.9b.

Referring to Figure 7.9b, a concrete pressure vessel **20** of the nuclear reactor **1** includes a cavity **21** which is disposed on a vertical axis and which has a lower portion lined with a steel shell **22** and an upper portion lined with a steel lining **23**. The steel shell **22** is tapered conically at the bottom and is connected to the line **6** while at the top, the steel shell **22** protrudes into the cavity **21**.

The upper end of the steel shell **22** is likewise tapered conically and is connected to a tubular wall **25** which is arranged concentrically with the shell **22**. The wall **25** is conically tapered at the bottom and is connected to the line **4**. In addition, the tubular wall **25** is provided with openings **26** in the vicinity of the upper end. A horizontal plate **30** is tightly connected to the wall **25** below these openings 26 and rests on a web structure 31 (Figure 7.9e) to which the plate 30 is fixedly joined by welding. The plate 30 has openings 32 of oval shape between the webs of the web structure 31. A heat exchanger subassembly 27 is mounted in each of these openings 32.

Referring to Figure 7.9c, each of the subassemblies 27 consists of an upper tube sheet 40 and a lower tube sheet 41 between which a large number of straight heat exchanger tubes 42 extend. A spherical head or shell 44, 45 extends over each of the tube sheets 40, 41. A guide tube 46 is welded to the upper tube sheet 40 and extends over the major part of the length of the heat exchanger tubes 42 in surrounding relation and terminates in spaced relation to the lower shell 45. The guide tube 46 has an outer flange 47 intermediate the ends as well as several passage openings 48 between the outer flange 47 and the upper tube sheet 40.

As shown in Figure 7.9d, the outer flange 47 has an oval shape and supports a cylindrical sleeve 50 (Figure 7.9c) which is tightly connected to the flange 47. The sleeve 50 is located on an axis which is parallel to the vertical axis of the subassembly and is equipped with an expandable sealing means such as a stuffing gland 51. The flange 47 is drilled-out beside the guide tube 46 to the inside diameter of the sleeve 50. An outlet line in the form of a pipe section 52 is tightly connected via a U-shaped pipe 53 to a nozzle 54 fastened to the spherical head 45. This line 52 extends through the sleeve 50 and the stuffing gland 51.

Referring to Figures 7.9b and 7.9c, the outlet lines 52 of the subassemblies 27 are connected via connecting pipes 55 to a tube sheet which is arranged in the vicinity of the steel lining 23 and to which the line 13 (Figure 7.9a) is connected. The line 12 is welded, likewise in the vicinity of the lining 23 to a tube sheet (not shown) from which connecting pipes which act as inlet lines 56 lead to and are tightly connected to nozzles 49 arranged on the heads 44 of the subassemblies 27.

During operation of the thermal power plant, gas, e.g., helium, which comes from the reactor 1 and is expanded in the gas turbine 3, flows as the primary medium via the line 4 into the space below the subassemblies 27. The gas then enters into the bundles formed by the tubes 42 of the individual subassemblies 27 through the openings formed between the lower end of the guide tube 46 and the tube sheet 41.

Thereafter, the gas flows along the tubes 42 into the vicinity of the upper tube sheet 40 and exits into the cavity 21 lined with the lining 23 through the passage openings 48. From this cavity 21, the now cooled-down gas passes through the openings 26 into the annular space between the tubular wall 25 and the steel shell 22 and then flows out through the line 6.

After being cooled further in the cooler 7 and being compressed in the compressor 9, the gas is delivered as a secondary medium into the shells 44 of the upper tube sheets 40 via the line 12 and the connecting lines 56. The gas then flows through the tubes 42 while absorbing heat and is thereupon collected in the shells 45 of the lower tube sheets 41. The preheated gas then is exhausted via the nozzles 54, the U-pipes 53, the pipe sections 52 and the connecting pipes 55 back to the nuclear reactor 1 via the lines 13.

Different thermal expansion of the tubes **42** and the U-pipe **53** causes relative displacements of the pipe sections **52** and the sleeve **50** which are taken up by the stuffing gland **51**. The sleeve **50** and the stuffing gland **51** may also be replaced by expansion compensators in the form of bellows.

Protective Tubes for Liquid Sodium Heated Water Tubes

J. Essebaggers; U.S. Patent 4,140,176; February 20, 1979; assigned to U.S. Department of Energy describes a heat exchanger in which the water tubes are heated by liquid sodium which minimizes the results of accidental contact between the water and the sodium caused by failure of one or more of the water tubes. A cylindrical protective tube envelopes each water tube and the sodium flows axially in the annular spaces between the protective tubes and the water tubes.

Figure 7.10: Liquid Sodium Heated Water Tubes

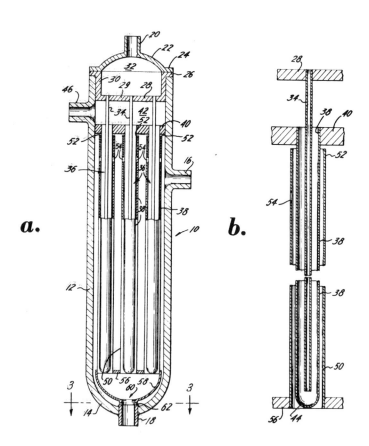

(continued)

Figure 7.10: (continued)

c.

(a)   Elevational view partly in section of a steam
        generator.
(b)   View partly in section of one of the protective
        tubes shown in Figure 7.10a with the water
        tube within it and the associated tube sheets.
(c)   View partly in section taken substantially along
        the line 3-3 of Figure 7.10a.

Source: U.S. Patent 4,140,176

Figure 7.10 shows a heat exchanger indicated generally as **10**, having a generally cylindrical outer shell **12** with a closed lower end **14**. The outer shell **12** includes a sodium inlet **16** and a sodium outlet **18**, which is positioned centrally in the closed lower end **14** of the outer shell **12**. Heat is supplied by hot liquid sodium which enters at the inlet **16** and which flows downwardly to eventually exit at the outlet **18**.

Water to be heated to steam or steam to be superheated enters at a water inlet **20** which is located centrally in a dome-like upper cover **22** which serves to close the upper end of the cylindrical shell **12**. The cover **22** has at its periphery an annular horizontally extending flange **24**. The flange **24** and the top of the cylindrical shell **12** clamp between them an annular radially extending lip **26** of an upper tube sheet **28**.

The upper tube sheet **28** is generally flat, and includes a generally horizontal circular flat portion **29** which is connected to the lip **26** by an annular vertically extending cylindrical sidewall portion **30**. The cover **22** and the tube sheet **28** define a chamber **32**.

Extending through the upper tube sheet **28** and secured to it are a plurality of inner tubes **34**, each of which is a component of one of an equal number of bayonet tube assemblies **36**. Each of the bayonet tube assemblies **36** has, in addition to its inner tube **34**, an outer tube **38** each of which extends down from a main tube sheet **40**.

The main tube sheet **40** is below the upper tube sheet **28** and defines with the upper tube sheet **28** and cylindrical sheet **12**, a chamber **42**.

Each of the outer tubes **38** is closed at its lower end **44** but the inner tubes **34** are open at both ends. Each inner tube bottom is a little higher than the corresponding closed end **44** of its associated outer tube so that water either in its liquid phase or its gaseous phase (steam) coming in the inlet **20** and filling the chamber **32** will flow downwardly through the inner tubes **34** to impinge against the closed lower ends of the outer tubes **38** to reverse direction and flow upwardly in the annular spaces between the inner tubes **34** and the outer tubes **38**.

It is during this upward travel that the water is heated. Thus, the outer tubes 38 are the heated tubes. In a steam generator, liquid water is heated and converted to steam. In a superheater, gaseous water (steam) is heated further. In either case, steam will collect in the chamber 42 and leave the heat exchanger 10 through a steam outlet 46 in the side of the heat exchanger 10 between the upper tube sheet 28 and the lower tube sheet 40.

As is known, the failure of a heat exchange surface can result in a sodium-water reaction which can cause considerable damage. In order to minimize this damage, each of the bayonet tube assemblies 36 is provided with a protective tube 50. Each protective tube 50 is cylindrical in configuration and is larger in diameter and coaxial with one of the heated outer tubes 38. Each protective tube 50 extends upwardly to its upper end 52 which is slightly below the main tube sheet 40. Each protective tube 50 has an orifice 54 in its sidewall below the tube sheet 40.

Liquid sodium entering the orifices 54 will flow in the annular spaces between the protective tubes 50 and the outer tubes 38. Some of the sodium will flow upward through these spaces and over the tops 52 of the protective tubes but most of it will flow downward to leave the protective tubes at their bottoms. As shown best in Figure 7.10b, each of the protective tubes has an open bottom 56. The sodium flowing through the annular spaces between the protective tubes 50 and the heated outer tubes 38 supplies heat to water flowing (as liquid or gas) in the annular spaces between the inner tubes 34 and the outer tubes 38.

In its flow upward between the orifices 54 and the tops 52 of the protective tubes 50, the sodium is cooled sufficiently that it will not excessively heat the main tube sheet 40. This eliminates the necessity of a cover gas immediately below the tube sheet 40.

The bottoms of each of the protective tubes 50 is secured to a lower tube sheet 56 which at its periphery is sealing-secured to the periphery of a generally bowl shaped inner shell 58 which covers the closed lower end 14 of the cylindrical outer shell 12.

The lower tube sheet 56 does not extend outward to the outer shell 12 and, therefore, the inner shell 58 does not contact at its periphery the outer shell 12. The inner shell 58 has at its bottom a centrally located open neck 60 which extends more or less vertically and has a number of small drains 62.

With this arrangement, a body of slow moving sodium which flows out of the protective tubes 50 at their tops 52, flows downward between the spaces between the protective tubes 50 and then, at the lower portion of the heat exchanger, outward, because of the lower tube sheet 56, to flow down and inward in the space between the inner shell 58 and closed lower end 14 of the outer shell 12.

The sodium then flows through the drains 62 and out of the steam generator 10 through the sodium outlet 18. The drains 62 are of a size and number to assure that this body of sodium will move at a very slow flow rate so that the amount of sodium following this route will be much less than the sodium which flows down through the protective tubes 50. It prevents the creation of a large temperature gradient across the protective tubes 50.

The arrangement has the advantage of more or less uniform flow of sodium axially along each of the outer tubes **38**. This allows for a better prediction of the hydraulic and thermal performance of the heat exchanger than would otherwise be possible. These parameters are not as easily predicted in exchangers when the liquid sodium flows over baffles placed along the length of the water tubes.

In the event of a failure of one of the outer tubes **38**, any sodium-water reaction will be limited because of the inclusion of protective tubes **50**. No adjacent bayonet tube assemblies will be damaged. It is contemplated that the protective tubes **50** will have thicker sidewalls than the outer tubes **38**. This is so because the protective tubes **50** have as their main function, the ability to stand up under a sodium-water reaction and its concomitant rise in pressure. The outer tubes **38** must be limited in thickness to assure a good heat transfer if the heat exchanger **10** is to operate efficiently.

Another advantage of the arrangement is that the products of any sodium-water reaction will flow downwardly more quickly than in a heat exchanger utilizing baffles. This means that any detection device downstream of the bayonet tubes will detect the sodium-water reaction more quickly so that remedial action can be taken.

Another advantage of the arrangement is that it prevents a slow leak in an outer tube **38** from causing erosion of an adjacent tube. Without the protective tubes **50**, a small leak in one of the outer tubes **38** would result in a small stream of products of a sodium-water reaction which would be directed outward from the tube to impinge against one or more adjacent tubes to erode them and eventually result in their failure.

In the heat exchanger **10**, any slow leak in any of the outer tubes **38** would result in an impingement against the associated thick-walled protective tube **50**. Further, the adjacent outer tubes **38** are protected from erosion by their associated protective tubes **50**.

Design for Mutually Reactive Transfer Fluids

R. Dickinson; U.S. Patent 4,090,554; May 23, 1978; assigned to The Babcock & Wilcox Company describes a heat exchanger in which mutually reactive heat transfer fluids are utilized. Tubes are interposed coaxially with other tubes so as to form a limited volume, wherein these fluids can mix and react in a controlled manner, permitting detection of leakage across the boundaries separating the reactive fluids before a major chemical or physical reaction can occur.

Figure 7.11a shows a heat exchanger **10** comprising a cylindrically shaped shell **12**, oriented with its longitudinal axis in a vertical plane, closed at its lower and upper ends by a lower hemispherical head **14** and an upper hemispherical head **16**, respectively. In the lower head **14**, a nozzle **52** provides an inlet for a tubeside fluid **61** that enables the fluid to flow into a lower chamber **46** formed by the internal surface of the lower head **14** and by a first outer tubesheet **22** disposed perpendicularly to the longitudinal axis of the shell.

Located at the lower end of the heat exchanger is a first inner tubesheet **32** spaced toward the center of the shell parallel to and in close proximity with the first outer tubesheet **22** forming a lower inner chamber **42** between the first inner and outer tubesheets. A second inner tubesheet **34**, transversely oriented with respect to the longitudinal axis, is spaced on the opposite side of the longitudinal center of the shell from the tubesheet **22** and **32**; a second outer tubesheet **24** is located further beyond the longitudinal center, parallel to and in close proximity with the second inner tubesheet **34** so as to form a second inner chamber **44** within the upper portion of the illustrated heat exchanger. The tubesheets **22, 24, 32** and **34** are integrally attached to the shell by welding or other means to provide a leakproof relation.

A plurality of openings in the lower-outer tubesheet **22** receives the first ends of a bundle of tubes **20**. The tube **20** extends, in parallel orientation with the longitudinal axis of the heat exchanger, coaxially nested within one of larger diameter tubes **30**. The first end of each tube **30** is received in a hole in the first inner tubesheet **32**.

In between the inner tubesheets **32, 34**, each outer tube **30** coaxially circumscribes an inner tube **20** forming a concentric volumetric gap **50** in the space intermediate with each coaxially arranged inner and outer tube combination. The opposite end of each outer tube **30** is received in an opening in the upper inner tubesheet **34**. The inner tubes **20** extend beyond the upper inner tubesheet **34** and are received in tube receiving holes of the upper outer tubesheet **24**.

Figure 7.11: Heat Exchanger

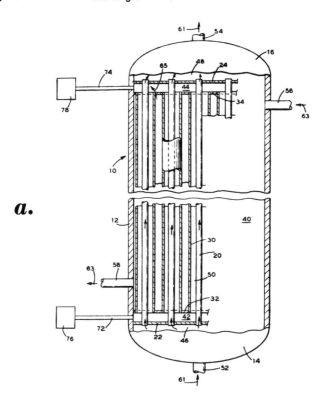

(continued)

Figure 7.11: (continued)

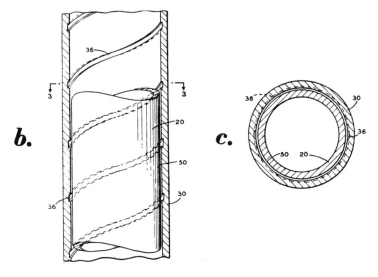

b.   **c.**

(a)   Elevation view, partly in section, of a heat exchanger.
(b)   Sectional side view of a part of a coaxial tube combination of Figure 7.11a.
(c)   Cross-sectional view taken along line **3-3** of Figure 7.11b.

Source: U.S. Patent 4,090,554

The tubeside fluid **61** flows from the lower chamber **46** through the inner tubes **20**, into an upper-outer chamber **48** formed by the upper-outer tubesheet **24** and the upper hemispherical head **16**. In the upper head, a nozzle **54** enables the tubeside fluid **61** to flow out of the upper chamber **48**.

A nozzle **56** on shell **12** provides an inlet for a shellside fluid **63** to a shellside chamber **40** formed by the shell and the outside surface of the outer tubes **30** in the space between the inner tubesheets **32** and **34**. A second nozzle **58** on the shell **12** provides an outlet for the shellside fluid **63** from the shellside chamber **40**.

As shown in Figures 7.11b and 7.11c the inner surface of outer tube **30** has formed therein one or more spiral grooves **36**. Each spiral groove is continuous over the length of the tube in between the inner tubesheets **32, 34**. The continuous groove **36** need not be spiraled and in differing embodiments could be formed on the outside surface of the inner tube alone or on both the inner surface of the outer tube and the outer surface of the inner tube. Typically, the volume of the spiral grooves **36** is significantly greater than the volumes of gaps **50**.

The gaps generally assure fluid communication between leaking inner tubes, the spiral groove and the inner chambers. Since some inner and outer tube contact may be experienced due, for example, to the minimal radial gap clearance or to tube vibration, spiral groove 36 assures fluid communication of the leakage with the inner chambers 42, 44. Nozzles 72, 74 on shell 12 (Figure 7.11a) provide a means of connecting appropriate devices 76, 78, e.g., pressure transducers, to detect reactive conditions in the inner chambers.

As shown in Figure 7.11a, the ends of inner tubes 20 are hermetically sealed at the outer tubesheets 22, 24 by welding, rolling or by other widely known sealing means. In the preferred case, the outer tubes are not hermetically sealed to the inner tubesheets 32, 34. This permits limited amounts of shellside fluid as shown by the shellside fluid flow lines 65 to enter and fill inner chambers 42 and 44, gaps 50, and spiral grooves 36 thereby establishing continuous fluid communication between the two inner chambers. Alternate means, such as drilling a small hole or holes through the inner tubesheets 32, 34 or notching the tube receiving holes therein, may be used to permit controlled and limited shellside fluid 63 communication from chamber 40 to the inner chambers 42, 44 and subsequently, to the spiral grooves 36 and the gaps 50 (Figure 7.11b and 7.11c).

The volumes of the inner chambers 42, 44 are carefully controlled by the spacing of the tubesheets to minimize the amounts of shellside fluid 63 contained therein relative to that of the main body of shellside fluid within chamber 40 resulting in a higher ratio of tubeside to shellside fluid than would result for a given leak rate of tubeside fluid directly into the shellside volume 40. The inner chamber volumes and the volume of the grooves and gaps must be limited to values which permit detection of leakage and remedial action prior to the occurrence of heat exchanger damage. Pertinent parameters for determining the volumes of the inner chambers, the gaps and the grooves include the pressure and temperature of the fluids 61, 63, their chemical reaction rates, the nature of the reaction products formed, the thermodynamic nature of the reaction, and the structure of the tubes, i.e., materials of construction and tube dimensions.

It appears that grooves having either a depth or width of less than sixty mils would not be suitable for the purposes of this process. In general, the total limited volume of the inner chambers, the spiral grooves and the gaps should not be greater than 10% of the shellside volume 40 in order to assure ratios of tubeside to shellside fluid resulting in reactions which are readily detectable. The resulting reactions due to the leakage can be calculated by assuming, for example, varying leakage rates into the fixed limited volume described above. Alternately, a required limited volume can be determined based on the pressures attained by varying the limited volume for a maximum permitted or anticipated leak rate.

In a preferred design, the higher pressure fluid is tubeside fluid 61. Typically, a heat exchange apparatus which can utilize the process is a liquid metal steam generator wherein a liquid metal, e.g., sodium is the shellside fluid and water or steam at a higher pressure is the tubeside fluid. Leakage across inner tube 20, or the outer tubesheets 22, 24 results in the mixing and reaction of the higher pressure tubeside fluid 61 with the shellside fluid 63 within the limited volume defined by chambers 42, 44, spiral grooves 36 and gaps 50, and in the formation of sodium hydroxide and gaseous hydrogen.

Within the limited volume, the high concentration ratio of tubeside to shellside fluid relative to shellside volume 40 results in a higher and more easily detectable pressure excursion therein. The leakage and reaction of water or steam with the limited volume of liquid sodium can be detected by monitoring pressure or hydrogen concentration changes within the inner chambers 42, 44.

PIPELINES

Insulated Pipelines

A process described by *J.S. Best; U.S. Patent 3,990,502; November 9, 1976; assigned to The Dow Chemical Company* concerns an arrangement of elements for controlling heat flow between a member and its environment. For example, there has been a problem of providing adequate thermal protection for heated members, such as an oil pipeline or building structures in Arctic regions where undue melting of the permafrost could have a major impact on the stability of the environment and possibly the security of such heated members.

The process provides various combinations of insulating, thermal absorbing and/or dissipating elements to take care of heated members in environments having varying thawing indexes, e.g., from the North Slope of Alaska, having a mean annual temperature of about 10°F to Valdez, Alaska, having a mean annual temperature of about 32°F. The combination of elements can include plastic foam or equivalent insulation adjacent the heated member and, in combination therewith, a heat sink and/or a ventilating or thermal absorbing cell, the latter allowing free air passage or its equivalent between other elements in the system.

The heated member, such as a pipeline carrying oil or the like at a temperature significantly above that of the adjacent earth, can be located below or above the ground and employ an arrangement of insulating, thermal absorbing and/or dissipating elements. Combinations of such insulating and thermal absorbing and/or dissipating elements can be used effectively in supporting other heated members, such as heated buildings and the like on permafrost. Bleeding of excess heat from members can also be affected by such elements. Such elements can also act to maintain a relatively stable temperature differential between cryogenic materials and the supporting earth, and generally to control heat flow between a member and its environment.

Figures 7.12a through 7.12e represent an oil pipeline **10** extending from the North Slope of Alaska, having an air-freezing index of about 8,500 degree days each year, to Valdez, Alaska, having an air-freezing index of about 3,000 degree days per year. A "degree day" as used herein represents 1 day with a mean air temperature 1 degree below or above freezing. The "freezing index" merely represents the number of degree days below freezing during a year and is commonly used to calculate the depth of ground freezing during the winter. The "thawing index" on the other hand, is merely the number of degree days above freezing during a year. For example, a freezing index of 10 degree days may result when the mean air temperature is 31°F for 10 days or when the mean air temperature is 22°F for 1 day.

Figure 7.12: Insulation of Pipelines in Permafrost Regions

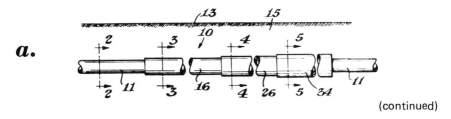

a.

(continued)

Figure 7.12: (continued)

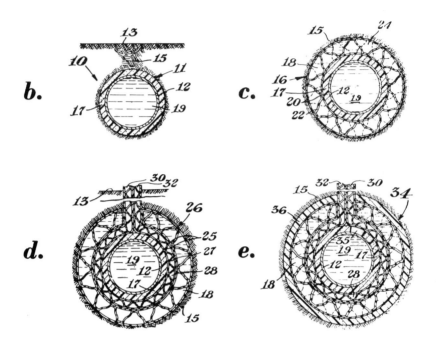

   (a)   Underground pipeline, with portions broken away, running through
           a permafrost region either continuously or only in sections where
           circumstances dictate, and extending left to right from an extremely
           cold region to a generally milder region
(b)-(e)   Partially exaggerated cross section of various arrangements of insulating,
           thermal dissipating and/or absorbing elements for different situations
           in various permafrost regions taken along lines 2-2 through 5-5 of
           Figure 7.12a

Source: U.S. Patent 3,990,502

In a specific example the following situation could exist: pipeline **10** includes
standard 48" diameter steel oil pipe **12** having a $\frac{1}{2}$" to $\frac{3}{4}$" wall thickness; the
pipeline is located underground so that the center line of pipe **12** is about 6 ft
below surface **13** of the permafrost soil **15**; hot crude oil at a temperature of
about 180°F is the fluid **19** carried by the pipe **12**.

Referring to Figures 7.12a and 7.12b, pipeline section **11** comprises pipe **12** surrounded about its circumference by insulation layer **17**. The insulation layer **17**
can be a closed cell urethane foam, or its functional equivalent, having a K factor
of about 0.24 Btu/ft^2/hr/°F/in thk, and is about 4" in thickness. For example,
a styrene-maleic anhydride foam could be used as it has good solvent resistance

to hydrocarbons. The environment for which this pipeline is adapted can be one such as is found on the North Slope of Alaska described above. Since in this location the conductance and volumetric heat capacity of the permafrost below the thawing temperature is so large insulation of the type described is adequate to prevent any significant disruption of the permafrost layer due to the heat dissipated from the hot oil **19** in the pipeline through the insulation layer **17**. However, even in this region if the water content of the soil is too great, and/or does not have a sufficient structural integrity it may be necessary to employ a pipeline section **16**.

Heat absorbing and/or dissipating elements are added to the insulation system, as shown in pipeline section **16** of Figure 7.12c to meet conditions found in Alaska south of the North Slope, e.g., at the 66° north latitude, which has a mean annual temperature of about 15°F. Here the air-freezing index is somewhere between 6,500 to 7,000 degree days F and the air-thawing index is around 2,500 degree days F. The pipeline section **16** is the same as that pipeline section **11** previously described only with a closed annulus heat sink in the form of a thermocell or heat sink **18** being located about insulation layer **17**. The thermocell can comprise various configurations for containing or enveloping a heat sink material. The thermocell **18** comprises inner and outer cylindrical skins **20** secured on opposite nodular ends of a core structure **22**, which core structure, for example, can have a shape such as that shown in U.S. Patent 3,277,598.

The core structure primarily is one which separates the 2 skins and permits fluid flow therethrough and, as such, can also be of bent corrugated metal, a granular or other particulate fill or other configurations and various materials such as also shown in U.S. Patents 3,086,899 and 3,190,142. The member containing the heat sink material should have sufficient impermeability to contain the heat sink material in its fluid state so that substantially none is lost in such state.

The skins **20** can be adhered by adhesives, welded, heat-sealed, or otherwise secured with the core **22**. The thermocell can be made of a plastic material such as polyethylene or rubber-modified polystyrene, but can be formed of other polymeric, metallic, organic or other synthetic or natural substances having sufficient strength and impermeability to satisfy the requirements of such a thermocell. A liquid **24** enclosed within the thermocell **18** can be a saline, glycol or other solution sufficient to give the thermocell heat sink a freezing point slightly less than the transition temperature of the surrounding permafrost (usually about 32°F), as, for example, 30°F.

The liquid **24** incorporates a freezing point suppressant in water that in solution acts in a eutectic manner in the range of temperatures below 32°F. One such material can be a frozen solution containing less than 5% sodium sulfate. Specifically, a 3.84% solution by weight of sodium sulfate in water has a freezing point of approximately 30°F. Likewise, a 1% propylene glycol solution, by weight, has a freezing point of approximately 30°F. When the liquid is frozen, the heat required to melt the solution is great. The total heat of fusion of the sodium sulfate solution, e.g., is available within a few degrees below 32°F, thus allowing reverse cycling of heat flow at less than 32°F, but stopping heat flow at heat source temperatures above 32°F. The heat sink in this particular instance had a thickness of about 12" from skin to skin and is substantially filled with liquid **24**. In designing the thermocell **18** care should be taken to allow for expansion and contraction of the liquid **24** as the temperature changes.

The presence of the heat sink **18** substantially eliminates fluctuations of the heat loss from the pipe **12**. Thus, the underground pipeline **16** is kept in a near constant temperature environment thereby reducing the expansion and contraction effects, and, therefore, the need for expansion and contraction joints in the pipeline.

The presence of the heat sink around an insulated pipe buried in permafrost will increase the amount of heat transferred from the pipe to the soil over a 1 year cycle. By controlling the heat transfer from the pipe to the soil such that the permafrost is not thawed, the available seasonal low temperature of the air during the winter cycle is more effectively utilized. The sink system keeps the permafrost in the frozen state throughout the year and the latent heat of fusion of the water in the soil is not required. The heat exchange between the pipe and the atmospheric air is maximized since the permafrost is not allowed to go through the thawing and freezing cycle. The net result is that the average effective temperature differential is greater, therefore, more heat can be dissipated.

For a milder area such as might be found at Fairbanks, Alaska, for example, which has a mean annual temperature of about 25°F, a modified insulated pipe section **26** such as shown in Figure 7.12d can be employed. Here the freezing index is about 5,500 degree days F and the thawing index is about 3,000 degree days F. The insulated pipe section **26** is like the pipeline section **16** only in between the insulation layer **17** and the thermocell **18** is formed a ventilating annulus or thermal bleed **28** to permit air circulation between the two. Since the amount of thawing which the permafrost experiences is greater here than at the colder latitudes, the thermocell **18** need not be designed so that it alone is adequate to prevent significant thawing of the adjacent permafrost soil due to the hot oil **19** in the pipeline **10**. Of course, the size of the thermocell could be increased to satisfy the additional demands made by the warmer climate. However, it is found more effective, practical and economical to provide ventilation in the form of air duct or annulus **27**.

This is not to say, however, that air is the only method that can be used to transfer the heat flow through the insulation layer to the atmosphere. To take additional heat from the pipe section **26** out to the atmosphere and to ease the load upon the thermocell **18**, the annulus **28** can have a structure not unlike that of the thermocell **18**, i.e., having skins **25** like skins **20** and the core structure **27** like core structure **22** secured together in a similar manner, only at one end thereof, preferably the upper end so as to have the least effect on the soil, providing an inlet **30** and outlet **32**, which can be reversed from that shown, whereby air from the atmosphere can have passage therethrough. In certain integral constructions, it may be possible for the thermocell **18** and annulus **28** to share a common skin.

The core is designed so that through passageways are a natural result of construction. It has been found that a bleed-off of this nature can remove from about 50 to 100 Btu per hour per lineal foot of pipe section **26** located at Fairbanks, Alaska.

During the summer there is the natural flow of heat from the adjacent soil to thermocell **18** which means the thermocell **18** will have enough capacity to absorb this heat and store it while remaining at 30°F and, therefore, not affect the permafrost in the adjacent area. Thermocell **18** also has to have enough capacity to handle warm air coming through annulus **28**. In permafrost regions the heat absorbed by the thermocell **18** during the summer can be dissipated during the winter. In the winter the annulus **28** is removing heat from the thermocell **18** as well as from heat passing through insulation layers from the pipe. Likewise, in the winter the thermocell **18** is losing heat through the soil to the atmosphere. Thus thermocell **18** is primarily needed for the summer months to store both the artificial heat from the pipeline and natural solar heat at 32°F or less until such heat can be dissipated during the winter months.

In achieving stabilization of the permafrost the process thus takes into account both artificial heat, i.e., the heat from the product being carried in the pipeline, and natural heat, i.e., solar radiation penetrating the soil. It recognizes that artificial heat can be controlled. That is, the heated pipe can be insulated and a heat sink or a ventilating or radiating annulus can be used to absorb and dissipate the artificial heat until the atmospheric temperature cools enough, at which time the accumulated heat can be dumped. The arrangement also recognizes that the natural heat source is balanced by the seasons. Thus, the natural heat source varies between hot (generally above 32°F) and cold (generally 32°F or less) while the artificial heat source is continuously hot. Natural heat source thaws from the top of the permafrost soil downward, while the artificial heat source thaws from within the permafrost outward.

In accomplishing stabilization of the permafrost, the process uses both energy absorption and energy transfer. It is thus recognized that the entire job for a greatly varying permafrost region cannot be satisfactorily accomplished with insulation alone or even necessarily together with a heat sink at all locations. But with a combination of these elements, together with added elements, such as an annulus where necessary, a practical balance providing a protective arrangement in all permafrost regions is achieved.

At warmer locations, as for example, as found at 64° north latitude in Alaska where there is a mean annual temperature of about 27°F with a freezing index of about 4,000 to 4,500 degree days F, more insulation is required as illustrated for the pipeline section **34** of Figure 7.12e. Pipeline section **34** is like previously described pipeline section **26** only with an added insulating foam layer **36** about the outer circumference of the thermocell **18**, the insulation layer **36** being of the same type and having properties like that of the insulation layer **17**. This is a preferred combination of insulation and heat dissipating elements to prevent significant thawing of permafrost in an economical structure of this somewhat milder permafrost region.

Pipeline section **34** functions in the same manner as pipeline **26**, but instead of having a thermocell forming a heat sink of a depressed temperature solution such as a glycol solution at 30°F, it includes, alternatively and optionally, a heat sink filled with water, thereby having a freezing point temperature of about 32°F. In this type of arrangement an insulation layer **36** is provided between the permafrost and the thermocell thereby reducing the interface temperature between the permafrost soil and the system. Otherwise, a direct interface relationship between the thermocell and the permafrost soil would result in some melting of the adjacent permafrost soil which, over the years, could result in settling of the pipeline. The temperature drop through the insulation layer **36** reduces the interface temperature between the insulation layer **36** and the permafrost soil **15** to less than 32°F. The insulation layer **36** may, in the winter time, slow down the dissipation of heat from the thermocell, but this is not a great amount and is more than compensated by the fact that during the summer the heat flow from the natural environment is slowed to the thermocell **18**.

Where the situation changes to one having a mean annual temperature of about 32°F, wherein the thawing index is substantially greater than the freezing index as, for example, at Valdez, Alaska, where the thawing index is about 30,000 degree days and the freezing index is only about 1,500 degree days, a simple pipe insulation arrangement as shown for pipe section **11** in Figure 7.12b can be used. The insulation in this instance, however, is for the opposite effect than that earlier described for the soil in this location is not permafrost. Here the insulation serves only to prevent the oil from cooling too significantly during the winter and causing significant desiccation and other variations in the adjacent soil.

So at both extremes of temperature, from that experienced, e.g., at the North Slope of Alaska to the southern tip of Alaska, and corresponding other places of the world, such as found in Canada and Siberia and even certain latitudes in Japan and the United States, by applying combinations of insulated pipelines sections, abovedescribed, damage to the adjacent soil and to the pipeline can be prevented no matter what the weather.

In the example of Figure 7.12e a section **35** of insulation layer **17** can optionally be removed from adjacent the pipe **12** so that a hot spot is formed. This hot spot will accelerate the flow of cooling fluid, in this case air, about the annulus since it will be substantially warmer than the air from the atmosphere flowing through the annulus. The wide difference in temperature between the hot spot and the atmospheric air thus greatly accelerates heat dissipation. The hot spot, however, is not so large to exceed the capacity of the thermocell **18** adjacent thereto to absorb the heat escaping therefrom. This is an optional advantage which can be included where circumstances permit. Alternatively, a high heat conductive material such as steel or aluminum could be located in open section **35** and extend upwardly adjacent the ground surface.

By conductance the heat would flow through the pipe wall into the high conductance material acting as a thermal bleed dissipating the heat near the surface of the ground. To control where the major heat dissipation takes place, the high conductance material can be insulated such that the major heat loss takes place in the active surface layer of earth above the pipeline.

Heat Transfer Composition for Piping

W.A. Weidenbenner and I.J. Steltz; U.S. Patents 3,972,821; August 3, 1976; and 3,908,064; September 23, 1975; both assigned to Amchem Products, Inc. describe a heat transfer product which is in the form of a heat curing plastic composition, preferred for use as an elongate extrusion or ribbon stripably adhered to a backing release sheet. The plastic composition is an intimate mixture consisting essentially of particulate solids having a thermal conductivity of at least about 100 Btu/in/hr/ft²/°F in an amount of from 15 to about 60%; epoxy resin in an amount from about 15 to 40%; a high temperature curing agent which is inactive or nonreactive at ordinary room temperatures (about 70°F), preferably dicyandiamide, in an amount of from about 1 to 15 parts per hundred of the epoxy resin; nitrile rubber in an amount of from about 2 to 15%; polybutene in an amount of from about 2 to 15%; and a cold flow control agent, preferably asbestos fiber, in an amount of from about 3 to 10%. The percentages are by weight based upon the weight of the composition.

The preferred form of the product can be readily applied to any shape or contour of piping, even in congested spaces or where the application surface cannot be seen (as below or behind piping). The product can be formed and packaged in unit lengths corresponding to lengths of piping or tracing so that there is not required calculation or conversion from units of volume to units of length in estimating quantities. In fact, it can be formed around and in contact with a unit length of piping or wire heat tracing ready for use.

The following materials in the stated proportions (percent by weight, based on the weight of the entire composition) are used in an illustrative embodiment.

Compound	Percent by Weight
Epoxy resin (epoxide equivalent of 185-192, from epichlorohydrin and bisphenol A)	28.51
Nitrile rubber crumbs (medium high acrylonitrile content)	7.89
Graphite (electrode grade, 1% on 20 mesh, 20% through 200 mesh	16.45
Graphite (electrode grade, 1% on 100 mesh, 82% through 200 mesh)	27.41
Polybutene (MW, 1,900-2,500; viscosity at 210°F, 14,660 SSU)	12.72
Asbestos fiber (7 R chrysotile)	5.92
Dicyandiamide	1.10

The foregoing materials are mixed in a sigma blade mixer with incremental additions of components selected between liquid (epoxy resin and polybutene) and solids to maintain a doughy consistency. The dicyandiamide is added last after the temperature of the mixture has been checked to insure it being not over 180°F, and care being taken that the temperature of the mixture not rise above 180°F during mixing in of the dicyandiamide.

The doughy mixture is then extruded through a die having a rectangular cross section of about 3/8" x 7/8" onto a moving strip of silicone release paper about 1 3/4" wide. The resulting tape is wound about a cardboard mandrel to form a flat coil. The material as applied remains doughy and tacky indefinitely at room temperature. After application it need not be cured to serve its heat transfer function. However, in use it will become heated and eventually cured. The rate of cure depends upon temperature, ranging from about 48 hours at 212°F to about 1/2 hour at 400°F.

Underground Heat Pipe

According to a process developed by *E.R. Perry and M. Rabinowitz; U.S. Patents 4,042,012; August 16, 1977; and 4,142,576; March 6, 1979; both assigned to Electric Power Research Institute, Inc.* the heat transfer of a heat pump using the ground as a heat source or sink is improved by surrounding the underground heat pipe with soil containing a plurality of water-soaked absorbent particles to provide a jacket of high thermal conductivity. A preferred form of absorbent particles is a hydrophilic polymeric gel material. To minimize loss of water from the particles, after soaking with water, they may be coated with a water-impermeable film and then mixed with the soil.

In an alternative example, the particles are formed of flexible balloon-like bags filled with water without an absorbent core. Other ways to prevent loss of water from the soil surrounding the heat pipes include laying a water-impermeable film above the pipes, or completely surrounding them.

Recently, various hydrophilic polymeric gel substances have been developed with extremely high water-holding capacities. One such product is described in "Super Slurper-Compound with a Super Thirst", *Agricultural Research*, June 1975. It is a hydrolyzed starch-polyacrylonitrile graft copolymer. One use described for this material is to increase the water-holding capacity of sand to enhance the top growth of crops such as oats. The article states that the sand, by itself, retains only 24 g of water compared with 317 g of water by the sand-gel mixture at a concentration of 1 part gel to 250 parts sand. Such gels are said to absorb as high as 1,000 to 2,000 times their weight of water.

Another type of synthetic hydrophilic gel (Viterra) is suggested to be used as an additive to the soil to assist transfer of water and nutrients to a growing plant. Product literature suggests that the Viterra hydrogel can retain more than 20 times its dry weight of water. Another such product (Imbiber Beads) has a holding capacity of 27:1.

All these polymers have the capacity to take in a large quantity of water without becoming dissolved. The water actually penetrates the polymer network causing the size of the particle to increase, but in so doing, no large pockets of water are formed which might later leak out. The water is actually entrapped by the molecular structure of the polymer. It is extremely difficult to squeeze out entrapped water from the polymer. However, water can be evaporated from the polymers, and the starch-based copolymer is biodegradable.

PLASTICS AND TEXTILE PROCESSING

Cyclic Heating and Cooling of Processing Equipment

R.E. Hinkle; U.S. Patent 4,071,075; January 31, 1978; assigned to American Hydrotherm Corporation describes a heat exchanger system for processing equipment to be heated and cooled. The heat exchangers are used for separately heating and cooling pressurized water, i.e., water under a pressure of at least about 30 psig. The process includes at least two heat recovery vessels where pressurized water is stored in such vessels during initial phases of one cycle for use during initial phases of a subsequent cycle whereby heat transfer fluid at an intermediate temperature level in one of such vessels is passed through the processing equipment prior to the passage of heated or cooled pressurized water from one of such heat exchanger through such processing equipment.

In accordance with the process, the efficiency of heat recovery is substantially improved (40 to 45%) by such two-stage change of water whereby water in the processing equipment is first displaced by tempered water from a first vessel and then replacing the tempered water after exchanging heat with the equipment, with water which is fully heated or cooled with such tempered water being stored in a second vessel, as distinguished from no recovery system.

Referring to Figure 7.13a, the process and apparatus for heating and cooling processing equipment includes user equipment, generally indicated as **10**, such as platen press for decorative laminates, a pump **11**, heat recovery vessels **12** and **13**, hot accumulator tank **14** and a cooler **15**.

Figure 7.13: Heat Exchanger System for Processing Equipment

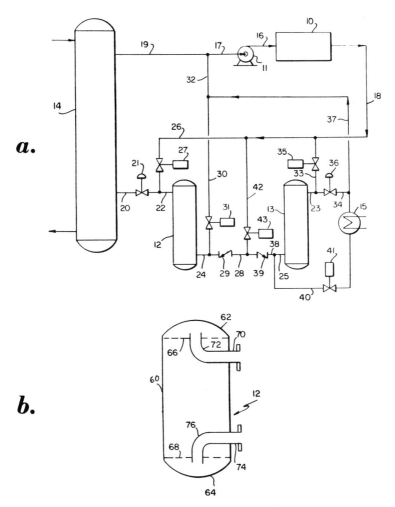

(a)   Schematic flow diagram
(b)   Schematic cross section of a heat recovery vessel

Source: U.S. Patent 4,071,075

The user equipment **10** is connected to the discharge side of the pump **11** by a conduit **16** with the suction side of the pump **11** being connected to a conduit

17. The downstream side of the user equipment 10 is connected to a conduit
18. The outlet of the hot accumulator tank 14 is connected by a conduit 19
to the conduit 17 with the inlet to the hot accumulator tank 14 being connected
to a conduit 20 under the control of valve 21.

The heat recovery vessels 12 and 13 are provided with upper conduits 22 and
23 and lower conduits 24 and 25, respectively. The upper conduit 22 of the
heat recovery vessel 12 is connected to the conduit 20 and to a conduit 26
under the control of valve 27 with conduit 26 being connected to the conduit
18. The lower conduit 24 of the heat recovery vessel 12 is connected to a conduit 28 under the control of a one-way valve 29 and to a conduit 30 under the
control of valve 31 with conduit 30 being connected by a conduit 32 to the
conduit 17.

The upper conduit 23 of the heat recovery vessel 13 is connected to conduits 33
and 34 under the control of valves 35 and 36, respectively. The conduit 33 is
connected to conduit 18 with conduit 34 being connected to a conduit 37 which
is in fluid communication with the downstream side of the cooler 15 and the
conduit 33. The lower conduit 25 of the heat recovery vessel 13 is connected
to a conduit 38 under the control of a one-way valve 39 and to a conduit 40
under the control of valve 41 with the conduit 40 being in fluid communication
with the inlet to the cooler 15. A conduit 42 under the control of valve 43 is
connected to the conduits 28 and 38 and to the conduits 18 and 26.

In operation, assuming initiation of a heating cycle in tempered water has been
previously stored in the vessel 13, in a first stage of a heating cycle, the suction
side of the pump 11 is placed in fluid communication with the upper portion of
the heat recovery vessel 13 via conduits 23, 34, 37 and 17 by opening the valve
36. The discharge side of the pump 11 is in fluid communication with the user
equipment 10 by the conduit 16 with the downstream side of the user equipment
being in fluid communication with the lower portion of the heat recovery vessel
13 via conduits 18, 42, 38 and 25 by opening valve 43 whereby tempered water
in the upper portion of the vessel 13 is caused to be displaced by cooler water
flowing upwardly within the heat recovery vessel 13 since the user equipment had
been operating within final stages of the cooling cycle.

The volume of the heat recovery vessels 12 and 13 may vary between 75 to
125%, but preferably is substantially about equal to the volume of the heat transfer conduits within the user equipment 10 and the conduits to and from such
user equipment. Thus, after a corresponding volumetric replacement, the valve
36 is closed and the valve 21 is opened to permit initiation of a second stage of
the heating cycle whereby hot water from the accumulator tank 14 is introduced
via conduits 19, 17 and 16 into the user equipment 10. Tempered cold water
is withdrawn from user equipment 10 via conduits 18, 42, 28 and 24 and is
introduced into the lower portion of the heat recovery vessel 12 to displace upwardly hotter water from the upper portion of the heat recovery vessel 12 into
the accumulator tank 14. Second stage heating is effected for a time sufficient
to replace the pressurized water in the heat transfer conduits of the user equipment as well as the associated conduits. It will be appreciated that heat-up times
are improved by passing the water stored in vessel 12 to hot accumulator tank
14 via passage through the user equipment.

The final stage of the heating cycle is effected by closing valve 43 and opening valves 27, 36 and 35 whereby hot heat transfer fluid is withdrawn from the accumulator tank 14 by conduit 19 and combined in conduit 17 with recirculating heat transfer medium in conduit 32 and introduced by conduit 16 into user equipment 10. The heat transfer fluid withdrawn from the user equipment 10 in conduit 18 is split with a portion being passed to hot accumulator tank 14 by conduits 26 and 20 with the remaining portion by-passing accumulator tank 14 by being passed by conduits 33, 34 and 37 to conduit 32 as the recirculating heat transfer fluid. The amount of heat transfer fluid by-passing the hot accumulator tank 14 increases as the desired temperature level is reached with concomitant reductions in the flow of fluid from accumulator tank 14.

After a time period determined by the capabilities of the user equipment 10 with regard to the materials being treated, the heating cycle is stopped and first stage of the cooling cycle is initiated. Accordingly, valve 21, 35 and 36 are closed and valve 31 is opened to permit hot heat transfer fluid from the user equipment 10 to be introduced by conduits 18, 26 and 22 into the upper portion of the heat recovery vessel 12 to displace downwardly thereby tempered colder heat transfer fluid therein, such dispersed fluid being passed to the user equipment 10 by conduits 24, 30, 32, 17 and 16.

As discussed with respect to first and second stages of the heating cycle, after a volume of heat transfer fluid is displaced equal to from 75 to 125%, preferably substantially equal to the volume of fluid in the user equipment 10 and related conduits, the second stage of the cooling cycle is initiated by closing valves 27 and 31 and by opening valves 35 and 41. Cooled heat transfer fluid is passed by conduits 37, 32, 17 and 16 to user equipment 10 with tempered hot fluid introduced into the upper portion of heat recovery vessel 13 to downwardly displace cold water which is passed to cooler 15 by conduit 40.

After a similar volumetric change, the final stage of the cooling cycle is effected by closing valve 35 and opening valve 43 whereby cooled heat transfer fluid is passed to user equipment 10 from cooler 15 by conduits 37, 32, 17 and 16 with tempered heat transfer fluid being returned to cooler 15 by conduits 18, 42, 38 and 40. After a time period similarly dictated by process requirements, the cooling cycle is discontinued and the heating cycle initiated as discussed.

As will be appreciated by one skilled in the art, when introducing a hot heat transfer fluid into a recovery vessel, the hot fluid is introduced into the upper portion thereof to thereby downwardly displace relatively cooler heat transfer fluid, whereas when introducing cool heat transfer fluid into a heat recovery vessel, the cooled fluid is introduced into the lower portion thereof to thereby upwardly displace warmer water. It will be further appreciated that mixing in the heat recovery vessels of the heat transfer medium at various temperature levels should be minimized. In this regard, one aspect of the process is concerned with minimizing the mixing of heat transfer fluid at various temperature levels.

Referring to Figure 7.13b, there is illustrated a heat recovery vessel generally indicated as 12 formed by vertically-disposed drum 60 enclosed by top and bottom walls 62 and 64, respectively. Proximate to the top and bottom walls 62 and 64, there are positioned horizontally-disposed perforated distribution plates 66 and 68, respectively. A horizontally-disposed conduit 70 is mounted on the drum 60 at a point below and proximate to the upper perforated distribution plate 66 and is formed with an elbow 72 extending upwardly through and terminating above the upper perforated distribution plate **66**. A horizontally-disposed conduit **74** is mounted on the drum **60** at a point above and proximate to the lower perforated distribution plate **68** and is formed with an elbow **76** extending downwardly through and terminated below the lower perforated distribution plate **68**. It will be readily appreciated that the introduction and withdrawal of a fluid from such a vessel will minimize convection currents therein. It will be further understood that the vessels are completely filled in operation since the process and apparatus relates to the use of water as the heat transfer fluid under pressures of at least above about 30 psig.

The time of a complete cycle (heating, cooling and molding) will vary depending on platen size, materials being treated, the number of layers of materials, etc. Generally, cycle times are about an hour's duration and can be as low as 20 minutes for laminating a plastic to a plywood substrate.

Continuous Transfer of Heat to Moving Solid Material

G.R. Coraor, H.L. Jackson and F.W. Mader; U.S. Patent 3,998,588; December 21, 1976; assigned to E.I. DuPont de Nemours and Company describe a process for continuously transferring heat to a moving band of solid material, e.g., a textile, a film, particles, or any shaped or unshaped article, the band preferably comprising a disperse dyeable synthetic polymer, e.g., a polyester or polyamide. The process comprises passing the band through a substantially enclosed purging-drying region containing the superheated condensable vapor of a liquid having an atmospheric pressure boiling point of less than 120°C, immediately thereafter passing it through a substantially enclosed heat transfer region containing the superheated condensable vapor of a fluorocarbon having a fluorine to carbon atom ratio of at least 1.5, a solubility parameter of not more than 6.5 and a molecular weight of at least about 300.

The pressure in the heat transfer region is no greater than that in the purging-drying region. Immediately thereafter the band passes through a substantially enclosed purging region containing dry steam, the pressure in the purging region being no less than that in the heat transfer region, and thereafter recovering the band of material. The process is characterized by efficient heat transfer resulting from rapid convection movement of superheated fluorocarbon in the heat transfer region and by minimum loss of such vapor to the atmosphere.

Infrared Radiator and Cooling Technique

According to a process described by *O. Rosenkranz, H. Goos and K.-H. Seifert; U.S. Patent 4,076,071; February 28, 1978; assigned to Heidenreich & Harbeck Zweigniederlassung der Gildemeister AG Germany* workpieces of a synthetic plastic material are heated to enable further molding to be effected. Heating is effected by infrared radiators and, in order to ensure that localized overheating is avoided, the surface of the workpiece which is exposed to radiation is cooled during the heating process.

Infrared radiation is eminently suitable for reheating. It enables the plastic material to be heated not only at its surface but simultaneously over its entire depth in a few seconds. Through the absorption characteristic of thermoplastics, a more intense heating occurs in the surface layers which face the radiation source. As can be seen from Figure 7.14a a temperature distribution develops which is determined by the radiation temperature and the specific absorbing properties of the plastic material. The more intense the irradiation, the higher is the temperature peak at **2**. Depending on the material that is to be processed, this can cause an undesired change in or damage to the affected layer, as is shown in Figure 7.14a at **5** and **6**, since the thermal conductivity of thermoplastics is relatively low and the temperature balance within the wall is not achieved immediately, but occurs only after some time, as is evident from the lines **1'**, **1''**, **1'''** which show the temperature distribution at predetermined time intervals after the irradiation process has been completed. This phenomenon is of particular significance in the case of relatively large wall thickness, for example, of more than 3 mm. The improvement of the properties of the material in thermoelastic processing increases considerably with an increase in the extent by which the material can be stretched, for example, stretching by more than 10 times is possible in the case of saturated polyethylene terephthalate (PET) and materials which behave in a similar manner.

In order to achieve specific properties, more especially in the case of pressure tanks, it may be necessary to start with a blank having a wall thickness that is more than 3 mm.

As is known, PET is highly transparent in its amorphous and orientated state. Through crystallization PET becomes opaque and brittle. The rate of crystallization reaches its maximum at temperatures around 150°C.

Figure 7.14: Plastic Molding Technique

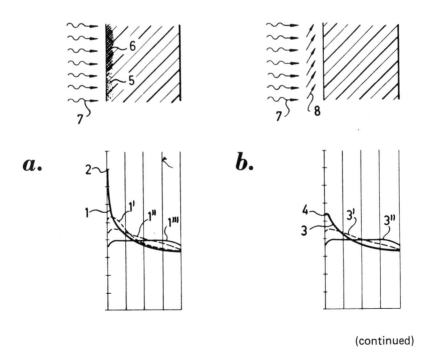

(continued)

Figure 7.14: (continued)

c.

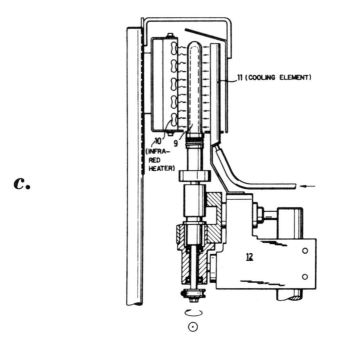

11 (COOLING ELEMENT)

10
9
(INFRA–
RED
HEATER)

12

(a)   Illustrates infrared radiation of a synthetic plastic material without
      simultaneous cooling and the resulting temperature distribution
      curve
(b)   Illustrates radiation with simultaneous cooling and the resulting tem-
      perature distribution curve
(c)   Apparatus for use in carrying out the process

Source: U.S. Patent 4,076,071

During the heating process by infrared radiation, in which the thermoelastic processing stage is reached, the critical temperatures in the vicinity of the surface
can be present for such a length of time that an at least partial crystallization
as indicated at 5 and 6 occurs on the surface of the originally amorphous workpieces. The consequences of this are as mentioned above, and further processing
by orientation is no longer practicable to the desired or otherwise possible extent.

For example in the case of polyvinyl chloride (PVC), polyacrylonitrile, polystyrene (PS) and materials which behave in a similar manner during processing, the
overheating in the irradiated surface layer brings about a release of dissolved substances and/or, by chemical changes, a release of constituents of the macromole-
cules. This can damage the material and/or lead to the formation of gas bubbles
which impair the appearance of the workpiece that is to be produced.

In order to prevent the abovementioned consequences of the temperature peak
2 in the irradiated surface layer, the peaks are, according to the process, reduced
by cooling the surface layer during the heating process. In Figure 7.14b, the
infrared radiation is designated by 8. The temperature peak 4 of the temperature
profile 3 is now considerably lessened; the period before which an approximate
equalization of the temperature profile 3 is achieved is likewise shorter (lines 3',
3"). Consequently, the undesired changes in the surface do not occur. In the
case of PET, it is now possible to reheat semifinished products or blanks that
have a fairly large wall thickness so that they reach the thermoelastic processing
temperature within a few seconds.

Biaxial orientation of blown hollow bodies is becoming of increasing importance,
but it is only an economical process, involving large quantities of materials, if an
improvement of the properties of the material, i.e., higher strength, reduced gas
permeability and better transparency, can be obtained. These properties are necessary in the case of pressure bottles, e.g., for drinks containing carbon dioxide
and for aerosols. PET is now being used as a particularly suitable thermoplastic
material for this purpose.

The starting material for the process comprises cold premolded blanks 9 with
closed bottoms, which can have been manufactured according to any one of a
number of known methods. For the purpose of being heated, the blanks 9 are
disposed in a heating chamber which contains a plurality of infrared radiators 10,
which are arranged on one side or on several sides of the chamber. Preferably,
the blanks 9 pass continuously through the heating chamber which is illustrated
in Figure 7.14c, while being rotated simultaneously about their own axes. A
cooling device 11 for the blanks is fastened to a support 12 on which the
blanks are mounted and by means of which the blanks are guided past the
radiators 10. During the heating process, the blanks are fanned by the cooling
device 11, which is stationary in relation to the axes of rotation of the blanks.

Cooling of Coarse-Grained Mixing Materials

*H. Bremer; U.S. Patent 3,903,957; September 9, 1975; assigned to Gunther
Papenmeier KG Maschinen-und Apparatebau, Germany* described an apparatus
and process for cooling of fine- to coarse-grained mixing materials, e.g., dry-
blended PVC agglomerate or granulate, etc., whereby the individual particles of
mixing material are moved along an essentially closed circular or similar path
and at the same time are passed against cooled surfaces, in conjunction with mixing within a cooling mixer.

This process is carried out with a cooling mixer which has at least one body with
circular base rotatable around its axis and disposed in the inside space of the
container, whereby at least one of the rotating bodies has a surface area closed
upon itself in a known manner, and is disposed parallel, but eccentrically to the
middle axis of the container or while enclosing an angle with the axis in such a
way that a gap is formed between the rotating body and the container wall or
the container bottom or a further rotating body, through which gap the mixing
material is moved.

The surface of the rotating body, advantageously, is movable toward the container bottom, whereby contact pressure means can be provided for the rotating
body opposite the container wall. Effectively, an adjusting and regulating arrangement for the contact pressure means is provided. In this way an adaptation of

the pressure exerted on the particles of the mixing material is made possible, which is an advantage, especially in case of substances with a variable mechanical loading capacity.

The cooling mixer shown in Figure 7.15a has a container **10** with a double side wall **11** through which a coolant flows, and with a double-walled bottom **12** through which coolant also flows, and through which the drive shaft **13** for a mixing tool **14** is guided centrally. On drive shaft **13**, which is driven by a motor disposed below the container **10** and which shaft extends to a location near the upper edge of the container, there sits a bipartite carrier arm **15**, which carries a rotating body **16**. The two parts **17, 18** of the carrier arm **15** are connected with one another on a hinge joint, and their combined length at mutual alignment is sufficient so that even when the surface **19** of the rotating body **16** fits against the container wall **11**, the arms cannot be aligned; they are angular, i.e., their disposition is at an angle.

The two parts **17, 18** of the carrying arm are biased by a suitable pressure agent, such as a spring **20** (Figure 7.15b) or a hydraulic piston **21** (Figure 7.15d) in the direction of their mutual alignment and thus the rotating body **16** is pressed against the inside wall **11** of the container. The rotating body preferably has coolant passages, e.g., a double wall, and coolant flows through it. Its surface **19** is adapted to the contour of the container wall **11** and of the container bottom **12**, especially in their region of transition **22**.

Figure 7.15: Process for Cooling Granular Materials During Mixing

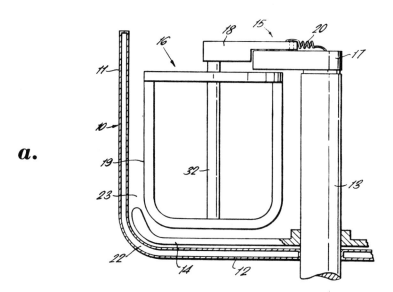

a.

(continued)

Figure 7.15: (continued)

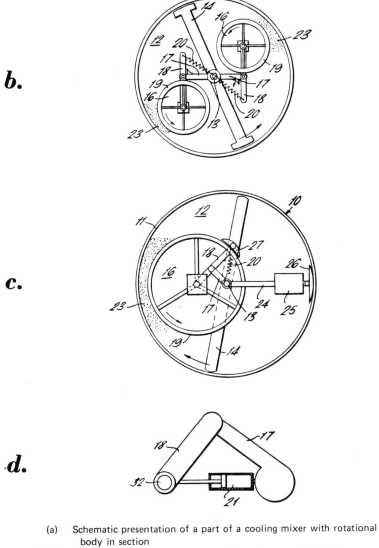

b.

c.

d.

(a)   Schematic presentation of a part of a cooling mixer with rotational body in section

(b)(c)   Views in schematic presentation with various rotating bodies as seen from above

(d)   Contact pressure means for the rotating bodies and an alternative to the one in (b)

Source: U.S. Patent 3,903,957

During rotation of the mixing tool **14** in the container **10**, the rotating body **16** is moved simultaneously through the mixing material via the drive shaft **13** and carrying arm **15**, and rolls off during this movement in relation to the container wall **11**. At the same time the mixing material is drawn into the gap **23** formed between the rotating body **16** and the container wall **11**, and is compacted between the surface **19** of the rotating body **16** and the container wall **11**. Subsequently, whenever the rotating body **16** is again turned farther, the just compacted mixing material is again relaxed and at the same time broken up (loosened up) again by the mixing tool. This breaking up is desirable to enable an easy movement of the rotating body **16**.

At the same time the elastically malleable particles of the mixing material at these surfaces are easily pressed flat and as a result the surface needed for the heat transfer is greatly multiplied. The contact pressure, while being determined by the construction of the device, the rotational speed, the quantity of the mixing material and other characteristics, is also determined by the level of the pressure exerted by the spring **20** or the piston **21**. This pressure can be adapted, sensitively possibly by adjustment of the pressure agent, to the pertinent requirements.

Figure 7.15b shows a top view of a horizontally arranged, schematically shown cooling mixer with two rotating bodies **16**. Their basic structure and arrangement corresponds essentially to the one according to Figure 7.15a. The mixing tool **14** is only schematically indicated and can have hook-like ends. The spring **20** differing from the embodiment according to Figure 7.15a is attached directly on the drive shaft **13**. The mixing material compacted in the gap **23** is indicated by dark coloring.

Figure 7.15c is a top view schematically showing a further embodiment of a cooling mixer **10** which has only one rotating body **16**, the diameter of which is about as large as half the diameter of the mixing container **10**. While this was not necessary for the example according to Figure 7.15b, in this case the carrier arm **15** must be disposed below the rotating body **16**. Since only one rotating body is provided, a counterweight **25** is disposed on a second arm **24**, mounted on the drive shaft **13**, which counterweight rotates together with rotating body **16**. Effectively a stripping element **26**, which glides along the container wall **11**, is disposed at the end of this arm **24** or of the weight **25**, and frees the wall of adhering particles of mixing material. The outer part **18** of the carrying arm **15** likewise carries a stripping element **27**, the surface area **19** of the rotating body **16** moving past it during rolling off movement and thus freeing it of adhering particles of mixing material.

In both cases, according to Figures 7.15b and 7.15c, cooling of the rotating bodies **16** can be accomplished by a fed-in coolant. A hydraulic piston can also be used instead of the spring **20** as a pressure means according to Figure 7.15d. It is possible to operate without a pressure means whenever the outward pressure exerted during rotation of the rotating bodies as a result of the centrifugal force suffices. In such case, adjustment of the pressure to the mechanical loading capacity of the mixing material can be accomplished simply by regulation of the rpm. It can possibly even be effective to provide a spring **20** acting in reverse direction, e.g., between the drive shaft **13** and the rotational axis of the rotating bodies **16**, in order to counteract the centrifugal force.

In order to increase or decrease the quantity of mixing material drawn into the gap **23**, the rolling-off movement of the rotating bodies **16** on the container wall **11** can be replaced by a forced rotational movement of the rotating bodies whereby the driving means can consist of customary toothings, etc. For purposes of cleaning and servicing, the drive shaft **13** can be removable in a known manner. Instead of or in addition to a mixing tool **14**, an arrangement for blowing in and possibly also for sucking off of air or inert gas can be provided, as a result of which the loosening up of the mixing material and the cooling time can be influenced favorably.

Ultrasonic Welding of Thermoplastics

A process described by *C.A. Johnson; U.S. Patent 4,088,519; May 9, 1978; assigned to Home Curtain Corporation* relates generally to the field of ultrasonic welding of synthetic resinous planar materials using a rotating pattern roll and a relatively fixed horn, and more particularly to a means and method for maintaining uniformity of welds during substantially continuous operation.

The process involves controlling the dissipation of heat within the pattern roll, so that thermal expansion is relatively uniform as the roll heats with continued use. The roll is not maintained at any particular temperature, but is forced to warm slowly by dissipation of heat accumulated in the roller through the use of a fluid coolant, the coolant absorbing considerable heat in the process. In this manner, the roller may be continuously operated for long periods of time at temperatures well above initial starting temperature.

Structurally, the result is accomplished using a dual flow rotary union in one end of the roll, connected in series with a small pump and a coolant reservoir. Using a roller of the abovedescribed size, a flow of as little as one gallon per minute provides adequate stabilization. During operation, the roll is approximately half full of coolant, and the inner surface of the longitudinal cylindrical wall thereof is continuously contacted as the roller rotates.

OTHER APPLICATIONS

Bauxite Extraction Process

A process described by *F. Kampf and H.-G. Kaltenberg; U.S. Patent 4,055,218; October 25, 1977; assigned to Vereinigte Aluminum-Werke AG, Germany* relates generally to systems wherein a fluid which forms incrustations undergoes temperature change.

One feature of the process resides in a method of effecting a temperature change of fluids which form incrustations wherein a fluid medium to undergo temperature change is conveyed along a predetermined path. The temperature of the fluid medium is changed along a first segment of the path which is maintained substantially at a first temperature at which the fluid medium forms incrustations at a first rate. The first segment of the path has a first length. The temperature of the fluid medium is also changed along a second segment of the path which is maintained substantially at a second temperature at which the fluid medium forms incrustations at a second rate. The second segment of the path has a length such that the ratio of the first and second lengths substantially equals the ratio of the aforementioned first and second rates at which incrustations form.

The process further provides an arrangement for effecting a temperature change of fluids which form incrustations which includes means defining a flow path for a fluid medium and means for conveying the fluid medium along the path. First temperature changing means, e.g., a heat exchanger, extends for a first distance along the path and is adapted to be maintained substantially at a first temperature at which the fluid medium forms incrustations at a first rate. Second temperature changing means, e.g., a heat exchanger, extends for a second distance along the path and is adapted to be maintained substantially at a second temperature at which the fluid medium forms incrustations at a second rate.

The first and second distances for which the respective first and second temperature changing means extend along the path are in a ratio which substantially equals the ratio of the first and second rates at which incrustations form.

Of special interest is the extraction of bauxite with sodium aluminate lye, particularly the continuous extraction of bauxite with sodium aluminate lye.

An apparatus for the extraction or decomposition of bauxite is, in general, optimally utilized when the various parts thereof which come into consideration become incrusted after the same operating period, i.e., when cleaning becomes necessary for each of the parts concerned at about the same time. However, for a given operating period, the intensity of incrustation normally depends upon the particular temperature involved. In accordance with the process, account is taken of the difference in growth of the incrustations at different temperatures by providing for the heat exchange surfaces of those heat exchangers which tend to be more strongly incrusted to be of larger dimensions than would be selected conventionally.

For example, if the formation of incrustations at 170°C amounts to 1.5 times that which occurs at 140°C, then according to the process, the length of a heat exchanger operating at 140°C would be selected so as to be about 1.5 times the length of a heat exchanger operating at 140°C. Hence, both heat exchangers would become fully incrusted after the same operating period and maximum utilization of the apparatus may be insured since the heat exchange occurring in the two heat exchangers would begin to decrease at about the same time due to the incrustations.

In this manner, it becomes possible to insure that despite the different intensity of incrustation in the various heat exchangers, the final temperature achieved in each heat exchanger may be maintained substantially constant for a maximum period of time so that, as a result, the periods of uninterrupted operation may be maximized.

Fluidized Bed Cooler for Ore Beneficiation

A process described by *W.E. Dunn, Jr.; U.S. Patents 3,960,203; June 1, 1976; and 4,081,507; March 28, 1978; both assigned to Titanium Technology NV, Netherlands Antilles* relates to a fluidized bed cooler capable of effecting fluidization and cooling of finely divided particles introduced at high temperatures without causing oxidization of those particles.

In the process, the thermal expansion of a volatile liquid to its gaseous state is utilized to fluidize the bed. Simultaneously, the heat of vaporization of such liquid is employed to absorb heat from the bed. In most instances occurring in industry, the particles are at high temperatures due to previous processing and can be introduced into the bed at temperatures high enough to volatilize the liquid introduced into the bed and because of film-boiling, without wetting the surface of the particles. If this is not the case, the particles will become wetted as by nucleate boiling, will agglomerate and more conventional means will have to be used.

However, in most cases, the particles desired to be cooled are hot enough to evaporate substantial quantities of water at the maximum evaporation rate without being cooled on their surfaces to such an extent that their surfaces become wetted. Because of the very large difference in temperature between the particles and the water, boiling takes place in the film-boiling regime and heat is removed through a vapor blanketing film by conduction and radiation.

In this process, preferably the ore being cooled has a temperature of 200° to 1100°C, and most preferred, 350° to 1050°C. The liquid preferably is water, but it can be any nonoxidizing liquid depending upon the temperature of the ore.

In its simplest form as shown in Figure 7.16a, the apparatus can consist of a tank 1 into which the very hot, finely divided fluidized particulate material 6 flows from a feed inlet tube 2. Immersed in this material is a pipe 3 having a plurality of perforations 5. A nonoxidizing fluid, e.g., water, is passed through the perforations into the reactor. As the water contacts the hot surfaces of the particles, it vaporizes into steam which serves to fluidize the bed and cool the material in the bed at the same time. As the material is cooled, it is withdrawn continuously through a discharge outlet 4, and the steam which developed passes upwardly and out of the vessel through an outlet tube 8.

Figure 7.16: Fluidized Bed Cooler

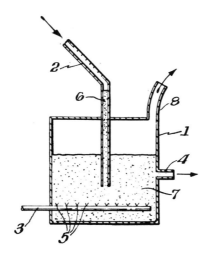

a.

(continued)

Figure 7.16 (continued)

b.

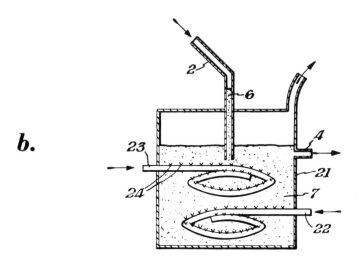

Source: U.S. Patent 3,960,203

Depending on the heat of the material and the volatility of the liquid, it may be necessary to initially fluidize the bed in a conventional manner before introducing the liquid. Accordingly, in Figure 7.16b there is shown another design in which tank **21** has disposed along its bottom portion a pipe **22** with a plurality of orifices in the surface of the pipe.

Similarly disposed is a second pipe **23** with a plurality of orifices **24** through which a volatile liquid can be injected into the tank. When a hot, finely divided material is introduced into the tank, the inert gas is supplied through the orifices of pipe **22** in sufficient volume to fluidize the bed. Once the bed is fluidized, the volatile liquid is supplied through pipe **23**, and as it comes into contact with the fluidized particles, the evolution of its gas will maintain fluidization and the flow of inert gas supply can be shut off or reduced.

Example: In a steel vessel having a diameter of 2.5 ft and a height of 7 ft with an outlet at the 3 ft level, there is placed a 2 inch steel pipe formed into a circle with an outside diameter of 30 inches within the tank. 30 lb/min of partially beneficiated finely divided ilmenite at a temperature of 1000°C is then flowed into the tank. Water at a temperature of 70°C is then introduced into the pipe at a rate of 0.57 gal/min. As the water exits from the orifices in the pipe, the heat of the ore vaporizes the water instantly to steam which fluidizes the bed as it expands and rises to escape. Ore is continuously removed from the bed at the same rate of 30 lb/min and leaves the bed at a temperature of approximately 350°C.

Partially beneficiated ilmenite ore can be produced by passing chlorine gas through a mixture of finely divided ilmenite ore under reducing conditions at a temperature of about 900° to 1050°C as described in U.S. Patent 3,699,206.

Surface Temperature Control in Carbon Black Reactor

T.A. Ruble; U.S. Patent 4,059,145; November 22, 1977; assigned to Sid Richardson Carbon & Gasoline Co. has developed a method and apparatus for controlling surface temperature of a body such as a carbon black reactor without subjecting the body to high pressure fluids.

In the process, a first heat exchange liquid having a relatively high boiling point at approximately atmospheric pressure is circulated in direct contact with the outer surface of the body being cooled to absorb heat from the body. Such first heat exchange liquid (Therminol 77) can preferably operate at temperatures in the range of 300° to 700°F at approximately atmospheric pressure without boiling.

The first heat exchange fluid is then circulated to a second heat exchanger containing a second heat exchange liquid having a lower boiling point and the heat is transferred from the first liquid to the second to cause the second liquid to boil and produce vapor. Preferably the second heat exchange fluid will be water and its vapor will be steam. Escape of the steam generated in the second heat exchanger, or boiler, is regulated by a valve or other pressure control means so that the pressure in the second heat exchanger can be controlled. As the steam pressure increases, the temperature of the boiling water also increases.

Conversely, as the steam pressure is reduced, the temperature of the boiling water will fall correspondingly. The temperature of the first heat exchange liquid circulating between the body being cooled and the second heat exchanger will similarly rise and fall along with the steam pressure in the second heat exchanger. As the boiling water in the heat exchanger increases in temperature, the temperature differential between the water and the first heat exchange liquid will decrease and less heat will be transferred from the liquid during its passage through the second heat exchanger.

Consequently it will return to the first heat exchanger and contact the body being cooled at a higher temperature, removing less heat from the body being cooled so that the body's temperature will also rise. Conversely, lower steam pressure in the second heat exchanger will reduce the temperature of the circulating first heat exchanger liquid and remove correspondingly more heat from the surface of the body being cooled per unit of time. By controlling the pressure of the steam in the second heat exchanger in response to the temperature of the body being cooled, completely automatic temperature control may be maintained.

Figure 7.17 illustrates a preferred example in which the temperature of the outer metal shell of a furnace-type carbon black reactor is regulated utilizing the method and apparatus of the process. A conventional carbon black reactor **10** includes an outer metallic shell **12** lined with suitable refractory material **14**. The refractory material is of normal thickness, as indicated at **14a**, in the combustion zone and a portion of the reaction zone, but there is a reduced thickness, as shown at **14b**, throughout the remainder of the reaction and quench zones.

The portion of the reactor shell **12** for which the thinner refractory lining **14b** is provided is surrounded by a first heat exchanger **16** covered by suitable insulating material **18**. The first heat exchanger **16** is formed by a metal shell **20** concentric with the reactor shell **12** and partitions **22** which extend radially between the reactor shell **12** and the concentric heat exchanger shell **20** to provide a plurality of fluid passageways **24**.

Figure 7.17: Carbon Black Reactor

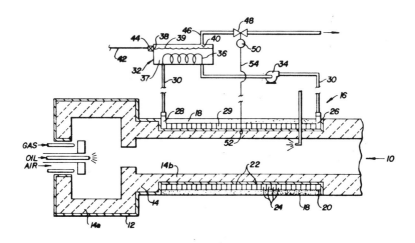

Source: U.S. Patent 4,059,145

Many forms of passageways are available from the extensive art of heat exchange. Preferably, the partitions **22** may be formed by a single member extending radially and spirally around the shell **12** so that fluid passageways **24** are formed in a continuous spiral. An inlet **26** and exit **28** from the heat exchanger **16** are provided so that a first heat exchange liquid **29** may enter the heat exchanger through inlet **26**, circulate through the heat exchanger in contact with the reactor shell **12** to remove heat therefrom and exit from the heat exchanger through outlet **28**.

Suitable conduit means **30** are provided for circulating the first heat exchange liquid from the first heat exchanger **16** to a second heat exchanger **32** and back from the second heat exchanger **32**, through circulating pump **34** to the inlet **26** to the first heat exchanger **16**. In the second heat exchanger **32**, suitable heat exchange means are provided, as indicated diagrammatically by the heat exchange coil **36**, for transferring heat from the first heat exchange liquid **29** to the second heat exchange liquid, preferably water **37**, maintained in the second heat exchanger **32** at a level covering the coil **36**. Heat transferred from the first heat exchange fluid to the water will cause the water to boil, producing steam **39** which is present in the second heat exchanger **32** above the level of the water **37**.

The second heat exchanger **32** is provided with an inlet **38** and an outlet **40** so that water may be supplied to the heat exchanger **32** and the generated steam vented therefrom. Water is supplied through a conduit **42** by means of a valve **44** which is preferably of a type which will automatically maintain a predetermined level of water in the heat exchanger 32. As water in the heat exchanger is converted to steam and vented, additional makeup water will be supplied through valve 44, always maintaining the level of water in the exchanger 32 above the heat exchange coils 36.

Steam generated in the exchanger 32 is vented through outlet 40 into a suitable conduit 46 through which it is preferably conveyed to another location where it is used to perform useful work. Means are provided for controlling the pressure in the second heat exchanger 32 to thereby control the temperature of the second heat exchange liquid (water 37). Preferably such means comprise valve 48 in the steam conduit 46 which may be adjusted to regulate the back pressure of steam in the second heat exchanger 32 to allow such steam to escape as will vary the pressure to maintain the predetermined temperature on the reactor shell.

Operation of the valve 48 is controlled by a conventional valve operator 50 in response to the temperature sensed by thermocouple 52 located at the inner surface of the carbon black reactor shell 12 and connected to the operator 50 by a suitable electrical connection 54. The valve operator means 50 operates to partially close valve 48 and increase the pressure of the steam in heat exchanger 32 in response to a decrease in the temperature of the reactor shell, as sensed by the thermocouple 52, and to open the valve 48 to decrease the pressure in the second heat exchanger 32 in response to an increase in the temperature of reactor shell 12 above its desired point.

This will automatically maintain the temperature of the reactor shell within a desired range since an increase in the steam pressure in heat exchanger 32 will increase the temperature of the water 37, resulting in less heat being removed from the circulating first heat exchange liquid 29, which will then return to the first heat exchanger 16 at a higher temperature. The resulting drop in temperature differential between the reactor shell 12 and the first heat exchange liquid will result in less heat being removed from the reactor so that its temperature will begin to increase.

When the increased temperature reaches the desired maximum, the operator 50 will partially open the valve 48 to decrease the pressure in the second heat exchanger, resulting in lower temperatures for the water 37 and first heat exchange liquid 29 so that the reactor shell 12 will be cooled at an increased rate to a lower temperature and start the control cycle over.

Caustic Cooling System

A process described by *P. Grosso; U.S. Patent 4,077,463; March 7, 1978; assigned to Kaiser Aluminum & Chemical Corporation* relates to an improved cooling system employed for cooling of hot, aqueous caustic solutions, such as obtained in the evaporative concentration of electrolytically-produced dilute caustic solutions. In the production of caustic solutions by electrolysis of aqueous brines, generally a relatively dilute caustic solution is obtained. This aqueous caustic solution frequently has only a 15 to 20% caustic content and is unsuited for most uses. Thus, it needs to be concentrated to a commercially acceptable concentration, usually up to a 40 to 55% caustic level. Evaporative concentration is the most commonly employed process for the removal of a desired portion of the water content of the solution and this is generally accomplished by heating the dilute caustic solution to about 100° to 150°C. The concentrated, hot caustic exiting from the evaporator is generally cooled to a temperature at which it can be handled, but more importantly, cooling is applied to reduce the solubility of caustic-soluble impurities. At lower temperatures, a significant quantity of the dissolved impurities will precipitate and can then be removed from the caustic solution by conventional solid-liquid separation methods.

Cooling of the hot caustic solution has been generally accomplished in agitated, open vessels or tanks equipped with coils through which a cooling medium, usually water, is conducted. For optimum operating efficiency, several series-connected tanks or vessels have been used and the caustic solution temperature has been gradually reduced from vessel to vessel until it reached the desired range.

In this process, a closed-loop pressurized cooling system is provided for the cooling of hot, aqueous caustic solutions containing dissolved brine impurities which, upon cooling, tend to deposit on the surfaces of the cooling system. The system consists of a shell and tube heat exchanger or similar device through which the hot caustic solution is circulated at a relatively high flow rate to obtain a predetermined temperature drop across the heat exchanger tubes. Since the system employs high flow rates and cooling is accomplished by constant recycle of the caustic solution through the heat exchanger, scaling will be significantly reduced with corresponding improved operational efficiency. Due to the use of a closed system, carbonate formation is avoided, resulting in improved caustic quality and yield.

Example: A conventional caustic cooling system, consisting of 7 series-connected agitated open vessels equipped with cooling coils, was first employed for cooling hot, aqueous caustic (NaOH) solution discharged from a multieffect evaporating apparatus. The aqueous caustic solution had an NaOH concentration of about 45 to 55% by weight, had a temperature in the range of about 85° to 95°C, and contained 2.7% by weight dissolved NaCl.

The hot caustic solution was charged to the first vessel of the cooling train at a rate of 504 l/min (133 gal/min) and cooling water was conducted through the coils at the rate of 2,274 l/min (600 gal/min). The cooling water temperature at entry in the coil was about 29°C, and it was discharged from the coil at a temperature of about 33°C. The caustic solution discharged by overflow from the first vessel to the second vessel of the cooling train after an average residence time of about 1 hour had a temperature of about 74°C. Cooling continued in the series-connected system until the caustic solution temperature was reduced to about 26°C.

The first vessel of the train had to be bypassed after about 24 to 36 hours of operation due to the heavy scaling of the cooling coils, causing decreased cooling efficiency indicated by reduction of T. The value of ΔT is calculated by using the formula $T_1 - T_2 = \Delta T$, where T_1 is the temperature of the hot caustic charged to the cooling vessel and T_2 is the temperature of the caustic discharged after the predetermined residence time.

In order to maintain the efficiency of the cooling system and assure the running of the plant, an additional open vessel had to be provided in lieu of the bypassed vessel which underwent cleaning.

To indicate the improved efficiency of the improved cooling system, hot caustic solution discharged from a multieffect evaporator was also cooled in a closed-loop, pressurized circuit. The circuit consisted of a shell and tube heat exchanger, the shell being of approximately 3.65 m in height and having a diameter of approximately 0.61 m. The shell housed approximately 64 tubes of about 5.1 cm outer diameter (16 bung) and of 91.4 cm length, providing a cooling surface of about 37.1 m².

Hot caustic of approximately 72°C temperature was continuously pumped through the heat exchanger at a rate of about 15,141 l/min (4,000 gal/min) and at a pressure of about 1.44 atm (6.5 psig). The ΔT, i.e., the temperature difference between the temperature of the caustic entering the heat exchanger and the temperature of the caustic discharged from the heat exchanger, was kept at about 0.56°C (1°F) by the rapid throughput rate. A portion of the cooled caustic was continuously discharged into a second closed-loop circuit, while the major portion (approximately 97% of the total volume) was recycled through the first shell and tube heat exchanger.

Cooling of the caustic solution was continued in the first circuit until a drop in ΔT indicated that efficiency of cooling was decreased due to scaling of the internal surfaces of the tubes. This was observed after a 168 hour operation period. The volume of hot caustic cooled in the closed-loop, pressurized system without requiring bypass and descaling of the shell and the tube heat exchanger amounted to 600 tons per day. By comparing these results with the cooling efficiency of the abovedescribed conventional system, it became clear that the improved cooling system operates far more efficiently than the conventionally-employed open vessel cooling trains.

In addition to the advantages shown above, the scaled surfaces of the shell and tube heat exchanger can be cleaned more rapidly than the coils and walls of the open vessel system. This is due to the pressurized flow of cleaning solution through the tubes, which at relatively high Reynolds numbers, e.g., above about 2,500, causes a turbulent flow in the system which aids the dissolution of the precipitated salts.

As far as the purity of the cooled caustic solution is concerned, the carbonate content of the caustic removed from the closed-loop cooling system was found to be essentially the same as the carbonate level of the hot, incoming stream. This, in addition to maintaining a constant quality, improves the yield of caustic production due to the elimination of undesired carbonate formation.

Utilization of Waste Energy from Cooking Equipment

According to a process described by *V.D. Molitor; U.S. Patent 4,125,148; November 14, 1978; assigned to Stainless Equipment Company* a gaseous source of waste energy, including heat, is passed through a gas to liquid heat exchanger and then a chiller to agitate a water bath and produce evaporation of water for cooling. The heated transfer liquid is passed through a heat exchanger to heat makeup air or the like, while, alternatively, chilled water from the chiller is passed through the heat exchanger to cool the makeup air. Supplemental heating, as by a furnace, or supplemental cooling, as by refrigeration, may be utilized when called for.

Separate heat exchangers, as for separate areas, are alternatively supplied with heated transfer liquid or chilled water, or one or more supplied with heated transfer liquid and one or more others supplied with chilled water.

The source of waste energy, including heat, is normally fumes and heated air from cooking equipment passed through a grease extraction ventilator. Such a grease extraction ventilator may include water contact means, the water of which may be circulated through the heat exchanger when cooling only, by circulating cooled water produced by the chiller, is desired. Other sources of waste energy may be suitable, such as heated air which has risen to the upper portion of a large room or enclosure, such as an auditorium, theater, meeting hall or the like.

The heated air removed from such a position is adapted in part to be recirculated, but all of it may be used to furnish heat for heating fresh makeup air. Two forms of a specialized double compartment chiller are described, as well as alternate ways of returning circulated water back to the chiller.

Temperature Control for Electrocoating Bath

A process described by *E.A. Russell, R.L. Knipe and J.E. Leaver; U.S. Patent 4,026,775; May 31, 1977; assigned to Kaiser Aluminum & Chemical Corporation* is directed to an improved method of controlling the temperature of a high temperature electrocoating bath, and, in particular, to heating an electrocoating bath to an operating temperature between about 90° and 135°F prior to start-up.

In accordance with the process, the electrocoating bath is passed through an indirect heat exchanger wherein the temperature of the heat exchange surfaces does not exceed 180°F, preferably less than 150°F. In one form of the process, the heating and cooling systems are integrated to provide one operative system. It is preferred to use tap water as the heat exchange fluid because it is readily available and because it facilitates the integration of the heating and cooling systems. The temperature of the cooling medium should be below about 90°F.

Figure 7.18a is a schematic diagram of the electrocoating tank and the bath treating facilities. Electrocoating bath is continuously withdrawn from the electrocoating tank and transferred to a storage or makeup tank. Bath from the storage or makeup tank is withdrawn and pumped through a heat exchanger to control the temperature at the desired operative levels and then to an ultrafilter wherein low molecular weight materials are removed from the bath. The permeate or ultrafiltrate from the ultrafilter, which contains low molecular weight materials, such as organic amines and the like, can be discarded to the drain as shown or further treated to recover the low molecular weight species therein. The treated bath can be transferred back to the holding and makeup tank, returned directly to the electrocoating tank or proportioned therebetween.

In Figure 7.18b, the features of the process are shown in greater detail. As shown in this figure, tap water (or other cool heat exchange fluid) is introduced into line 10 which is split into conduits 11 and 12 containing valves 13 and 14, respectively. These conduits rejoin into a conduit containing a check valve 16 which is in fluid communication with conduits 17 and 18. Conduit 17 directs the tap water directly to the heat exchanger 20. Conduit 18 directs the tap water to pump 21 which is in fluid communication with heater 22.

Figure 7.18: Electrocoating Process

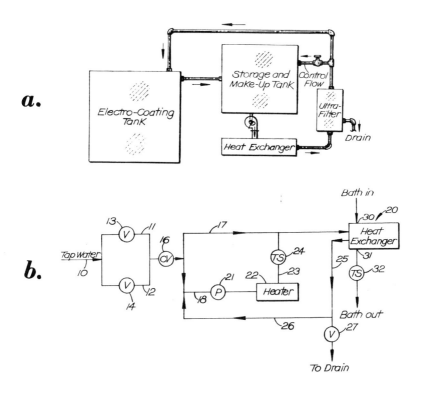

(a) Schematic drawing of the overall bath treatment facilities

(b) Schematic drawing of the temperature control system of the process

Source: U.S. Patent 4,026,775

The heated heat exchange fluid passes through conduit **23** which contains temperature sensing element **24** which is utilized to sense the temperature of the heat exchange fluid and thereby control the temperature at the desired level. Conduit **23** is in fluid communication with conduit **16** so as to direct heat exchange fluid to the heat exchanger **20**. Heat exchange fluid exits the heat exchanger **20** through conduit **25** which can direct the expended heat exchange medium to the drain or return to the pump **21** through conduit **26** depending upon whether valve **27** is open or closed. The electrocoating bath enters the heat exchanger **20** at inlet **30**, passes in an indirect heat exchange relationship with the heat exchange medium and then out of the heat exchanger through outlet **31** where the temperature of the electrocoating bath is sensed by temperature sensor **32**. The indirect heat exchange device can be of any convenient design, such as a conventional shell and tube-type heat exchanger.

For improved bath temperature control, it is preferred in the process to automatically control the heating and cooling system. This can be conveniently accomplished by utilizing the temperature sensed by sensor 32 for control purposes.
Thus, for heating when the temperature of the bath is below a desired level,
valves 13 and 27 will be automatically closed, valve 14 automatically opened and
pump 21 and heater 22 automatically activated, all in response to the temperature sensed by sensor 32. For cooling when the bath temperature is above a
desired level, valves 13 and 27 would automatically be opened, valve 14 automatically closed and pump 21 and heater 22 automatically deactivated in response to the temperature signal.

The system shown in Figure 7.18b was designed to employ an automatically
adjustable valve 13 to provide closer control of the bath temperature by providing a continuous operation rather than an on-off type cooling system. The
valves 14 and 27 can be conventional open-closed, solenoid-operated valves.
With the use of an automatically adjustable valve 13, the temperature sensor 32
generally will provide two signals, one an on-off or open-closed signal to operate
valves 14 and 27, pump 21 and heater 22 and a second signal which adjusts
valve 13 so as to provide the coolant flow necessary to control the bath temperature at the desired level. This latter signal, in accordance with conventional
practice, will be compared by suitable means with a signal representing the desired signal and by control means responsive to the error signal generated by the
comparator control position of valve 13.

Preferably, sensor 32 comprises two sensing elements, one which generates the
on-off signal or open-closed signal and the other the signal for controlling valve
13. A suitable on-off, open-closed sensing element is Temperature Controller
Model T654A1602 (Honeywell). Adjustable valve 13 can be a Silphon valve
(Fulton Valve Corporation). The solenoid valves are of a conventional nature.

During the start-up of electrocoating operations, the bath temperature will be
below the desired level for high temperature operation. In that instance, the
temperature sensor 32 will direct a suitable signal to close valve 13, open valve
14, and actuate the pump 21 and the heater 22. The tap water or other heat
exchange fluid will then pass through valve 14, pump 21, and heater 22, where
it is heated and then passed to the heat exchanger 20. The temperature of the
heated fluid is controlled so that the heat exchange surfaces in the heat exchanger 20 do not exceed 180°F, preferably less than 150°F. The check valve
located in line 15 precludes back flow of water into the supply system. Little
or no fluid is introduced into the system except that needed for makeup. Upon
discharging from the heat exchanger, the expended heated fluid then returns to
the pump 21 through conduit 26.

As soon as the temperature of the electrocoating bath reaches the desired operating level, the electrocoating process is started. The bath immediately begins
to heat up due to the normal heat input of the electrocoating process. When
the temperature reaches a predetermined level, the temperature sensor 32 generates a second signal which activates the cooling system. Valve 14 closes and
valve 13 opens. Pump 21 and heater 22 are shut off. Valve 27 to the drain is
opened. The cool tap water or other cool heat exchange medium then passes
directly to the heat exchanger through conduit 16 and the expended heat exchange fluid is discarded through conduit 25 and valve 27. The volume of tap
water or other heat exchange medium through valve 13 is automatically con-

trolled by suitable means in accordance with the temperature sensed by temperature sensor **32**.

If for some reason the electrocoating process is shut down, e.g., for repairs and the like, the system will automatically switch off the cooling system when the temperature of the bath is reduced below a predetermined minimum level and the heating system will be activated. The heating system thereby maintains the temperature of the bath at the desired levels during the downtime so that no delays occur during start-up and most importantly no significant amount of scrap is generated because efficient electrocoating begins immediately.

The following is given as an example of the process utilizing the system shown in Figure 7.18b. Prior to start-up of the electrocoating operations, the bath is at ambient temperature and tap water at about 60°F is directed to a heater wherein the temperature is raised to about 110°F. The heated water is passed through a shell and tube-type heat exchanger and in an indirect heat exchange relationship with the electrocoating bath to heat the bath. When the temperature sensor determines that the bath temperature is at an operating level at about 108°F, the heating system is deactivated and the electrocoating process is started.

Shortly thereafter, due to the heat generated by the electrocoating process, the bath temperature rises. When the temperature rises 1°F above the operating level, the cooling system is activated and tap water is directed to the heat exchanger for cooling purposes.

Close control of the electrocoating bath temperature is essential because the rates of the various electrocoating phenomena, such as electrophoretic deposition, electrocoagulation, electroendosmosis and the like, are all temperature dependent. Small changes in temperature can generate significant changes in the electrocoating process. It is preferred to maintain bath temperature variation during operations to less than 2°F from a desired temperature.

Olefin Polymerization

In the catalytic liquid phase polymerization of olefins to polyolefins, such as polyethylene, polypropylene, polybutylene, etc., the polymerization reaction is strongly exothermic. Unless the heat generated is removed the temperature of the reaction mass rises above acceptable limits. Accordingly, the reactors are conventionally provided with cooling plate coils which extend through the reactor and which also function as baffles. These coils are cooled with water. However, they are not adequate to prevent reaction temperatures from exceeding acceptable limits. Accordingly, further exothermic heat is removed from the reaction mass by external pump-around cooling which comprises continuously pumping the reaction mass from the reactor through the tubes of an external heat exchanger (cooler) and back to the reactor.

The olefin monomer is polymerized in the reactor in a liquid state and under a positive pressure to retain it in that state. As the polymer is formed it becomes dissolved in the liquid monomer to form a highly viscous reaction mass, the viscosity of which may vary from about 1,000 to 15,000 cp depending on the ratio of polymer to monomer maintained in the reaction mass, which may vary from 10 to 35% by weight.

This very high viscosity of the reaction mass and its nature, i.e., polymer dissolved in liquid monomer, together present the several serious problems in pump-around cooling unique to this reaction system.

According to a process described by *C.D. Miserlis and P.J. Lewis; U.S. Patent 4,089,365; May 16, 1978; assigned to The Badger Company* the problems of fall-out of polymer on heat transfer surfaces with its attendant disadvantages, of high pressure drop across the pump-around cooler with its attendant disadvantages, of the relatively long time required to dry out the system after the reaction is killed with its attendant disadvantages, and of leakage across the head baffles are sharply reduced.

This is achieved by use of an external, single-pass, pump-around tubular cooler, rather than a multipass cooler, having an increased number of tubes of smaller diameter, together with a pump of increased capacity.

The number of tubes of the single-pass cooler, the internal tube diameter and the pump capacity are preferably sufficient to achieve a coefficient of heat transfer of between 10 to 20 Btu/ft^2/hr/°F and a velocity of between 0.3 to 0.7 gallon per minute per tube.

The increased number of tubes in the single-pass cooler varies directly with the polyolefin production capacity of the plant using as a base for calculation between 2,000 and 3,000 tubes per 15,000,000 pounds per year of polyolefin plant capacity.

The number of tubes is preferably increased to a number equal to about the same total number of tubes in all the banks or passes of the multipass cooler.

The length of the tubes is preferably about the same as the length of the tubes in each bank of the multipass cooler so that the length of travel of the reaction mass is reduced by a factor equal to the inverse of the number of passes of the multipass cooler.

The internal diameter of the tubes is reduced (L/D factor is increased, L being length of tubes and D being internal diameter) to increase velocity and heat transfer coefficient. The internal tube diameter is preferably reduced to between $\frac{1}{4}$ and $\frac{3}{4}$, more preferably between $\frac{1}{4}$ and $\frac{5}{8}$ inch.

The pump capacity is increased to also increase velocity and coefficient of heat transfer. Preferably the pump capacity is increased by a factor equal to the number of passes of the multipass cooler to thereby substantially increase the mass flowing through the cooler tubes by the same factor.

It has been discovered that by the use of a single-pass tubular cooler that polymer does not drop out at the ends of the tubes and in the exit head, and that the system can be dried in 24 hours, only $\frac{1}{7}$ or $\frac{1}{8}$ the time required with multipass coolers.

Further, the pressure drop is reduced by the inverse of the number of passes of the multipass coolers with substantial savings in power, the danger of leakage past the head baffles is eliminated and the mass flowing through the cooler is increased.

Heat Wrap Control for Corrugated Paperboard Processing

W.A. Massey; U.S. Patent 4,095,645; June 20, 1978; and D.J. Evans; U.S. Patent 3,946,800; March 30, 1976; both assigned to Molins Machine Co., Inc. describe a heat wrap control system for providing uniform heat transfer to a moving web while automatically compensating for nonlinearities in the amount of web wrap effected in response to web speed. A set point means selectively provides a set point heat signal indicative of the amount of heat to be transferred to a moving web. A speed sensing means provides a reference speed signal indicative of web speed.

Control means operatively associated with the set point means and speed sensing means controls the wrap of the web about the preheater drum to effect uniform heat transfer to the web. The control means includes linear means for controlling the web wrap as a linear function of web speed.

Figure 7.19a shows a preheater which may be used in conjunction with a single facer or double facer machine. For the purpose of this process, the preheater is assumed to be operatively associated with a single facer machine. The preheater as shown in Figure 7.19a includes two preheaters 10 and 12 in tandem with a guard 14 therebetween. Since the preheaters are identical, only preheater 10 will be described in detail.

The preheater 10 includes end frames 16 which rotatably support a drum 18 for rotation about its longitudinal axis. The journals which support the drum 18 may be hollow so that steam or some other fluid such as heated oil may be introduced into the drum 18 to heat the same. Other means for heating the drum such as electrical resistors may also be utilized.

Associated with the drum 18, there are provided idler rollers 22 and 24 pivotable about the periphery thereof. The idler rollers 22 and 24 have a thread-up position as shown in Figure 7.19a. Each of the idler rollers may be moved to position A which designates the minimum or zero wrap position. Also, the idler rollers may have any one of a variety of intermediate positions down to position B which is the maximum wrap position. As web speed decreases, the amount of wrap also decreases to maintain constant heat transfer to the moving web. The moving web is indicated as 26 in Figure 7.19a.

Figure 7.19: Heat Wrap Control System

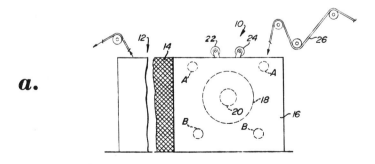

a.

(continued)

Figure 7.19: (continued)

b.

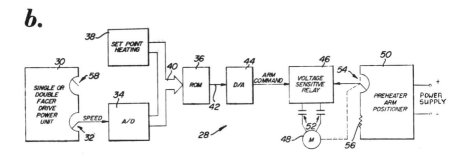

(a)   Side elevation of a preheater
(b)   Block diagram of the heat wrap control system

Source: U.S. Patent 4,095,645

The position of the idler rollers is varied by means of a reversible motor which drives the rollers through appropriate gearing as described in U.S. Patent 3,946,800.

Referring to Figure 7.19b, there is shown a heat wrap control according to the process designated generally as **28**. A single facer drive power unit **30** includes a motor operated potentiometer **32**. The motor operated potentiometer **32** provides a reference speed signal indicative of web speed.

The motor operated potentiometer **32** is connected to an analog to digital converter **34**. The analog to digital converter **34** converts the reference speed signal provided by potentiometer **32** to a digital multibit signal for purposes of addressing a read-only memory **36** as described more fully hereinafter.

A set point heating circuit **38** comprising thumb wheel switches or the like is selectively adjusted by the operator to provide a set point heat signal which is a digital multibit signal also used for addressing read-only memory **36**. The digital signals provided by set point heating circuit **38** and analog to digital converter **34** are combined to form a single digital word at the input **40** to the read-only memory. The read-only memory **36** contains digital information for compensating for the nonlinear relationship between the web speed and the amount of web wrap effected by adjustment of the position of the idler rollers. In effect, the read-only memory is utilized to introduce offsetting nonlinearities to linearize the relationship between web wrap and web speed.

Information accessed in the read-only memory appears as a digital signal at the input **42** to digital to analog converter **44**. The digital to analog converter generates an arm command signal for purposes of controlling a voltage-sensitive relay **46** associated with reversible motor **48** and preheater arm positioner **50**. The reversible motor **48** drives the idler rollers to effect uniform heat wrap control as a linear function of web speed. The voltage-sensitive relay includes normally open contacts **52** connected to the reversible motor. The reversible motor is also mechanically coupled to motor operated potentiometer **54** in the preheater arm positioner.

The preheater arm positioner includes a trimmer resistor **56** in series with motor operated potentiometer **54** and the power supply. By the foregoing arrangement, motor operated potentiometer **54** is mechanically driven by motor **48** to provide a voltage that is the desired function of the position of the idler rollers.

The direction of rotation of motor **48** is dictated by the relationship of the signals communicated to voltage-sensitive relay **46** by digital to analog converter **44** and motor operated potentiometer **54**. The potentiometer **58** in the single facer driver power unit **30** provides a manual adjustment of the voltage across resistor **56** to establish the correct relationship between wiper voltage and single facer speed.

In operation, the single facer drive power unit **30** sets the speed of the single facer machine, hence, web speed. The motor operated potentiometer **32** generates a reference speed signal indicative of web speed. The reference speed signal is used to control reversible motor **48**. The set point heating circuit **38** and analog to digital converter **34** address the read-only memory **36**. The output of the read-only memory, therefore, is a function of the web speed, as indicated by the reference speed signal, and the desired amount of heat to be transferred to the web as indicated by the set point heat signal. The output of the read-only memory introduces a nonlinearity in the relationship between idler roller position and web speed by means of the arm command signal.

Any difference between the arm command signal and the signal provided by motor operated potentiometer **54** causes voltage-sensitive relay **46** to close one of the contacts **52** to activate reversible motor **48**. Reversible motor **48** causes the idler rollers to move between positions A and B or vice versa. As the idler rollers move, motor operated potentiometer **54** is driven by motor **48** to null out the difference in signals appearing at the inputs to the voltage-sensitive relay. The nonlinearity introduced by the read-only memory offsets the nonlinearity between web wrap and position of the idler rollers, particularly in the low wrap region. As a result, web wrap is made a linear function of web speed to enable uniform heat transfer to the web over the full range of control.

Each of the electrical components described above are relatively inexpensive and readily available. The read-only memory may be an integrated circuit programmed to provide the offsetting nonlinearities referred to above. The voltage-sensitive relay may be a solid state relay. Equivalent devices may be substituted without exceeding the purview of the process.

OTHER PROCESSES

Fluidized Bed Reactor

A process described by *A.B. Steever and W.W. Jukkola; U.S. Patent 3,983,927; October 5, 1976; assigned to Dorr-Oliver Incorporated* is directed to a special arrangement for a heat exchanger structure provided to control the operating temperature of the fluidized bed of particulate solids in a fluid bed reactor.

The heat exchanger coil assembly of the process comprises a plurality of vertically-oriented coils or platens in which the lower return bends of the coils are protected from the erosive conditions within the reactor by bed material and the lower return bends have positioning means secured thereto which cooperate with tuyeres fixed to the constriction plate and extending into the fluid bed reaction chamber.

More specifically, the lower return bends of the heat exchanger coils are covered by a layer of quiescent or static bed material which protects them from the erosive conditions that prevail in the volume of the reaction chamber occupied by the fluidized solids. The static layer of solids is established on the top surface of the fluid bed constriction plate and extends upwardly to just below the level of the tuyere ports, the tuyeres extending into the reaction chamber a substantial distance. The particulate solids below the level of the tuyere ports are essentially undisturbed by the fluidizing gas issuing through the ports. In this static layer there is little or no movement of gas or solids. The lateral positioning means for the heat exchanger are plates, welded or otherwise secured to each of the bottom return bends, which extend into close proximity with tuyeres adjacent each return bend.

In one application of the fluidized bed reactor of this process, the reactor is employed to burn coal as a fuel and thereby generate hot gases capable of driving a gas turbine, which, in turn, drives a generator for the production of electric power. The use of combustion gases from the burning coal for driving gas turbines has the disadvantage that the combustion gases often contain large amounts of sulfur compounds which make the gases highly corrosive when contacting the turbine blades. Further, the discharge of noxious gases containing sulfur compounds into the atmosphere is undesirable from the environmental point of view. The combustion gases also contain substantial amounts of solids which are erosive when they impinge upon the turbine blades.

Accordingly, a substantial amount of gas cleaning apparatus must be interposed between the fluidized bed reactor and the gas turbine and traversing this apparatus results in considerable pressure drop and, hence, loss in energy. The apparatus described herein tends to minimize these problems, first, by lowering the sulfur content in the combustion gases, and second, by reducing the dust content of the hot gases. The first objective is attained by proper temperature control in a fluidized bed with appropriate bed solids, while the second objective is realized by using hot dust-free air as one component of the hot gases employed for driving the gas turbine.

Quartz Pebbles Imbedded in Polyester Film

According to a process described by *J.C. St. Clair; U.S. Patent 3,933,195; January 20, 1976* a high efficiency liquid-liquid heat exchanger is made by imbedding 1 mil polyester plastic film in a mass of quartz pebbles. The quartz pebbles are 0.125 to 0.25 inch in diameter and are placed in 0.25 to 0.5 inch thick layers between the plastic sheets. The two liquids flow on alternate sides of the sheets and the flows of the liquids are given a 90° angular spiral flow in relation to each other by strips of plastic cemented between the sheets. In this way a stream tube, or small division of the main flow of one of the liquids, is heated by short elements of a large number of stream tubes of the other liquid and the effects of uneven placement of the pebbles and the resulting channeling of the liquids are overcome.

Heat transfer coefficients as high as several hundred Btu/°F/hr/ft² of plastic surface have been obtained with very low pressure drops. The plastic sheets and quartz pebbles are low cost and the heat exchanger is easily assembled. The heat exchanger can be operated, if desired, at relatively high flow rates and pressure drops, and higher heat transfer rates obtained.

Corona Discharge Cooling Technique

J. Yamaga and M. Jido; U.S. Patent 3,938,345; February 17, 1976; assigned to Agency of Industrial Science & Technology, Japan describe a method for cooling an article by corona discharge caused by placing the article opposite a high-potential electrode and applying high electric potential between the article and the high high potential electrode. The method is characterized by interposing between the article subjected to cooling and the electrode, a corona focusing ring possessed of a central hole defining a centrally-inclined upper face such that when high electric potential is applied between the electrode and the corona focusing ring and the article to generate corona discharge therebetween, the avalanche of ions generated by the electrode in consequence of the corona discharge is focused in the direction of the central hole by the inclined upper face of the corona discharge ring to be released through the central hole onto the article.

Consequently the impinging ions effect local cooling of the article. Since the avalanche of ions is concentrated as described above, this method permits effective cooling of a local part of the article.

When this cooling method is utilized for cooling the edge of a cutting tool in the course of a cutting operation, for example, it is expected to lengthen the service life of the cutting tool to a great extent, decrease the frequency of cutting tool replacement required because of worn edge and improve the productional efficiency to a marked extent.

Localized Corona Discharge

A process which is described by *K.G. Kibler and H.G. Carter, Jr; U.S. Patent 4,015,658; April 5, 1977; assigned to the U.S. Secretary of the Air Force* is concerned with an apparatus for increasing heat transfer from an object at a temperature above ambient by directing a localized corona discharge toward the heated object. Even more particularly, the process relates to an apparatus for increasing heat transfer from an object at above ambient temperature by applying a voltage (+dc or ac) of about 5,000 V between a needle and a small conducting screen near the needle point and placing this assembly near the heated object in air or any other gas. If the object of interest is initially at a temperature lower than ambient, the apparatus may be used to raise the object temperature to ambient.

Thus, the mechanism for cooling by these methods is a phenomenon usually known as the electric, or corona, wind. A nonuniform electric field, such as produced by a charge needle and grounded plane, generates air motion away from the charged point by several electric forces. The significant feature of the effect is that all physical phenomena responsible for the wind occur within usually a millimeter of the charged point. It is therefore not necessary to impose a very high voltage across an air gap of typically centimeters. Rather, the charged probe can be miniaturized, i.e., made physically smaller, operated at considerably lower voltage, placed in close proximity to the heated sample if desirable, and electrically screened to largely shield the heated object and its environment from the electric field.

The referred to miniaturization requires that the electric field be concentrated along the axis of the probe and in the direction of the object to be cooled. An essential feature of the design of the process is that the proper field distribution is obtained by using a grounded screen which subtends a relatively small angle at the probe tip and which is rigidly supported on the probe axis by a thin cylindrical housing of nonconducting material provided with slots to allow the entrainment of air or other cooling gases. The construction provides for a concentrated directional cooling, i.e., the structure is designed to concentrate the amperage or electric current flow in the desired direction, e.g., toward the sample being cooled.

Previously the direction of current flow has usually been dispersed outward from a probe type over a wide angle. Since under the theory of ionic drag the airflow moves in the direction of current flow, the airflow has not been concentrated.

With the process the sides of the probe are an insulator in the form of a tube, in this case cylindrical and longitudinally slotted, and the screen is small. Thus, the current flowing from the probe to the screen is concentrated in the direction of the screen and air currents resulting from ionic drag flowing from the screen to the sample are directionally concentrated toward the sample. If the screen were greatly enlarged or if the probe had screen all around or if no screen is used, air movement is more dispersed and is not the concentrated directional airflow provided by this process.

It is noted that in the prior art, 15 to 60 kV were required for usable cooling. In this process, typically 5 kV are needed although higher voltages can be used including those of the prior art. The advantages are smaller, more compact, less expensive power supply and interconnecting cables, less chance for dielectric breakdown (insulation failure), and greater safety for personnel.

Steam Absorbent Solution

G.C. Blytas; U.S. Patents 4,094,355; June 13, 1978; and 4,102,388; July 25, 1978; both assigned to Shell Oil Company describes a process in which thermal energy derived from a heat source, preferably a low grade heat source, is concentrated and transferred to another medium. If the heat source is steam, it may be used directly, exchanged indirectly with liquid water to make steam, or exchanged or combined with low grade steam to increase the heat content of the low grade steam. If the heat source is other than steam, it may be heat-exchanged with liquid water to produce steam. In any event, even if low grade, the heat source must be steam or be suitable for converting liquid water into steam.

Accordingly, the process involves transferring thermal energy comprising contacting steam and a water-absorbent liquid or solution in a contact zone under conditions to absorb water, form a solution or dilute the solution, and raise the temperature of the solution; and transferring heat from the diluted or higher temperature solution by indirect heat exchange to a medium to be heated, thereby raising the temperature of the medium. The medium to be heated is a matter of choice, but liquids, such as water, are normally preferred. In its preferred form, the process comprises a multistage method of transferring thermal energy. It involves contacting steam and a water-absorbent solution in a first contact zone under conditions to absorb water, dilute the solution, and raise its temperature, the solution being in indirect heat exchange with an aqueous stream. The temperature of the aqueous stream is raised, preferably converting at least the bulk of the aqueous stream to steam.

Diluted solution is then passed from the first contact zone to a second contact zone. The aqueous stream of raised temperature produced in the first contact zone is passed to a second contact zone wherein the stream is contacted with the diluted solution from the first contact zone, under conditions to absorb from water and further dilute and raise the temperature of the solution. This causes the transfer of heat from the further diluted or heavily diluted solution to a medium to be heated.

Preferably, the diluted or heavily diluted solution is regenerated by evaporating the diluted or heavily diluted solution, and then returning the concentrated solution to the contact or first contact zone. The initial heat source is a preferred source of heat for this evaporation, and the process additionally provides a variety of energy-saving features. Use of the process makes possible the recovery of thermal energy at, e.g., 100°C etc., and the concentration and upgrading of the energy to provide, in some examples, the heating of a desired medium to 250°C or higher.

In Figure 7.20a, a thermal energy source **1**, such as a hot waste stream, a geothermal source, or stack gas, enters combination boiler-generator-preheater **2** at **3**. For the purpose of illustration, steam at 150°C will be employed as the thermal energy source.

Figure 7.20: Heat Recovery Process

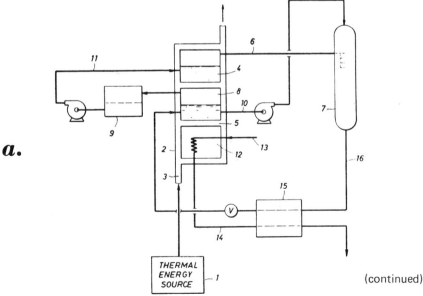

a.

(continued)

Figure 7.20: (continued)

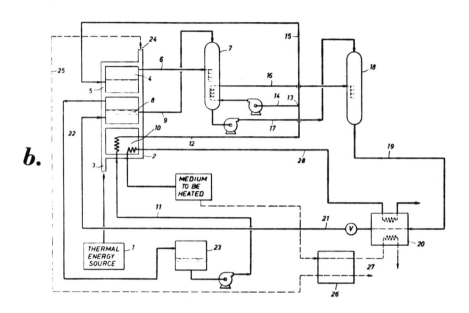

b.

Source: U.S. Patent 4,094,355

Again, although unit 2 is a preferred heat recovery configuration, other units may be used. For example, separate generator and evaporator units may be used. The advantage of unit 2 lies in its excellent heat recovery characteristics. A steam source may be used directly, if contaminants in the steam are not a problem. Moreover, it is within the scope of the process that multiple different temperature sources of heat may be used, so that, e.g., low temperature steam might be supplied to unit 7 to supplement the heat supply. Unit 2 comprises, as shown, a three-sectioned heat exchange apparatus. In section 4 a liquid water supply is converted to steam by indirect heat exchange with the steam passing through the passages 5 of unit 2. Steam at, e.g., 150°C and 65 psia is passed through line 6 to contactor 7 at a rate of 61 lb/hr.

Concomitantly, the steam in passages 5 of unit 2 provides heat for evaporation of water from a regenerating absorbent solution of glycerin in section 8. Section 8 is operated at 150°C and condenser 9 is operated at about 5 psia, so that an absorbent glycerin solution containing, e.g., about 3.5% water by weight is provided in line 10 at 1,000 lb/hr. Water from condenser 9 is passed through line 11 to boiler 4. At the same time, the steam in passages 5 provides heat in section 12 for preheating a medium to be heated in line 13. The medium to be heated in line 13 is preferably water, as indicated previously.

Steam from line 6 and the glycerin solution in line 10 are combined or contacted in contactor 7. Because the vapor pressure of the diluted absorbent solution leaving unit 7 is less than the steam pressure in line 6, the steam will condense or be absorbed, diluting the absorbent solution and raising the temperature of the diluted solution. For the conditions mentioned, the temperature of the solution will be raised to a calculated temperature of about 188°C, and the solution is diluted to approximately 9% by weight.

Concomitantly, water from line **13**, having been heated to approximately 150°C in section **12** of unit **2** is passed through line **14** and pumped in indirect heat exchange unit **15** with the diluted solution from line **16**. The solution in line **16** is pumped through exchanger **15** and returned to **8** for regeneration. Obviously, this heat exchange procedure may be carried out in unit **7** by suitable equipment modification, if desired.

Figure 7.20b illustrates the use of two contact zones in the process, numbers **1** through **8** representing units analogous to those of Figure 7.20a, and the operation of such units will not be detailed again except insofar as necessary to describe the use of a different absorbent solution.

In section **4** of Figure 7.20b, a liquid water supply is converted to steam by indirect heat exchange with the steam passing through the passages **5** of unit **2**. Steam at, e.g., 150°C and 60 psia, is passed through line **6** to contactor **7** at a rate of 125 lb/hr. Concomitantly, the steam in passages **5** of unit **2** provides heat for evaporation of water from a regenerating absorbent solution of LiBr in section **8**. Section **8** is operated at 150°C and about 5 psia, so that the absorbent LiBr solution containing, e.g., about 32.5% water by weight is provided in line **9** at 1,000 lb/hr. At the same time, the steam in passage **5** provides heat in section **10** for water in line **11**. The water in line **11** may be from any water source.

Steam from line **6** and the LiBr solution in line **9** are contacted in contactor **7**. Because, as indicated previously, the vapor pressure of the diluted absorbent solution leaving unit **7** is less than the steam pressure in line **6**, the steam will condense or be absorbed, diluting the absorbent solution and raising the temperature of the diluted solution. For the conditions mentioned, the temperature of the solution will be raised to a calculated temperature of about 200°C, and the solution is diluted to approximately 40% by weight. Concomitantly, water from line **11**, having been heated to 150°C in section **10** of unit **2** is passed through line **12** and may be divided in portions at **13** into lines **14** and **15**. The water in line **14** is pumped through contactor **7** in indirect heat exchange with the LiBr solution.

The temperature of the water is raised to approximately that of the LiBr solution, i.e., about 200°C, and at least the bulk of the water is converted to steam. The aqueous stream leaves contactor **7** through line **16**. Water in line **15** is circulated to section **4** of unit **2**.

From contactor **7**, diluted LiBr solution at, e.g., about 200°C is pumped through line **17** to a second contactor **18**. Concomitantly, at least the bulk of the aqueous stream in line **16** is passed into contactor **18**. Preferably, even if some liquid water is present in the aqueous stream, all of the stream is sent to contactor **18**. Accordingly, the aqueous stream from line **16**, at a pressure of about 210 to 220 psig, is contacted with LiBr solution from line **17** in contactor **18**, with resultant absorption of water, further dilution of the LiBr solution, and raising of the temperature of the solution to a calculated temperature of about 245°C. The concentration of the further diluted solution is about 43.5% by weight LiBr.

The heated dilute LiBr solution may be heat exchanged indirectly with a stream or medium to be heated in contactor **18** or may be passed through line **19** to separate heat exchanger **20**, which may be a direct or indirect heat exchanger, depending on the material to be heated. If the medium to be heated is a water stream, the heat exchange is preferably indirect, and the temperature of the water stream will be raised to approximately the temperature of the LiBr solution, i.e., about 245°C.

The now-cooled, heavily diluted LiBr solution from heat exchanger **20** is passed through line **21** back through an expansion valve to section **8** of unit **2** where it is concentrated for reuse.

The design shown permits additional heat recovery. More particularly, water vapor from section **8** at 150°C and 5 psia, may be passed through line **22** to condenser **23**, where the water vapor is condensed. The liquid water may then be pumped through line **11** thus providing water, as indicated previously, through section **10**, line **12**, etc., instead of depending on a separate source of water. Simultaneously, if steam is the thermal energy source provided by **1**, the steam exiting unit **2** at **24** may be passed through line **25** (dotted line) to preheater **26** wherein the steam is exchanged with a medium to be heated (source line of medium shown as dotted line). The steam or water exiting **26** may be exhausted to waste or used further, while the heated medium may be passed through line **27** (dotted line) to exchanger **20** or contactor **18** where it is heated, as discussed previously.

Preferably, however, the medium to be heated may be heat exchanged or preheated initially with the heat source, e.g., in a modified unit **2**, so that the medium to be heated will already have the heat source temperature before it is passed through line **28** and heated, e.g., in unit **20**. Again, multiple different temperature sources of heat may be used, so that, e.g., low temperature steam might be supplied to unit **7**, while higher temperature steam from an available source might might be supplied to unit **18**.

The following table illustrates calculated temperature, concentration and pressure values from various absorbents and heat sources.

	Absorbent Concentration (wt % of water)	Temperature Rise (ΔT) in Contact Zones (°C)	Pressures in Contact Zones (psia)	Condenser Pressure (psig)
Heat source temperature, 150°C; absorbent—50:50 CaBr$_2$:LiBr in 17 wt % of water				
First zone	24	68	60	5
Second zone	26.5	34.5	260	—
Heat source temperature, 180°C; absorbent—glycerin solution containing 1.5% water				
First zone	11	38	150	3
Second zone	14.1	18.5	210	—

SPECIALIZED APPLICATIONS

CRYOBIOLOGY, BLOOD TREATMENT

Biological Freezer

According to a process described by *T.E. Williams and T.A. Cygnarowicz; U.S. Patent 4,117,881; October 3, 1978; assigned to NASA* blood cells, bone marrow, and other similar biological tissue are frozen while in a polyethylene bag placed in abutting relationship against opposed walls of a pair of heaters. The bag and tissue are cooled with refrigerating gas at a time-programmed rate at least equal to the maximum cooling rate needed at any time during the freezing process. The temperature of the bag, and thus of the tissue, is compared with a time-programmed desired value for the tissue temperature to derive an error indication. The heater is activated in response to the error indication so that the temperature of the tissue follows the desired value for the time-programmed tissue temperature. The tissue is heated to compensate for excessive cooling of the tissue as a result of the cooling by the refrigerating gas. In response to the error signal, the heater is deactivated while the latent heat of fusion is being removed from the tissue while the tissue is changing phase from liquid to solid.

Referring to Figure 8.1a, there is schematically illustrated a biological freezer **11** of a conventional type, such as is available from the Linde Co. Freezer **11** includes a fluid refrigerant source **12**, preferably a source of nitrogen gas. Refrigerant from source **12** is supplied to a volume being cooled by valve **13** that is controlled in response to a mechanical output of controller **14**. Controller **14** is responsive to an error signal from differential amplifier **15** which in turn is responsive to a temperature-indicating signal and a signal indicative of a desired temperature versus time relationship for the volume of freezer **11**, as derived from gas temperature program source **16**. The temperature-indicating input signal to differential amplifier **15** is derived from a thermocouple **17** in the gaseous refrigerant flow stream, which thermocouple supplies an input signal to a conventional thermocouple reference circuit **18**. Thermocouple reference circuit **18** derives output signals that are directly proportional to the temperature of the gas in the volume where the material being frozen is located.

Figure 8.1: Process for Freezing Biological Tissue

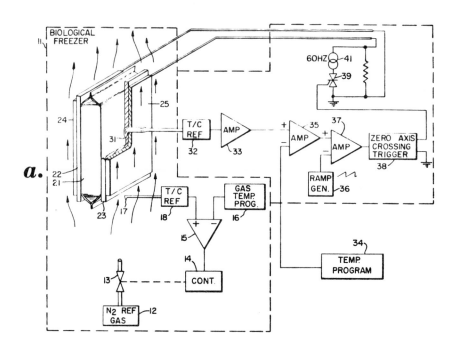

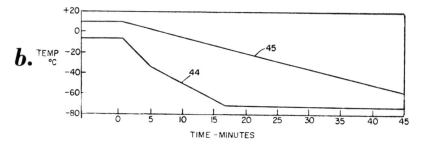

(a)   Partially schematic and partially perspective drawing of the process.
(b)   Graphs indicating the programmed time variations of the temperatures
       of blood being frozen and refrigerant gas supplied to the blood.

Source: U.S. Patent 4,117,881

Typically, in the prior art, blood cells, bone marrow and other similar biological
tissue were frozen, i.e., changed from the liquid to the solid state, by placing
the tissue in a polyvinyl container that was placed in the cooled volume of biological freezer **11**. Typically, in the prior art, the output of gas temperature
program source **16** had a linearly decreasing temperature versus time characteristic. Differential amplifier **15** responded to the outputs of reference circuit **18**

and source **16** to derive a control signal that caused the temperature of the refrigerating gas to follow the programmed value derived from source **16**.

It has been found that the temperature of the tissue being frozen follows the
gas temperature only until the fusion temperature of the tissue is reached. When
fusion begins, typically at −16°C, there is a step increase in the temperature of
the tissue to −13°C. It takes approximately four additional minutes to overcome
the rise in temperature at the fusion point of the tissue. Approximately ten
additional minutes are required before the temperature of the tissue is substantially the same as the temperature of the gaseous refrigerant.

In accordance with this process, these prior art problems are alleviated by deriv-
ing a signal that is substantially equal to the temperature of the tissue being
frozen and controlling the application of heat to the tissue in response to the
monitored temperature of the tissue. Also, the tissue is cooled with gas from
source **12** at a rate at least equal to the maximum cooling rate needed at any
time during the freezing process, a result that is attained by modifying the program of source **16**. The tissue is heated to compensate for excessive cooling of
the tissue as a result of the cooling by the refrigerating fluid. By virtue of a
high-resolution, fast-response time feedback loop between the temperature measuring device and the heater, the heater is deactivated while the latent heat of fusion
is being removed from the tissue while it is undergoing a change of state from
liquid to solid.

To these ends, the tissue is placed in a polyethylene bag **21** having parallel faces
that abut against opposed parallel walls of metal plates **22** and **23**. On the outer
walls of plates **22** and **23** are bonded electric heaters **24** and **25**; to provide a
better bond between the walls of the plates and the heaters, the walls are chemically etched. Heater plates **22** and **23** are arranged so that the gap between them
extends in the vertical direction and bag **21** has its longitudinal axis similarly disposed. Because bag **21** is mounted vertically, in line with refrigerant gas flow
to achieve maximum cooling of the tissue in bag **21**, there is a moderate hydrostatic pressure of the liquid tissue in bag **21** that tends to push the heater plates
apart. However, the plates **22** and **23** are sufficiently stiff, e.g., 0.75 mm thick
of hardened aluminum, so that negligible bending and a uniform thickness of the
tissue results, with uniform cooling of the tissue. Hardened aluminum is also
preferably employed for plates **22** and **23** because of its high thermal conductivity to provide a good heat transfer between heaters **24** and **25** and the aluminum plates and the tissue in polyethylene bag **21**.

During the freezing process, the entire heater assembly, including plates **22** and
23 and heaters **24** and **25**, is held together at the edge of the plates by a fiber
glass channel.

In one configuration, the relatively high thermal conductivity of bag **21** is
achieved by fabricating the bag of polyethylene with a wall thickness of 0.05
mm. The thickness of the bag, between its parallel faces that abut against plates
22 and **23** is relatively small, such as 4 mm, while the planar faces of the bag
can be any suitable size. Bag **21** can be of a type commercially available from
Union Carbide, and is provided with a self-sealing port so that cells can be injected and withdrawn from the bag under sterile conditions. In development
work that was actually performed, bags having planar faces with areas of 13
by 18 cm were designed to hold up to 100 ml of blood cells; other bags having

areas of 6 by 18 cm and 30 by 30 cm were also tested. Because of the construction of the heaters, and the thermal properties of the tissue in bag **21**, as well as the thermal conductance properties of the bag, the temperature of the tissue within the bag is substantially equal to the temperature of the surface of the bag.

In one preferred configuration, each of heaters **24** and **25** includes a copper alloy foil sandwiched between two layers of Kapton film, achieving a total heater thickness of less than 0.2 mm. Such a heater design minimizes the thermal resistance path from the heaters to bag **21** and the cells within the bag and provides a fast response time because of the small thermal heat capacity of the heaters. The maximum power requirement for heaters **24** and **25** depends on the flow rate of liquid nitrogen past bag **21** and the maximum desired cooling rate of the tissue. It was experimentally determined that 280 watts is the maximum power required for the described configuration.

The described configuration of plates **22** and **23**, as well as heaters **24** and **25**, is advantageous because: (1) the thin section of tissue in the 4 mm gap between plates **22** and **23** minimizes thermal gradients within the tissue in container **21**; (2) a standard blood retainer bag configuration can be employed; (3) it is easier to analyze and adapt the system for one-dimensional heat flow through bag **21**; (4) it is easier to install the heater in the refrigerator; and (5) there is a compatibility with readily available, commercial gaseous biological freezers.

To monitor the surface temperature of bag **21** and thereby provide a signal substantially indicative of the temperature of the tissue within the bag, thermocouple **31** is centrally located against one face of the bag. To this end, a hole is drilled through plate **23** and heater **25** to enable the thermocouple to extend through the heater assembly into contact with the center of the face of bag **21**. The temperature-indicating output signal of thermocouple **31** is supplied to a negative feedback loop for controlling the amount of electric power applied to heaters **24** and **25**, and therefore the temperature of bag **21** and the tissue therein.

The negative feedback loop includes a thermocouple reference circuit **32**, of a known type, responsive to the signal derived by the thermocouple. The output signal of reference circuit **32** is a dc voltage directly proportional to the temperature monitored by thermocouple **31**. The output signal of circuit **32** is applied to the input of dc, operational amplifier **33** which derives an output signal that is compared with a time-varying signal derived from source **34**, which signal is indicative of the desired or set point time versus temperature trajectory for the tissue in container **21**. The comparison between the signals derived from amplifier **33** and source **34** is performed by differential amplifier **35**, which derives an error output signal having a magnitude directly proportional to the difference between the inputs thereof.

The magnitude of the error signal derived from amplifier **35** is converted into a rectangular wave having a duty cycle directly proportional to the error signal magnitude. To this end, the output of amplifier **35** is compared with the output of ramp generator **36** in differential amplifier **37**. Ramp generator **36** derives a periodic, saw-tooth wave having a relatively long period, such as 0.5 second. The ramp output of generator **36** has a linear amplitude versus time characteristic over the 0.5 second interval and virtually zero flyback time. Comparator **37** responds to the ramp generator output and the error signal derived from amplifier **35** to derive a signal indicative of the difference between the error signal and ramp. The output of amplifier **37** is applied to a zero axis crossing trigger circuit **38** which derives positive voltage pulses at each positive transition of the load current in response to the difference signal from amplifier **37** being greater than zero; in response to the output of amplifier **37** being equal to or less than zero, a zero voltage level is derived from trigger circuit **38**. Thereby, the number of pulses from the output of trigger circuit **38** is directly proportional to the deviation of the temperature monitored by thermocouple **31** from the desired blood temperature, over one cycle of ramp generator **36**.

The output signal of trigger generator **38** controls an electric switch, in the form of Triac **39**, that selectively connects heaters **24** and **25** to 60 Hz ac power source **41**. Of course, the 60 Hz source cuts off Triac **39** twice during each cycle of the power source that the Triac has been activated into the conducting state in response to the positive voltage pulses derived from trigger circuit **38**. Because power source **41** has a period much less than the 2 Hz output of trigger generator **38**, many cycles of the power source occur during each cycle of the output of the trigger circuit. In the specifically described configuration, there is an opportunity to cut off Triac **39** 60 times during each cycle of ramp generator **36** to provide a resolution of $1/60°C$ and a full power band width of $1/2°C$.

The circuit is capable of going from zero to full power and vice versa in the $1/2$ second period of ramp generator **36**. Thereby, the heat applied by heaters **24** and **25**, and plates **22** and **23**, can be cut off or cut on at a very rapid rate. Because of the linear relationship between the variable duty cycle output of trigger circuit **38** and the error signal derived from amplifier **35**, the amount of power supplied to the heater is proportional to the error signal.

Typical curves for the output signals at gas temperature program source **16** and blood temperature program source **34** are illustrated by wave forms **44** and **45**, respectively, in Figure 8.1b. It has been found that the refrigerant gas and bag surface temperatures actually follow the desired curves indicated in Figure 8.1b. It is noted from curve **44** that the temperature of the gas supplied to the volume being cooled causes the tissue in container **21** to be cooled at a rate at least equal to the maximum cooling rate needed at any time during the freezing process.

For example, initially the temperature of the gas is at approximately –5°C, in contrast to an initial temperature of the prior art method of approximately +20°C. Heat is supplied by heaters **24** and **25** and plates **22** and **23** to compensate for the excessive cooling of the tissue as a result of the cooling by the refrigerating gas. When the latent heat is being removed from the tissue in container **21**, at the time the liquid is being fused to a solid, the heater is deactivated because of the tendency for the tissue to increase in temperature at this time, as indicated by the output of thermocouple **31**. Thus, the step function increase which occurred in the prior art at the time fusion was occurring, no longer is observed.

Conservation of Isolated Organs

According to a process described by *J. Kraushaar and R. Voss; U.S. Patent 4,008,754; February 22, 1977; assigned to Messer Griesheim GmbH, Germany* organs or parts of organs are conserved by being rinsed with an inert gas or mixture until they are free of water and blood. Their vascular systems are filled at slight excess pressure of an inert gas or gas mixture, and finally the organs or parts are cooled at excess pressure in an inert gas atmosphere to a temperature below –100°C and are stored at this temperature.

The suitability of the process, for the conservation of organs or parts of organs, becomes evident from the following examples which refer to the conservation of two pig kidneys. Both kidneys originated from the same animal, and were taken out in the slaughterhouse by a butcher en bloc. The time between the death of the animal and the taking out of the organs could not, as in all similar cases, be established. Since, however, in the meantime, the pig had to be scoured, shaved, divided, and taken out, one has to assume from one-half to one hour, especially, since in the customary course of the slaughtering, unavoidable delays occur.

Both kidneys were then kept for two hours in ice, until the continuation of treatment. Thereafter, they were separated and the one (K 19) was rinsed free of blood by means of cool (+4°C) gravity perfusion (1,000 ml of Ringer solution + 2.5 ml Liquemin). In the case of the other (K 20), there was used under the same conditions 1,000 ml of so-called Collin's solution. The duration of the perfusion amounted for the one (K 19) 87 minutes, for the other (K 20) 57 minutes.

To examine the vascular system and for the conditioning of the organs, they were subjected only to a mechanical, pulsating permanent perfusion (K 19, 3 hours; K 20, 3.5 hours). The perfusion solution consisted thereby essentially of 10% rheomacrodex with Liquemin and OH-ion-excess.

The amount flowing through amounted for both kidneys to 180 ml/min. After 2.5 hours, the pressure, which had to be applied to maintain this flow, measured for K 19 84/62 torrs (perfusion No. PZ: 2.46), for K 20 95/75 torrs (PZ: 2.11). After that K 19 showed constant behavior, K 20, with respect to the pressure, a dropping tendency (up to PZ: 2.6). By perfusion number there is thereby to understand the relation from flow passing through per minute to the medium pressure, with omission of the dimensions.

After the machine perfusion, the vascular system of both kidneys was rinsed about 10 minutes via its arterial connection with helium under a pressure of 100 torrs until it was free of water; then the system was filled with the gas at a slight excess pressure, with the veins and arterial stump closed. Then, above the ureter, the kidney base was filled to 50 torrs with helium and closed.

The kidneys, thus prepared, were then introduced jointly into a container being impermeable to gas, and were then stored therein under nitrogen with a pressure of 2 atm. The whole system was cooled by dipping it into liquid air, where it was stored for more than 16 hours. Then the container was opened, the kidneys taken out, and thawed again by irradiation at invervals with microwaves. The thawing process took for K 19 about 1 hour, for K 20 55 minutes.

Subsequently, both preparations were subjected to a new machine perfusion. Its procedure was the same as the preceding one.

The amount flowing through amounted for both kidneys again to 180 ml/min. The respective pressure measured always after 60 minutes for K 19 54/34 torrs

(PZ: 4.09), for K 20 90/75 torrs (PZ: 2.18). K 19 showed again a constant behavior, K 20 a dropping tendency (to PZ: 2.70).

For the morphological investigation, both organs were perfused with about 5% formalin-solution, and placed in formalin. After a critical diagnosis, the following finding resulted.

Both kidneys were well preserved, K 20 somewhat better than K 19. The rind structures were preserved optimally. Some middle pieces had fresh epithelial necrosis and the formation of homogeneous cylinders. Besides, Henle's loops occurred in the marrow, which show the same kind of epithelial necrosis and cylinders. In spite of the long time of perfusion, the interstitial edema was limited. The glomerule and the containers were not altered (alternated).

On the whole, also for the pig's kidneys, there were ideal results. Under the circumstances the degree of necrosis of the severed kidneys was not bad for the prognosis taken.

Surface-Attached Living Cell Cultures

K.O. Smith; U.S. Patent 3,943,993; March 16, 1976 describes a process for cooling, storing and reviving surface-attached living cells which includes the steps of: (1) exposing living cells which are attached to a surface of a storage vessel to a storage medium which is relatively nontoxic to such cells and which is compatible during cooling to a temperature below 0°C, storage and subsequent warming processes; (2) cooling the storage vessel and the surface-attached living cells therein to a storage temperature below 0°C effective for arresting the metabolism of the cells; and (3) storing the vessel and the surface-attached cells at the storage temperature for an extended period of time.

In its more specific aspects, the processes and package of the process include the use of a storage medium which contains between about 5 and 40% of either dimethyl sulfoxide, glycerol or ethylene glycol. A cooling rate of about 1°C per minute is particularly effective and storage temperatures between about 0° and –200°C have been found to be highly useful. A particularly useful storage temperature is about –70°C.

One of the important objectives of this process is to preserve surface-attached living cells in situ (in their cultured position) and thus minimize cell manipulation, with the associated shock and damage to the cells. By the process, shaking, scraping, agitation and exposure to enzymes or other chemicals normally required for suspending the cells are avoided, so that revived cells require less time to recover from the shock associated with storage. In fact, cells stored according to the process frequently can be used for viral studies or other experimentation immediately after thawing.

Example: Living cells (hereafter called "cells") are made to detach from the surfaces upon which they have been cultured. This is done by scraping, mechanical agitation or by treatment with solutions of enzymes (such as 0.2% trypsin) or chelating agents (such as EDTA).

Suspended cells are counted in a hemacytometer, pelleted by low-speed centrifugation (approximately 2,000 *g* for 5 minutes), and resuspended in an appropriate cell culture growth medium to give a concentration of about 100,000 cells per ml. Lesser concentrations (25,000 to 50,000 cells per ml) may be used in the case of rapidly growing cells; greater concentrations may be used in the case of slow growing cells, or when a rapid outgrowth of cells to a high density is desired.

Cell suspensions are dispensed into the type or types of culture vessels most convenient for the purposes intended, in volumes appropriate to the size of the vessels. Examples of vessels are test tubes (about 1 to 2 ml cell suspension per tube), multidepression plastic trays (about 0.2 to 2 ml cell suspension per depression), glass or plastic bottles (5 to 100 ml cell suspension per bottle) or Petri-type plates (1 to 5 ml cell suspension per plate).

Cells in the culture vessels are incubated at a temperature optimal for the cell type (about 37°C for many mammalian cells and 20° to 29°C for fish and reptile cells). In the case of unsealed vessels, a gaseous atmosphere having a composition favorable for cell culture is provided (a favorable atmosphere usually contains about 5% CO_2 and a high humidity).

Cells are incubated for periods of time sufficient to allow cell attachment to the surfaces within the vessel, and to allow sufficient cell division to give a concentration of cells suitable for use in specific types of work (this may vary from a sparse, low-density outgrowth of cells to a crowded, continuous sheet of cells).

The growth medium is drained or aspirated from the cell culture vessel, and replaced with a "storage medium" containing about 10% by volume dimethyl sulfoxide prepared in cell culture growth medium. The volume of "storage medium" is variable, but at minimum, the volume must be sufficient to bathe the attached cells.

The "storage medium" is allowed to bathe the cells for about 15 minutes before proceeding with the cooling procedure.

The "storage medium" may then be removed by draining or by aspiration, in the case of glass culture vessels or substrates, leaving only that "storage medium" which naturally adheres to the cell surfaces. Alternatively, "storage medium" can be left in the culture vessels to bathe the attached cells. The latter is optional when cells are cultured and adhered on glass substrates but is preferred when the cells are adhered to plastic surfaces.

The groups of cell culture vessels are then placed in an insulated container or, alternatively, the vessels are individually wrapped with insulating material. The insulated vessels are placed in a refrigeration unit, such as a –70°C freezer. If the culture vessels contain a significant volume of the "storage medium", the vessels are oriented in the freezer so as to allow the "storage medium" to bathe the cells. The cell culture vessels containing the surface-attached cells are then cooled slowly (at about 0.5° to 1°C per minute) so as to reach a temperature below –40°C. Alternatively, but less preferred, cell cultures with thin layers of "storage medium" can be placed into a refrigeration unit without insulation and frozen more rapidly.

The frozen cell cultures are kept in storage at a low temperature indefinitely. Temperatures lower than –60°C are preferred for long-term storage. Temperature fluctuations in the refrigeration unit are to be avoided.

The cell culture vessels containing the surface-adhered cells are withdrawn from storage when needed, insulation (if present) removed from the vessels and vessels warmed rapidly to a temperature of about 37° to 41°C. This is done by exposing the vessels to warm water, a warm surface or moving warm air. Sudden or excessive movements of the vessels during the thawing process should be avoided to prevent pieces of ice and moving fluid from dislodging attached cells.

The thawed "storage medium" is drained or aspirated from the culture vessel, a cell culture growth medium is added gently in appropriate volumes (see above) and the cell cultures are incubated and/or used as desired. Ideally, cell culture vessels should not be agitated mechanically by bumping or by shaking excessively after growth medium is added, until the cultures have incubated for a few hours (12 to 24) and have resumed their normal metabolic activity.

Types of cells which have been successfully stored by this method include human embryonic brain cells, human embryonic lung cells, canine kidney cells, human amnion cells, monkey kidney cells and various human epithelial cell lines, both neoplastic and nonneoplastic.

Blood Oxygenator

Extracorporeal circulation is and has been a routine procedure in the operating room for several years. An important component in the extracorporeal blood circuit is a heat exchanger used to lower the temperature of the blood prior to and during a surgical procedure and subsequently rewarm the blood to normal body temperature. The cooled blood induces a hypothermia which substantially reduces the oxygen consumption of the patient. The published literature indicates that the oxygen demand of the patient is decreased to about one-half at 30°C, one-third at 25°C, and one-fifth at 20°C. Light (33° to 35°C), moderate (26° to 32°C), and deep (20°C and below) hypothermia are commonly used in clinical practice. Hypothermia is used to protect the vital organs including the kidneys, heart, brain and liver during operative procedures which require interrupting or decreasing the perfusion.

A number of different structural configurations for heat exchangers have been used in the extracorporeal blood circuit including hollow metal coils, cylinders and plates through which a heat transfer fluid (typically water) is circulated. A survey of a number of different types of heat exchangers used in extracorporeal circulation is included in the book entitled *Heart-Lung By-Pass* by P.M. Galletti, M.D. et al, pages 165 to 170.

A process described by *J.E. Lewin; U.S. Patents 4,138,464; February 6, 1979 and 4,065,264; December 27, 1977; the latter assigned to Shiley Laboratories, Inc.* relates to a heat exchanger for an extracorporeal blood circuit formed by an aluminum tube having one or more integral hollow ribs along its length and having substantially its entire exterior surface electrolytically oxidized to form a hard anodized coating. This tube in turn is formed in an overall helical configuration and mounted between an inner cylindrical column extending within the helically configured tube and an outer cylindrical shell. Both the column and the shell are sized such that peripheral portions of the rib are in contact with or are closely proximate to the exterior wall of the column and the interior wall of the cylindrical shell. The method employed for regulating the temperature of blood using this type of heat exchange element involves flowing a heat transfer fluid through the tube and hollow rib and flowing the blood in a counterflow direction over the exterior surface of the ribbed tube. The combination of the rib and the contacting surfaces of the cylinder and chamber confine the flow of blood substantially within paths of restricted area and extended length provided by the hollow ribbing.

The heat exchanger of the process provides several significant advantages. Thus, its performance factor is very high due to the long residence time of the blood, the high conductivity of the heat exchange tube, the direct contact of the blood with the hard anodized surface, the counterflow operation, and high flow rate of the heat transfer fluid through the ribbed tube.

Heat exchangers constructed in accordance with the process have the reliability necessary for routine use in open heart surgery and other procedures utilizing extracorporeal circulation. The anodized metal heat transfer fluid tube is an integral member which may be completely tested, both before and after assembly into the blood chamber, for leaks under substantially higher fluid pressures than are ever encountered in an operating room environment. The integral nature of the heat exchange tube also provides an important advantage in that only the ends of the tube pass through the wall of the blood-carrying chamber, thus minimizing the number of openings in the chamber which must be hermetically sealed.

Moreover, no connections need to be made to the tube within the blood chamber since a heat transfer fluid inlet and heat transfer fluid outlet are provided by the ends of the tube extending out from the chamber. Any leak at the connection of the heat exchanger tube and the fluid supply conduit will merely leak water or other heat transfer fluid external of the blood chamber.

The hollow ribbed heat exchanger tube may be mounted within a blood chamber separate from the blood oxygenator or may be incorporated integral with the blood oxygenator, e.g., in the venous side within the blood-oxygen mixing chamber or in the outlet side within the defoaming chamber. In typical units in which the heat exchanger is incorporated within the mixing chamber of a bubble oxygenator the flow of the blood and blood foam through the lengthy paths of restricted cross-sectional area contributes to the blood-gas transfer process.

The heat exchangers are sufficiently economical in terms of material and manufacturing costs so that it is disposed of after use, thus avoiding the problems and cost of sterilization in the hospital. In addition, the heat exchangers may be made biologically inactive and compatible with human blood. A significant advantage of this process is that the anodized exterior surface of the metal heat exchanger is compatible with human blood.

Thus, no other coating such as a plastic coating is required over this anodized tube so that the blood may flow directly in contact with this surface and achieve a very high performance factor.

AUTOMOTIVE APPLICATIONS

Fan Shroud Design

H.D. Beck; U.S. Patent 3,872,916; March 25, 1975; assigned to International Harvester Company describes an internal combustion engine, having a heat exchange cooling system, a fan for moving air through and a shroud and shroud exit section for controlling the air path. The shroud exit encloses the fan and includes throat (CF), radial expander (R) and radial flat (RF) sections whereby air is drawn through the heat exchanger axially and expelled radially along the exit sections. The fan has a projected axial width (AW) such that a general relationship exists with the shroud exit sections: CF = AW/3, RF = AW/3, and R = 2AW/3.

In Figure 8.2a, there is shown a conventional water-cooled heat-producing internal combustion engine means **10** forwardly carried on longitudinally extending parallel support means **12** of vehicle means **14**. As shown, vehicle means **14** is a tractor; however, this process can be applied to any type of vehicle employing a heat-generating internal combustion engine or any other portable or stationary device requiring an air-moving fan. Forwardly mounted is a water-cooling radiator means **16** employed to dissipate the engine-generated heat. Water flows between the water jacket on the engine and the radiator through a series of fluid-communicating means **18** and **20**. In this particular example, sheet metal means **22** encircles engine means **10** thereby forming the engine compartment area means **24**.

Figure 8.2: Fan Shroud Exit Structure

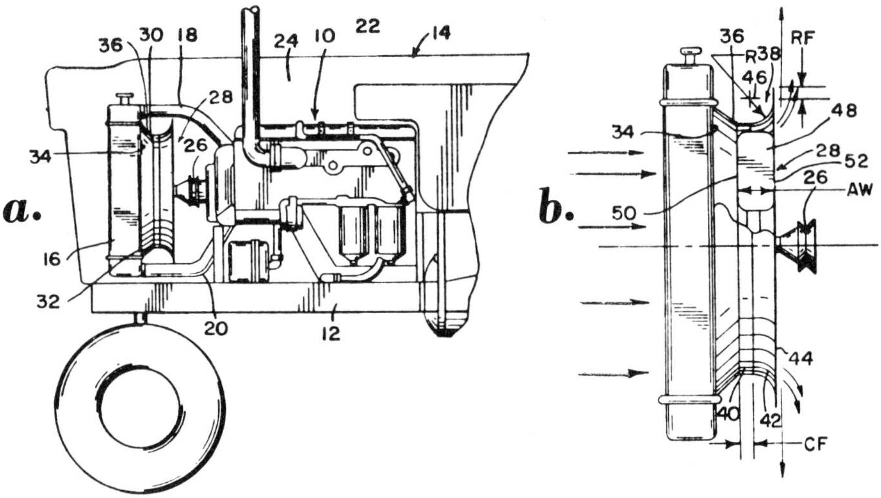

(a) Side elevation of an internal combustion engine showing the fan shroud attached to a vehicle.

(b) Fragmentary vertical section showing the relationship of the fan to the contoured exit section.

Source: U.S. Patent 3,872,916

Carried at the forward end of engine means **10** is a fan shaft means **26** whereby power is delivered to drive fan means **28**. As is apparent, the particular mode whereby power is transmitted is not critical and belts and pulleys could also be employed. As employed here, fan means **28** is a rotatable suction fan positioned opposite the radiator means **16**, and normally creating a flow of air or drawing in a stream of cooling air rearwardly through the radiator with a subsequent axial discharge thereof. This axial flow of air is directed to the fan means by a shroud means **30**. The particular shape of the forward section **32** is dependent upon the shape and design of the perforated heat exchanging design of the radiator. The nature of the connection between the leading edge of **32** and the rear face **34** of the radiator will be dependent upon the particular characteristics of these components, that is, some connections being provided with air gaps while others are substantially sealed over the entire circumference of the enclosure.

In the preferred form of this process, the entire perforated area is substantially sealed against the passage of air from any other direction except through the radiator. From the forward edges the shroud means **30** (be it a taper transition as shown or a box type) converges rearwardly to a circular rear section **36**.

Referring to Figure 8.2b, wherein is more clearly shown a shroud exit means **38** extending rearwardly and outwardly from shroud edge **36**, the connection between the shroud and the shroud exit can be achieved by any suitable means; however, it is desirable that such connection be relatively free of gaps or spaces which would allow the passage of air. Exit shroud means **38** includes a tubular means **40**, an arcuated means portion **42** and a flat flange portion means **44**. For the most part tubular means portion **40** forms the leading edge of the exit shroud means while arcuated means portion **42** still extending generally rearwardly simultaneously extends outwardly around an arch, the reference point of which is defined as point **46**. That is, arcuated section **42** has a general bell-shaped appearance, being a section of a transition surface or some approximation thereof.

In the preferred case arcuated section **42** is a section of a constant radius arch. Flat flange portion **44** forms the trailing edge of exit shroud means **38** and has a major plane perpendicular to that of tubular section **40**. For purposes of simplicity, tubular means **40** will be hereafter referred to as the cylindrical throat means, arcuated portion **42** will be referred to as the radial expanding means and flat flange portion **44** will be referred to as radial flat means. Overall the entire fan shroud exit means **38** has a horn-like configuration.

As previously stated, fan means **28** is rotatly carried adjacent the radiator means and operably to establish a flow of cooling air therethrough. Fan means **28** includes a plurality of fan blade means **48** (only one shown) as is well known in the art. As shown in Figure 8.2b, fan means **28** is surrounded by the contoured fan shroud exit section **38**. The enclosure of the fan means **28** within shroud means **30** is such that a front plane struck out by the leading edge **50** is coextensive and passes through the leading section of throat means **40** and a rear plane struck out by trailing edge **52** is about coextensive and parallel with the radial flat portion **44**. It should be noted, however, that there is a plus or minus error factor involved in both of these values of about 12% of AW. That is, the respective planes formed by the blade means can be within about 12% of optimum and still function satisfactorily within the scope of this process. Thus, within this range the direction air stream will still be substantially radial.

It has been determined, however, that best results are obtained when the front plane struck out by leading edge means **50** passes through the juncture point between converging shroud **36** and the throat section **40**. Even more determinative on the result is the relationship between the rear plane struck out by trailing edge means **52** and the radial flat portion **44**. Overall performance is achieved when the rear plane and the radial flat portion **44** are coextensive and parallel. Deviations from this orientation cause the air stream to change more rapidly than corresponding percentage changes in the front plane location.

The following relationship exists between these parameters: RF = AW/3, CF = AW/3 and R = 2AW/3 where RF is the length of the radial flat portion **44**, CF is the length of the cylindrical throat section **40** and R is a radius of the radial expanding section **42** or distance from the reference point to the transition surface and AW is the projected axial width of fan **28**.

Although not obvious from a simple consideration of the layout the cooling assembly embodies herein, horsepower savings and noise reduction are realized. The geometry of the shroud exit and positioning of the fan therein so effects the cooling characteristics of the assembly that fewer revolutions per minute of the fan are necessary to achieve the same temperature reduction of the coolant. Thus, decreasing the fan speed yields a reduction of fan-generated noise and power required to drive the fan. Because of the radial discharge of the air stream, dust and particle matter are swept away from the operator and not back on him. The same is true for the heat which has been passed to the air; it issues away from the operator station.

Internal Combustion Engine Coolant

J.T. Ciezko and C.L. Moon; U.S. Patent 3,939,901; February 24, 1976; assigned to White Motor Corporation describe an improved method and apparatus for increasing the heat transfer capabilities of an engine coolant system utilizing ethylene glycol-water mixtures while assuring effective degassing or deaeration of the coolant passing through the radiator.

It has been discovered that the practice of increasing the number of radiator core tubes, or otherwise increasing the heat transfer areas of the radiators such as by lengthening the tubes, has had retrogressive effects on the efficiency of engine coolant systems although cooling capacities have been increased by such practices. Put another way, the increase in required radiator size has been disproportionately large as compared to the resulting increase in radiator heat transfer effectiveness.

It has been found that the kinematic viscosity of a coolant composed of one-half ethylene glycol and one-half water, by volume, is roughly three times that of water at engine operating temperatures, e.g., about 200°F. This difference in viscosity results in significantly different Reynolds numbers when the respective coolants flow through a radiator core tube at the same velocity, with the Reynolds number of the ethylene glycol-water mixture being relatively low compared to that of the water. The Reynolds number of such a coolant is frequently below 2,100 resulting in the coolant passing through the core tubes exhibiting laminar flow characteristics.

The heat transferred to or from a cylindrical tube is generally proportional to an exponential power of the Reynolds number of the fluid flowing through the tube. Moreover, the magnitude of the exponent is larger when Reynolds numbers are high, and coolant flow is turbulent, than when the Reynolds number is below 2,100 and flow is laminar.

The range of Reynolds numbers from about 2,100 to 10,000 is a transitional range from laminar to turbulent flow and heat transfer characteristics of the fluid are generally thought of as indeterminant. However, when Reynolds numbers are 5,000 or greater, the flow through radiator core tubes has been found to be sufficiently turbulent to assure favorable overall heat transfer coefficients for the radiators when compared to the heat transfer coefficients of the same radiator with coolant flows establishing Reynolds numbers substantially less than 5,000.

It has been found that if the heat rejecting ability of a given radiator through which coolant passes in the laminar flow range is inadequate, very little improvement in heat transfer is realized by lengthening the core tubes. If the number of core tubes is increased, the coolant flow area is increased and the coolant flow velocity is reduced with the result that the Reynolds number is further reduced. This reduces the amount of heat transfer from a given core tube.

Experimentation has revealed that significantly improved heat transfer is obtained by passing the ethylene glycol-water coolant through core tubes at velocities which insure turbulent flow of the coolant in the core tubes. In particular, high heat transfer rates have been obtained when flow through the core tubes produces Reynolds numbers of 5,000 or more. The differences in heat transfer are sufficiently great that the size of many automotive radiators can actually be reduced without reducing the amount of heat which the radiator is capable of transferring from the coolant and without altering any remaining components of the cooling system such as the coolant pump.

The achievement of a coolant velocity through a standard size radiator core tube, i.e., a tube having a hydraulic radius of ¼" or ⅜" to achieve the desired minimum Reynolds number of 5,000 can be determined in terms of gallons per minute per tube. The number of core tubes required to reject the desired amount of heat from the coolant can then be determined empirically.

In a preferred case, an engine is provided with a multipass radiator in which the Reynolds number of the coolant flowing through the radiator core tubes is 5,000 or more in the normal operating speed range of the engine to assure substantially turbulent flow of the coolant in the core tubes and thus provide efficient heat transfer from the coolant without substantially affecting the capacity requirements of the system coolant pump. The radiator also maximizes heat transfer from coolant passing through the radiator core tubes by positively degassing the coolant.

In one preferred example of the process, a vertical two-pass radiator is provided having radiator inlet and outlet openings in a bottom tank. The bottom tank is partitioned to provide separate header chambers. First and second groups of core tubes extend from the respective bottom tank header chambers to a top tank. The top tank defines a coolant reservoir, a head space above the coolant level for gas collection and a partition or baffle plate which defines a header chamber into which all of the core tubes open. The header chamber is constructed and arranged so that coolant directed into the header chamber from the first group of core tubes is decelerated and exhibits laminar flow characteristics as it flows toward the second group of header tubes. The coolant is fractioned off from the chamber by the core tubes of the second tube group and as the coolant passes through the tubes of the second tube group it is accelerated to produce Reynolds number of no less than 5,000.

The baffle plate defines an enlarged header chamber portion adjacent the second group of core tubes having a cross-sectional area which is sufficiently large that the coolant in this portion of the header chamber is substantially quiescent. This construction therefore enables entrained gas or air to rise through the quiescent coolant to the top of the enlarged chamber portion. A vent tube communicates the enlarged chamber portion with the head space in the top tank reservoir so that detrained gas or air is exhausted from the enlarged chamber portion.

The enlarged chamber portion also functions to facilitate filling of the system when another vent tube is connected between the chamber portion and an elevated location in the coolant system at which air in the system would otherwise be trapped during filling. This air flows to the reservoir head space from the chamber.

In a preferred construction, a static pressure line extends between the top tank reservoir and the inlet of the cooling pump. The static line functions to minimize coolant pump cavitation by assuring a positive coolant pressure at the pump inlet, particularly when the engine is operating at high speed and coolant temperatures are high. The static line also functions to provide pressure communication between the top tank reservoir and the pump inlet so that gas or air detrained from the coolant is positively expelled from the header chamber portion into the coolant reservoir head space during filling.

In another example of the process, the radiator is constructed with horizontal core tubes and vertically extending side tanks. The radiator outlet and inlet openings are located in the upper and lower portions, respectively, of a first partitioned side tank. The opposite side tank defines the coolant reservoir and head space and includes a deaerating baffle plate which defines a header chamber for communicating the core tube groups. The baffle plate defines an enlarged header chamber portion adjacent its upper end for providing a region of substantially quiescent coolant from which gas or air detrained from the coolant is communicated to the reservoir head space.

The horizontal tube version of the radiator may also include a static pressure line extending from the reservoir to the coolant pump inlet for enabling positive deaeration of the header chamber portion and other functions mentioned above.

Electrostatic Field Applied to Cooling System

A development described by *O.C. Blomgren, Sr., O.C. Blomgren, Jr. and F.J. Lyczko; U.S. Patent 3,872,917; March 25, 1975; assigned to Inter-Probe, Inc.* relates to methods and apparatuses for increasing the coefficient of heat transfer between a solid surface and a heat transfer medium or between a heat transfer medium and a solid surface, or of increasing the coefficient of heat transfer of both surfaces of a heat exchange conduit or tube, one surface of which contacts a heat transfer medium of one temperature and the other of which contacts a heat transfer medium of another temperature.

The heat exchange apparatus includes a surface through which heat is to be transferred and conductors or probes that have a high-voltage low-current dc potential with respect to the surface and are spaced a distance from the surface such as to prevent arcing at a given potential while creating a strong electrostatic field between the surface and the conductors. A voltage source of 30 to 60 kilovolts produces the potential difference between the conductors or probes and the surfaces treated where a small ionic current in the low microampere range results in treating the surface with a low power economically. This arrangement results in a greatly increased rate of heat transfer through the surface by the electrostatic field surface bombardment.

In the process, the application of a high voltage dc potential to the conductors near the surface creates a strong electrostatic field that aids in the transfer of heat through the surface without requiring further apparatuses. For example, when conductors along the central axis of a condenser tube of a steam condenser or along the central axes of the cooling tubes of a radiator are energized with a high-voltage dc potential while the walls of the tubes are grounded, heat is transferred from the steam to the walls of the condenser tube and from the coolant to the walls of the cooling tubes at a greater rate than before the conductors were energized. Similarly, when probes spaced outside the cooling tubes are energized with a high-voltage dc potential, heat is transferred from the walls to the surrounding air at a greater rate than before the probes were energized.

Self-Cleaning, Water-Cooled Radiator

The water-cooled radiator employed on a construction vehicle, such as a log skidder, is subjected to plugging due to the ingress of debris. Such plugging is a particular problem with respect to radiators comprising closely spaced fins for the purpose of improving cooling capabilities and/or for reducing radiator core size. Proposed solutions to the plugging problem have included constructing the radiator with less than nine fins per inch and mounting screens in front of the radiator to screen-out debris prior to its ingress into the radiator. Such arrangements require close attention and periodic cleaning to maintain the cooling capacity of the radiator at an acceptable level.

J.M. Bailey; U.S. Patent 4,125,147; November 14, 1978; assigned to Caterpillar Tractor Company describes an economical and noncomplex method for maintaining a radiator free of debris.

The method comprises rotating an endless and perforated belt about the radiator and simultaneously twisting the belt 180° during each rotation to form a Mobius strip, whereby the front and back sides of the radiator are each exposed to both sides of the belt during rotation.

Figure 8.3a illustrates a conventional water-cooled radiator 10 having a self-cleaning screen assembly 11 of this process associated therewith. An engine driven fan assembly 12 of the blower type is adapted to be mounted in an engine compartment on the inboard or back side of the radiator to aid in blowing air through. Screen assembly 11 will function to pick up any debris prior to its ingress into the radiator and automatically discharge the same on the outboard or front side and exteriorly of a vehicle. Modifying the assembly **11** a suction fan could be used.

Figure 8.3: Water-Cooled Radiator

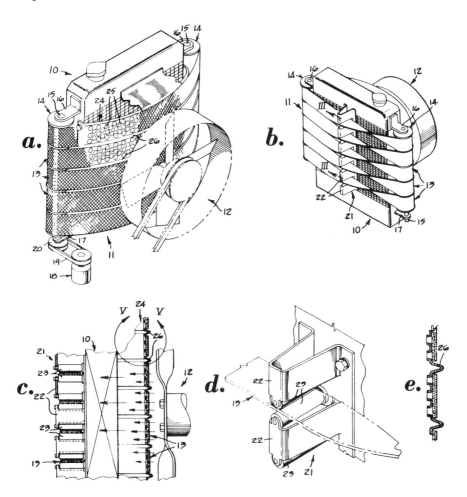

(a)   Air inlet side isometric view of a radiator having a self-cleaning screen assembly.

(b)   Air outlet side isometric view of the screen assembly.

(c)   Sectional view of the radiator and screen assembly, generally taken in the direction of arrows III—III in Figure 8.3b.

(d)   Enlarged isometric view of a pair of guide members employed in the screen assembly.

(e)   Enlarged sectional view, generally taken within circle V—V in Figure 8.3c.

Source: U.S. Patent 4,125,147

The screen assembly illustrated in Figures 8.3a through 8.3e comprises a plurality of horizontally disposed and perforated endless belt means or belts 13 which circumvent the radiator. Each belt may comprise a fine mesh screen construction, composed of rubber, rubberized fabric, high-strength stainless steel, plastic or any other suitable material which will flex and provide the other operational desiderata required. The lateral ends of the flexible belt engages and is guided by vertically disposed first guide means or rollers 14, mounted on either lateral end of the radiator. In particular, each roller is secured to a vertically disposed shaft 15 having its opposite ends suitably rotatably mounted in vertically spaced brackets 16 and 17, secured to a respective side of the radiator. Alternatively, a separate roller may be provided for each belt 13.

Figure 8.3a further schematically illustrates drive means for continuously rotating the belts during operation of the vehicle. Such drive means may comprise an electric, hydraulic or air-actuated motor 18. The motor may be suitably mounted on the frame of the vehicle to have its output shaft drive a belt 19, entrained about a pulley 20 secured on the lower end of shaft 15 to rotate a roller 14. Alternatively, a small windmill-type device, driven by the air stream emanating from fan assembly 12, could be employed as the drive means to rotate roller 14.

As shown in Figures 8.3b through 8.3d, belt means 13 are supported on the front side of the radiator by a vertically disposed first guide means comprising a bracket 21. The bracket is secured intermediate the lateral ends of the radiator and has a plurality of equally and vertically spaced guide members 22 extending forwardly therefrom. Each pair of vertically adjacent guide members has a roller 23 rotatably mounted thereon to define a guide opening for engaging and guiding a particular belt.

Each belt is preferably half-twisted, as clearly shown in Figure 8.3b, to form a Mobius strip. Such twisting will function to turn the belt completely over as it moves past the front side of the radiator to expose both sides thereof to the air stream forced therethrough by fan assembly 12. Thus, debris carried by the belts will be blown off the belt and dumped exteriorly of the vehicle. It should be further noted that the arrangement of guide means 21 may further induce a slight fluttering of the belts to further aid in the dislodgement of debris therefrom. However, the clearance defined between each pair of adjacent rollers 23 is sufficiently small to minimize vibrations imparted to the belt while yet permitting a slight fluttering due to its extended length between the guide means and each roller 14.

Referring to Figures 8.3a, 8.3c and 8.3e, each belt is guided on the cooling air inlet side of radiator 10 by a horizontally disposed third guide means comprising an arcuate guide plate 24. The guide plate is secured at its lateral ends to the ends of the radiator with the intermediate mid-portion of the plate being spaced at a maximum distance from the radiator. As shown, the guide plate may be stamped from a relatively thin sheet of metal to have a plurality of holes 25 and may have an upper plate and a bottom plate secured between it and the radiator to prevent the ingress of foreign materials. In addition, a plurality of vertically spaced and horizontally disposed guide ribs 26 are formed integrally on the plate with each pair of vertically adjacent ribs receiving a belt 13 therebetween.

As more clearly shown in Figure 8.3e, ribs **26** function to entrap and to prevent vertical movement of the belt. Holes **25** are formed sufficiently large to assure a relatively unrestricted flow of air and to provide a minimal metal contact with the belt means. If so desired, the metallic portions of the guide plate, engaging the belt, could be chrome-plated or provided with any other suitable low friction surface thereon (e.g., Teflon) to further reduce the coefficient of friction between the belt and the guide plate.

In view of the above description, it can be seen that the self-cleaning screen assembly of this process permits a relatively close spacing of the radiator fins together (e.g., more than nine fins per inch) and also assures that the radiator will function substantially up to its full cooling capacity, even when operating in heavy dirt and debris-laden environments. The half-twist imparted to each belt **13** aids in the cleaning function, as abovedescribed and also exposes a different side thereof to guide plate **24** upon each complete rotation of the belt means to reduce belt wear by approximately one-half. Also, by providing a substantially open structure at the air exit side, pressure loss normally caused by the screen is reduced by approximately 50%.

Air Brake Systems

J.D. McKenney and T.R. Rumsey; U.S. Patent 3,865,180; February 11, 1975; assigned to Royal Industrial Inc. describe a tank for conveying a hot, moisturized, pressurized fluid stream which is constructed for cooling the fluid stream and separating out the moisture. The tank is provided with an internal passageway for conveying the fluid stream therethrough and conveying the fluid stream adjacent the inner wall of the tank in a heat exchange relationship. The tank wall is constructed of a good thermal conductor with the associated fluid passage wall constructed of a good thermal insulator. With the passage of the fluid stream through the tank, the heat is transferred to the outer wall of the tank and radiated to the ambient air by means of fins, or the like.

The tank of the process is adaptable for the storage of various pressurized fluids wherein the pressurized fluid conveyed to the tank is hot and may contain moisture and is to be applied to a utilization system as a cool, dry, pressurized fluid. In a specific example, the tanks may be employed for an air brake system on trucks and trailers in lieu of the "wet" tanks.

A conventional air-controlled braking system is described in U.S. Patent 3,515,438. The tank of this process may be incorporated into such a pressurized air source in accordance with the general piping arrangement illustrated in Figure 8.4a. For this purpose, the tank **10** may be considered as the drying tank of the fluid-operated system and is connected by means of the dry tank **14** and the conduit **10a** to the air-operated brakes illustrated as a box **11** in Figure 8.4a. The drying tank **10** is coupled to receive the fluid or pressurized air conveyed thereto from a conventional compressor **12** which is coupled thereto by means of a conduit **13**. The drying tank **10** is coupled to the dry tank **14** by means of the conduit **15**. The dry tank **14** is coupled to an emergency tank **16** by means of a conduit **17** having a one-way check valve **18** connected in series between the tanks **10** and **16**. The tank **16** may be coupled to the fluid-operated controls for the brake system as more specifically described in U.S. Patent 3,515,438.

The important aspect of this control system is that the drying tank **10** will receive the hot, pressurized, moisturized air stream from the compressor **12** and
apply a cool, dry pressurized air stream to the air-operated brakes **11** as is desired.

Referring to Figures 8.4b and 8.4c, the specific structural organization of the
tank **10** may be examined. The tank **10** comprises the cylindrical shell **20** preferably constructed of a material having good thermal conductive properties such
as aluminum. The ends of the cylindrical shell **20** are enclosed by a pair of bell
heads **21** and **22** secured to the opposite ends of the shell **20**. The bell heads
21 and **22** are also preferably constructed of good thermal conductors compatible
with the material for the shell **11**. In one particular illustrated example of the
bell heads **21** and **22**, apertures for conveying the pressurized air through the
tank **10** are provided. For this purpose, the bell head **21** is considered the entry
end of the fluid system and the entry aperture **21a** is internally threaded and is
illustrated with the conduit or pipe **15** secured thereto for conveying the pressurized fluid into the tank **10** in a system of the type illustrated in Figure 8.4a.

Figure 8.4: Air Brake System

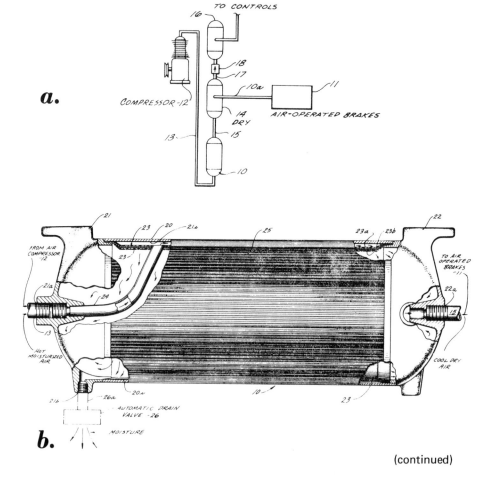

(continued)

Figure 8.4: (continued)

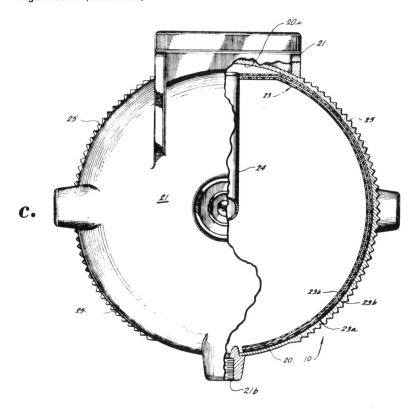

c.

    (a)   Schematic illustration of an arrangement of storage tanks utilized
          in a conventional motor vehicle braking system illustrating the
          use of the tank of the process in such a conventional braking
          system.
    (b)   Longitudinal elevational view, with portions broken away, and
          portions illustrated in cross section and elevation of a tank and
          an element illustrated in dotted outline.
    (c)   Left-hand elevational view of the tank of Figure 8.4b with portions
          broken away and the internal elements illustrated in cross sec-
          tion and elevation.

Source: U.S. Patent 3,865,180

The tank **10** includes a liner **23** secured adjacent the inner wall **20i** of the shell
20 at the joints between the shell **20** and the bell heads **21** and **22**. The liner
23 may be secured at the joints by welding, or the like. The liner **23** is preferably constructed of a thermal insulator for rejecting heat towards the wall **20i**.
The liner **23** illustrated in Figures 8.4b and 8.4c comprises a thin, flexible,
metallic substrate **23a** having insulative plastic coating **23b** coated on the opposite sides of the metal substrate **23a**. The plastic insulation preferably is a vinyl
coating. The liner **23** is constructed and defined to be resilient with respect to

any foreign objects such as ice that may be conveyed into the fluid conduit defined between the inner wall **20i** of shell **20** and the liner **23**. This resiliency and yieldability is provided so that no obstruction will result in the fluid passageway to prevent or restrict the passage of pressurized fluid or air therethrough. The liner movement is illustrated in Figure 8.4b in dotted outline. The pressurized fluid conveyed through the tank **10** is conveyed to the internal passageway through the provision of a fluid conduit **24** coupled to the entry aperture **21a** to receive the pressurized air from the conduit **15** and convey it upwardly as illustrated in Figure 8.4b for conveyance along the inner passageway described hereinabove.

For this purpose, the pressurized fluid will be conveyed into the passageway and circulated around and in contact with the inner wall **20i** of the shell **20** as it is conveyed longitudinally of the tank so as to be expelled or conveyed out of the tank through the exit aperture **22a** provided for the exit bell head **22**. The exit aperture **22a** is defined and substantially similar to the entry end of the tank **10** and has the exit conduit or pipe **10a** coupled thereto. The outer wall of the shell **20** is constructed and defined with a plurality of heat-radiating fins **25** extending longitudinally and on opposite sides of the tank **10** as best illustrated in Figure 8.4c. The fins **25** are good thermal conductors and provide a large surface area for radiating any heat that is derived from the pressurized stream by means of the inner wall **20i** to the ambient air.

The fluid conduit defined with the inner wall **20i** of the shell **20** has a cross-sectional area that is selected to optimize the heat transfer characteristic versus the pressure drop characteristic of the fluid conveyed to the tank **10**. To this end the fluid may be conveyed at the rate of 20 or 40 feet per second through the tank **10**. These rates of conveyance would provide sufficient velocity to effect the desired heat exchange to the shell **20** without a substantial drop in the pressure of the conveyed fluid. In a practical example, the space between the inner wall **20i** and the adjacent surface of the liner **23** has been selected on the order of 0.0030 to 0.0040 inch. This spacing would allow a velocity on the order of 40 feet per second through the tank to effect the desired heat exchange and thereby cooling of the fluid.

The liner **23** extends essentially around the entire inner wall **20i** of the shell **20** and as best illustrated in Figure 8.4c, has an open section adjacent the bottom of the tank **10** to allow any moisture, ice or foreign objects to collect along the bottom of the tank. The moisture, then, that is separated from the fluid stream as a result of the separating action afforded by the internal passageway will be collected at the bottom of the tank.

For the purpose of draining the moisture collected at the bottom of the tank **10**, the entry bell head **21** may be provided with a drain aperture **21b** coupled by means of a conduit **26a** to a commercially available automatic drain valve **26**. The automatic drain valve **26** is illustrated in dotted outline and is operative to remove the moisture collected at the bottom of the tank and eject it into the ambient air. It should also be noted that, although the liner **23** is illustrated and described as being a plastic insulative coating that it may be constructed of other materials that have thermal insulative properties and yet sufficiently flexible to allow it to yield to the fluid stream to prevent any restrictions or blockage in the internal passageway. For this purpose, the liner **23** may be constructed completely of plastic such as glass-reinforced nylon, polyester, polysulfone, or silicone.

The thermal properties that are important with respect to the liner **23** is that it have a high thermal insulative property relative to the thermal conductive properties of the shell **20** so as to reject any heat in the fluid stream conveyed between the shell wall **20i** and the liner. For this purpose it has been found that a ratio of 2,000 to 1 or similar very large differences is effective for producing the cooling action desired.

From the above, the passage of the fluid stream through the tank **10** should be evident. In summary, it will be noted that the hot, moisturized, pressurized air as it is received from the compressor **12** is conveyed by means of the entry aperture **21a** and the conduit **15** to the internal conduit **24** and delivered adjacent the inner shell wall **20i**. This pressurized fluid stream will be conveyed around the internal surface of the shell **20** and progress longitudinally toward the exit bell **22** of the tank **10**. In traveling from the entry bell **21** to the exit bell **22**, the heat of the fluid stream will be transferred to the outer shell wall and to the fins **25** to be radiated to the ambient air. The fluid stream is thereby cooled as it travels toward the exit aperture **22a** and the conduit **10a** to the air-operated brakes **11**.

During the interval that the pressurized stream is being cooled, the moisture is separated therefrom and will be deposited at the bottom of the tank and collected at that point. The automatic drain valve **26** or any other similar arrangement will be effective for removing the moisture from the tank and ejecting it into the ambient air. The tank **10** then will couple the cool dry air to the air-operated brakes to allow them to operate more efficiently as a result of the provision of such a tank **10** in an air-operated system.

AIRCRAFT APPLICATIONS

Electrostatically Cooled Brake

C.P. Han; U.S. Patent 3,952,846; April 27, 1976; assigned to The Bendix Corporation describes an apparatus for cooling a disc brake or the like wherein a plurality of electrically conductive probes are arranged in circumferentially spaced-apart relationship radially outwardly from frictionally engageable disc portions of the brake and adapted to direct a high-voltage, low-amperage electrical flow to the frictionally engageable disc portions which are of opposite electrical polarity and separated from the probes by a predetermined air gap. A suitable high-voltage electrical source is connected to the probes and the frictionally engageable disc portions to establish electrostatic discharge across the air gap.

Referring to Figure 8.5a, numerals **10** and **12** designate inboard and outboard annular wheel portions, respectively, of a conventional aircraft wheel generally indicated by **14**. The wheel portions **10** and **12** are provided with radially inwardly extending web or arm portions **16** and **18**, respectively, which terminate in associated hub portions **20** and **22**. The hub portions **20** and **22** are adapted to rotate on annular bearing members **24** and **26**, respectively, which are suitably mounted on a hollow stub axle **28** extending from an aircraft landing gear wheel support, not shown.

Figure 8.5: Electrostatically Cooled Brake

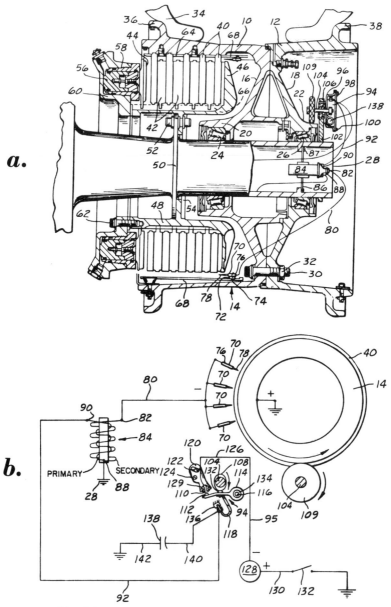

(a)   Schematic representation in partial section of a conventional
       aircraft wheel and multiple disc brake.
(b)   Schematic representation showing one form of the electrical
       components.

Source: U.S. Patent 3,952,846

The inboard and outboard wheel portions are fixedly secured together by a plurality of circumferentially spaced-apart bolts 30 extending through arm portions 16 and 18 and threadedly engaged by associated nuts 32. An inflatable tire 34 is retained on wheel 14 by wheel flanges 36 and 38 integral with wheel portions 10 and 12, respectively.

A plurality of interleaved annular rotor and stator discs 40 and 42, respectively, are adapted to be compressed between an annular pressure plate 44 and a backing plate 46 which backing plate 46 is integral with an annular torque member 48. The torque member 48 is fixedly secured to a mounting flange 50 integral with axle 28 and extending radially outwardly therefrom. A plurality of circumferentially spaced-apart bolt and nut assemblies generally indicated by 52 and plate member 54 serve to secure torque member 48 in position on flange 50.

The pressure plate 44 is engaged by a plurality of circumferentially spaced-apart fluid-pressure-actuated pistons 56 slidably carried in associated cylinders 58 in an annular carrier member 60 which is fixedly secured to torque member 48 by a plurality of circumferentially spaced-apart bolts 62 threadedly engaged with torque member 48. The cylinders 58 are connected via suitable passages, not shown, to a conventional controlled source of pressurized fluid, not shown, by means of which pistons 56 are pressurized.

The rotor discs 40 are slidably keyed for axial movement of a plurality of circumferentially spaced-apart slotted members 64 fixedly secured to inboard wheel portion 10. The stator discs are slidably keyed for axial movement to a plurality of circumferentially spaced-apart slotted members 66 fixedly secured to torque member 48.

A conventional annular heat shield 68 located radially outwardly from rotor and stator discs 40 and 42 is fixedly secured to inboard wheel portion 10 to reduce heat transfer radially outwardly from the discs 40 and 42 to the tire 34.

The brake is applied as a result of pressurization of pistons 56 which urge pressure plate 44 toward backing plate 46 thereby compressing rotor and stator discs 40 and 42 together to frictionally oppose rotation of wheel 14. The frictional resistance generated by the rubbing surfaces of rotor and stator discs 40 and 42 is transformed to heat which, in the case of high-capacity aircraft disc brakes, tends to raise the operating temperatures of the rotor and stator discs 40 and 42 as well as adjacent brake structure which operating temperatures cannot be permitted to exceed a predetermined limit if rapid deterioration of the discs 40 and 42 and resulting brake failure is to be avoided.

Referring to Figures 8.5a and 8.5b, the process includes a plurality of circumferentially spaced-apart electrodes or electrically conductive probes 70 as, for example, six or eight, only one of which is shown in Figure 8.5a, suitably secured by support members 72 fixedly secured to a suitable support such as heat shield 68 by rivets or other fastening members 74. The electrodes or probes 70 are each provided with an electrically insulated body portion 76 and a relatively sharply pointed electrical discharge member 78 extending therefrom in predetermined spaced-apart relationship to the adjacent radially outermost portion of the nearest rotor disc 40. The spaced relationship between rotor disc 40 and discharge member 78 is adequate to prevent electrical arcing therebetween. The electrodes or probes 70 are connected in parallel by a suitable wiring network including wire 80 to a high-voltage terminal 82 of the secondary coil of a conventional coil member 84 which is mounted within axle 28 on a bracket 86 attached to axle 28. A second terminal of the secondary coil as well as one terminal 88 of the primary coil of coil member 84 is grounded to axle 28. The second terminal 90 of the primary coil of coil member 84 is connected via wire 92 to one terminal of an electrical breaker assembly generally indicated by 94, a second terminal of which breaker assembly is connected via wire 95 to a suitable dc power source. The breaker assembly 94 is suitably mounted in a container 96 having a cap 98 removably secured thereto as by screws 100.

The container 96 is fixedly secured to a support member 102 which, in turn, is fixedly secured to axle 28. A shaft 104 rotatably mounted on antifriction ball bearing 106 suitably secured to support member 102 extends into container 96 and is provided with a cam member 108 fixedly secured thereto and rotatable therewith. The opposite end of shaft 104 is provided with driving wheel 109 fixedly secured thereto which frictionally engages hub portion 22 and is rotated thereby. The cam member 108 is engageable with breaker assembly 94 and serves to periodically open and close the same to thereby energize the coil member 84 as will be described.

Referring to Figure 8.5b, the abovedescribed electrical network and associated mechanical components are shown in schematic form in entirety. It will be noted that preferably the electrical polarity of the various component members including probes 70 is as shown. However, it will be understood that the polarity of the component members and, in particular, probes 70 may be reversed from that shown without disturbing the continuity of the electrical network. The breaker assembly 94 is conventional in that it includes stationary and movable contact members 110 and 112, respectively. The movable contact member 112 is fixedly secured to an arm 114 which arm is pivotally secured to a fixed support 116 and biased into engagement with cam member 108 by a flexible string member 118. The stationary contact 110 is secured to an adjustable support 120 provided with adjusting screw 122 and locking screw 124 to provide for gap adjustment between contacts 110 and 112.

The stationary contact 110 is connected via a wire or lead 126 preferably to the negative terminal of a suitable electrical power source generally indicated by 128 which may be an aircraft engine driven electrical generator or storage battery of suitable voltage as for exampe 24 volts dc. The stationary contact 110 is electrically insulated from adjustable support by suitable nonconducting sleeve means 129. A positive terminal of the source 128 is connected via wire or lead 130 and switch 132 to common ground.

Wire or lead 92 is electrically connected to movable contact member 112 via spring member 118 and arm 114 which are electrically insulated from cam member 108, support 116 and container 96 by suitable nonconducting means 132, 134 and 136. A capacitor 138 is connected via wires or leads 140 and 142 to lead 92 and ground, respectively.

It will be understood that the cam member 108, breaker assembly 94 and coil member 84 may be replaced with a conventional high-voltage transformer, not shown, suitably connected to be energized by source 128 and deliver a stepped-up voltage to electrodes 70. It will be recognized that a high-voltage transformer will supply a continuous voltage to electrodes 70 in contrast to the intermittent electrical energization produced by coil member 84.

In operation, application of the disc brake to retard rotation of wheel **14** during a landing or rejected take-off of a heavy, rapdily moving aircraft results in a rapid heating of the rotor and stator discs **40** and **42** with corresponding temperature increase thereof to the extent that progressive structural deterioration of the discs and subsequent mechanical failure of the same is likely to result. It has been observed that such a braking operation with subsequent aircraft taxiing can result in brake temperatures on the order of 2000°F which the conventional disc brake cannot tolerate for any significant time period without a drastic reduction in operating life of the rotor and stator discs or concomitant brake failure.

Assuming that the abovementioned overheated condition of the disc brake of Figure 8.5a is encountered, the switch **138** may be closed by the aircraft operator or by automatic temperature-responsive means, not shown, suitably connected to actuate the switch which establishes continuity to ground. As the wheel **14** rotates cam member **108**, the movable contact **112** moves into and out of engagement with stationary contact **110** resulting in cyclical energization of the primary coil of coil member **84** which, in turn, generates an electrical potential in the secondary coil of coil member **84** which secondary coil is provided with sufficient turns relative to the turns of the primary coil to establish a preferred potential of approximately 20 kV. The electrical charge thus imposed on probes **70** via wire **80** creates an electrostatic discharge between probes **70** and rotor and stator discs **40** and **42** adjacent to probes **70** which has the effect of increasing dissipation of heat from the rotor and stator discs **40** and **42** thereby cooling the same, causing a significant reduction in temperature thereof.

The reason or theory behind the effect created by directing an electrostatic discharge against a relatively hot object and the resulting cooling of the hot object is not fully understood. However, it appears that it may be due to the boundary layer of air adjacent the rotor and stator discs **40** and **42** which layer may normally act to impede dissipation of heat radially outwardly from the heated rotor and stator discs **40** and **42** particularly when the discs are extremely hot. The electrostatic discharge imposed on the boundary layer of air may act to change the characteristics of the latter thereby permitting heat to escape more readily from the rotor and stator discs **40** and **42**.

Aircraft Gas Turbine Engine

A process described by *D.S. Matulich and B.F. Saylor; U.S. Patent 4,102,387; July 25, 1978; assigned to The Garrett Corporation* relates to thermal control systems for aircraft and, more particularly, to a thermal control system for cooling bleed air from a gas turbine engine which utilizes a heat exchanger much smaller than that found in the prior art and yet safely maintains the air in the bleed air ducts at a temperature lower than that which might produce auto-ignition of aircraft fuel proximate to the ducts.

Thus, according to this process, a bleed air preconditioning system is utilized which preconditions the air in the desired manner with a substantially smaller and lighter heat exchanger without sacrificing safety of the aircraft. A first thermostat controls bleed air temperature during normal operation so that the desired thermal characteristics of the bleed air to be supplied to the air conditioning systems can be achieved. Under operating conditions which cause the temperature of the bleed air to exceed the capacity of the heat exchanger to maintain the desired temperature, a second thermostat, triggered at a predetermined higher temperature than the first thermostat, controls a valve upstream of the heat exchanger to limit flow of air to the heat exchanger to an amount which is within the capacity of the heat exchanger to cool to the desired safe temperature.

While this decreased flow of air might otherwise temporarily inhibit operation of the air conditioning systems of the aircraft, it permits the aircraft to fly safely through the transient high speed or "off design" condition. The second thermostat is pressure-biased so that its temperature-response characteristics are altered with changes in altitude to match the ignition characteristics which vary as a function of air density. This enables the system to raise the permissible temperature of the air at higher altitudes where the auto-ignition temperature of the fuel is higher.

The improved system of this process satisfies normal operating conditions and does not penalize the aircraft with excess weight in order to meet certain steady state performance objectives during transient or "off design" aircraft operating conditions.

Aircraft Instrument and Panel Cooling Apparatus

K.D. Groom; U.S. Patent 4,093,021; June 6, 1978; assigned to The Boeing Company describes a process which provides adequate cooling of an instrument panel and the instruments mounted in the panel.

The process involves supplying cooling air flow through a plurality of small, specially designed and located orifices in the panel to each instrument to be cooled, the plurality of orifices being located close to and surrounding the large apertures in which the instruments are mounted so that cooling air is directed along and parallel to the longitudinal sides of the instrument case. Air flowing through the orifices tends to attach to each instrument case, cooling it adequately, efficiently, and uniformly. Each instrument is served by its own tailored supply of air, rather than by air previously heated by other instruments, so that the location of an instrument on the panel does not affect its cooling. The amount of air supplied to each instrument is tailored and controlled by the number, size, and location of the orifices for each instrument.

At the flow rates used, and with the prescribed orifice sizing and location, the flow out of the orifices tends to coalesce into a sheet surrounding the instrument case and rapidly becomes turbulent. This improves the heat transfer between the air and instruments, increasing efficiency of the cooling. The structure is simple, requiring no specialized baffling or ducting.

There are two preferred designs. In the first, the panel is a single wall, and the air is drawn through the orifices from the space in front of the panel.

In the second case, the panel has a dual wall which forms a plenum chamber, orifices being in the rear wall so that the front wall utilization is not hampered.

FOOD SERVICE

Food Service Systems

A.E. Colato and J.L. Formo; U.S. Patent 4,005,745; February 1, 1977; assigned to Anchor Hocking Corporation describe an apparatus for storing, refrigerating and heating food items on a single tray in a single environment including a tray having at least one opening therein and a container together with means for positioning the container on the tray in alignment with the opening, a rack for the tray and a heater carried by the rack in alignment with the opening in the bottom of the tray when the tray is mounted on the rack, the heater being substantially in contact with the bottom of the container, which raises the container free of the tray to form an air barrier between the container and the edge of the opening in the tray, together with means for refrigerating the rack to refrigerate items on the tray, and means for selectively actuating the heater to heat food in the container while other food on the tray remains refrigerated.

Referring to Figure 8.6a, the food treating and storing assembly A includes a support in the form of a food storage and serving tray 14 which has a generally flat panel portion 16, obliquely disposed side and end walls 18 and 22 respectively.

Figure 8.6: Apparatus for Storing, Refrigerating and Heating Food Items

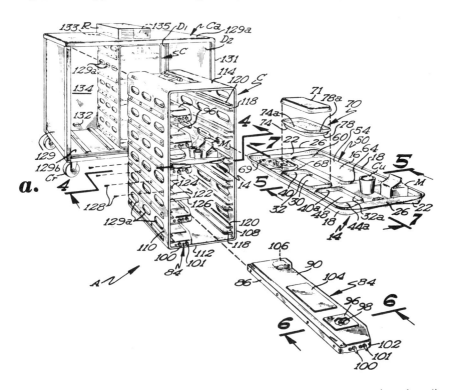

(continued)

Figure 8.6: (continued)

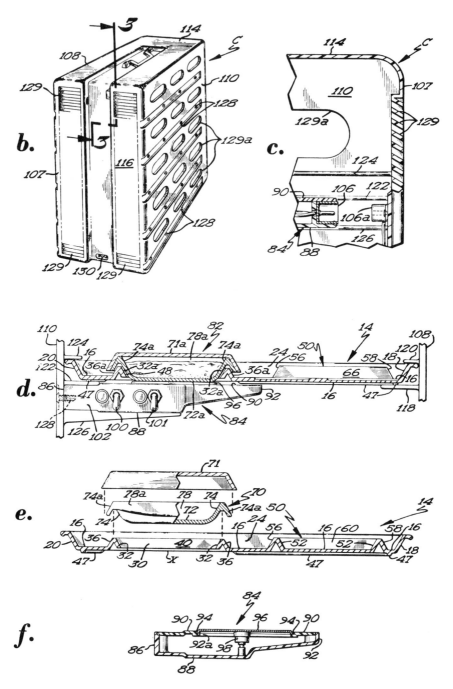

(continued)

Figure 8.6: (continued)

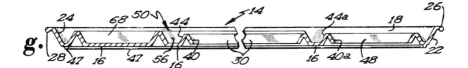

g.

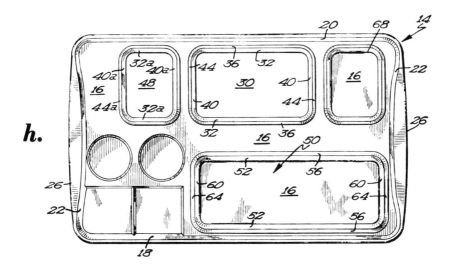

h.

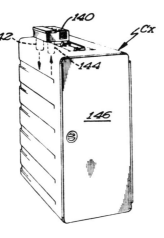

i.

(continued)

Figure 8.6: (continued)

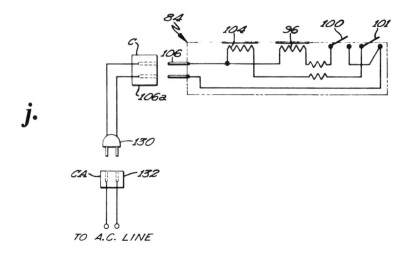

j.

(a)   Exploded perspective view of the apparatus for heating, refrigerating
            and storing food on the same tray.
(b)   Perspective rear view of the tray rack that is housed in the refrig-
            erated cart.
(c)   Section taken along line 3—3 of Figure 8.6b.
(d)   Section substantially along line 4—4 of Figure 8.6a with the tray
            supported in the rack and a covered dish upon the tray contact-
            ing the heater carried by the rack.
(e)   Section taken along line 5—5 of Figure 8.6a.
(f)   Section taken along line 6—6 of Figure 8.6a.
(g)   Longitudinal section of a tray through the openings therein.
(h)   Top plan view of the tray.
(i)   Perspective view of a tray rack substantially that of Figure 8.6a
            but which is a self-contained unit including refrigerating means
            for the rack.
(j)   Circuit diagram of the heater system and its connection with a
            power source.

Source: U.S. Patent 4,005,745

Hand grips **26** extend from the end walls **22**. The tray **14** has an opening **30**
with upwardly diverging side walls **32** and end walls **40** terminating in downwardly diverging wall portions **36** and **44** respectively which constitute a rib
formation (also see Figures 8.6e, 8.6g and 8.6h). The tray **14** may be molded
from a suitable plastic material such as styrene, styrene acrylonitrile, ABS or
the like as a unit.

A peripheral rib **47** is formed on the bottom surface of the tray **14** adjacent
the edges and sides except as at **X** (see Figure 8.6e) which allows insertion of
the tray without interference into contact with heating means.

A further opening **48** is formed in the panel portion **16** of tray **14** having wall
formations identical to those surrounding opening **30** with identical parts bearing

identical reference numerals accompanied by the letter a (see Figure 8.6h). The tray 14 also includes a rectangular section or food compartment 50 formed on the upper surface of the panel portion 16 which includes the upwardly diverging sidewalls 52 and end walls 60 which terminate in the downwardly diverging wall portions 56 and 64 respectively, which, in turn, terminate in the panel portion 16. The panel portion 16 forms the floor for the food section 50, the walls of which may be molded of plastic material. Various dishes and/or trays may be placed within compartment 50 which prevents movement of the dish, etc., upon the tray.

The tray 14 may also have a raised rectangular formation 68 which is identical to the raised rectangular formation 50 but of smaller area within which a dish or tray may be contained.

Used in conjunction with the opening 30 of the tray is a container in the form of a dish 70 having a bottom wall 72 from which extends upwardly diverging side walls and end walls 74 and 78 with corresponding downwardly diverging extensions 74a and 78a respectively. The dish 70 may generally be used for a hot meal entree such as meat, fish or the like and may be made of heat-resistant plastic such as melamine, a formaldehyde molding material, glass, ceramic, metal or the like. The divergent angles of the side and end walls of the dish 70 are substantially the same as that of the walls 32 and 40 bordering the opening 30 whereby the dish walls overlie the opening 30 with the dish nesting within the walls. The vertical dimension of the dish 70 is such that when it is within the opening 30 the bottom wall 72 of the dish extends below the panel portion 16 but above the lower edge of the rib 47 of the tray 14. The dish is provided with the tight-fitting cover 71.

Used in conjunction with the opening 48 of the tray 14 is dish 82 (see Figure 8.6d) constructed identical to the dish 70 except that it is shorter. The reference numbers are the same as with dish 70 except for the letter a following the number. The dish 82 has a vertical dimension such that when it is within the opening 48 the bottom of the dish extends below the panel portion 16 of the tray 14 but above the lower edge of the rib 47 as in the case of the dish 70. The remaining area of the tray is essentially flat for supporting items such as the cup Cu and the milk carton M.

The numeral 84 designates one of several identical tray supports, particularly Figures 8.6a and 8.6f. The tray support 84 includes the sidewalls 86 and 92 to which are connected the bottom wall 88 and the spaced top wall 90. Formed in the top wall 90 is rectangular opening 92a and on the periphery thereof is recess 94. Mounted in the recess 94 is a first conventional electric heating element 96 which extends slightly above the top surface of top wall 90. The recess 94 may be coated with silicone to insulate the heating element from the top wall 90 of the support 84. A conventional thermostat 98 is mounted upon the bottom wall 88 and extends into contact with the heater element 96. The element 96 is connected to a source of power as hereinafter described and is turned "off" and "on" by a conventional switch 100 mounted on the end wall 102 of the support 84.

A second conventional electric heater 104 is provided which is mounted identically to heater 96 and controlled by a thermostat as in the case of the heater 96. The heater 104 is connected to, and turned "off" and "on" by, the conventional switch **101**. The heaters **96** and **104** are conventionally wired to a male plug **106** mounted on the inner end wall of the tray support. The plug **106** of the tray support engages the receiver **106a** mounted on the rear wall **106** of the rack **C**, Figure 8.6c, when the tray support is inserted into operative position in the rack **C**. The heater **104** if used for an entree such as meat may have a control to start the heater at approximately 110 watts with a cutback to approximately 53 watts for holding. With the heater **96** for soup the same would be started at 44 watts with a cutback to 28 for holding.

The area of the heater in each instance is substantially that of the area of the bottom of the dish heated thereby, and as a result heat from the heater element does not go into the free air in the carrier but is substantially confined to the dish **70**.

The dimensions and the disposition of the sidewalls **32** and the end wall **40** of the tray and the walls of the dish are such that when the dish **70** is placed on the same as in Figure 8.6d and the tray **14** is placed with the ribs **47** on the flanges **118** and **122** of the rack **C**, the heater **96** of the support **84** is so positioned that it raises the dish upward slightly and substantially free of the sidewalls **32** and endwalls **40**. As a result an insulating air barrier is formed and there is essentially no heat conduction from the heated dish **70** to the tray via contact with the wall formations **32** and **40** which allows maximum refrigeration for food items not being heated.

The same is true of dish **71a** supported within opening **48** and the wall formations **32a** and **40a** and heater **96** whereby there is essentially no conduction of heat from the heated dish **71a** to the tray which allows maximum refrigeration for food items not being heated. Briefly put, with the dish raised slightly and out of contact with any part of the tray, particularly Figure 8.6d, the dish is in effect insulated from the tray due to the air space between the dish and the tray.

The letter **C** designates a tray carrier in the form of a rack for trays which includes the sidewalls **108** and **110** connected to bottom wall **112**, the top wall **114** and the inner end wall **116**. Secured to the inner surface of the wall **108** are pairs of spaced flanges **118** and **120** and in juxtaposition thereto on the wall **110** are pairs of spaced flanges **122** and **124**. Further provided is a major flange **126** connected to the wall **110** and underlying each flange **122**. The support **84** is positioned for support between the flange **122** and the flange **126** and secured by means such as screws **128** extending through sidewall **86** of the support **84** and through the wall **110** of the carrier. The top surface of the support **84** is substantially in the same plane of the flanges **118** and **122** and the rib **47** on the bottom of the tray will rest on and be supported by the flange **122** and the flange **118**.

As the tray **14** is placed on the flanges **118** and **122** the bottom of the dish **70** makes intimate contact with the heater unit **104**, for, as described the bottom wall of the dish extends below the surface of the opening **30** and as the tray is placed on the flanges and the dish contacts the heater the dish is raised from the tray to a position slightly above and free of the tray. The same is done with dish **82** upon the opening **48**. The rear wall **107** of the tray carrier **C** is formed with the louvers **129** and the sidewalls **108** and **110** are formed with the openings **129a** which allow circulation of the refrigerated air of the cart **Ca**, throughout the tray rack **C** to refrigerate the items on the trays **14**. The conventional male plug **106** of each tray **14** engages in a conventional female receptacle **106a** on the rear wall of the carrier **C** which hooks up the heaters **96** and **104**, Figure 8.6j.

The carrier or rack **C** has the conventional male plug **130**, Figures 8.6b and 8.6j, on the rear wall thereof and conventionally wired to the receptacles **106a** which engage the conventional female receptacle **132** on the rear wall **134** of the cart **Ca**. The rear wall is connected to the end walls **129** and **131**, the top **129a** and the bottom **129b**. The cart **Ca** is also equipped with a caster **Cr** at each corner of the bottom **129b**. Electric power is conventionally supplied to the receptacles **132** of the cart **Ca** provided for each of the racks **C**. The cart **Ca** carries the enclosure **R** in which is positioned conventional dry ice or the like which thermally communicates with and refrigerates the cart through openings **133** and **135**. Also mounted in the enclosure **R** is a conventional squirrel cage fan, not shown, to facilitate circulation. Sliding doors **D1** and **D2** are supplied for the cart to seal-off the cart when the tray racks **C** are enclosed within the cart.

Hinged doors may also be used on the cart **Ca**. A conventional electrical refrigeration unit may be carried by the cart and in communication with the interior of the cart for refrigerating the cart in lieu of the aforementioned refrigeration means. The rack **C** may be permanently mounted in the cart **Ca** and used relative thereto.

With reference to Figure 8.6i, there is illustrated a rack or carrier **Cx** which is identical to the carrier or rack **C** found in Figures 8.6a and 8.6b. However, with the carrier rack **Cx** there is mounted on the top thereof the enclosure **140** which is adapted to contain dry ice and the enclosure communicates with the interior of the carrier **Cx** by means of the conduit **142**. A further conduit **144** also communicates the enclosures **140** with the interior of the container or carrier **Cx**. The enclosure **140** also contains a conventional squirrel cage blower fan which directs cold air from the dry ice downwardly through the conduit **142** and the conduit **144** is used as a return. The blower fan is connected to an electrical source of power. The carrier rack **Cx** is also supplied with the door **146**. Thus, with a refrigerated carrier or rack **Cx** the cart **Ca** is not necessary.

Operation: Prepared food such as an entree in the form of a steak, which requires heating, is placed in the dish **70** and the dish placed upon the opening **30** with the cover **71** on the dish. Similarly prepared food such as soup, which requires heating, is placed in the dish **82** and the dish placed on the opening **48** with a cover **71a** on the dish.

Additional food such as a salad which requires cooling may be placed in the dish **69** which is positioned within the rectangular raised formation **68** to secure the dish against movement on the tray **14**.

Still further food requiring refrigeration may be placed in dishes positioned within the raised rectangular formation **50**. The milk carton **M** and the cup **Cu** are placed on the remaining portion of the tray, the milk being subject to and requiring refrigeration.

The tray **14** is then placed upon flanges **118** and **122** within the rack **C** along with other trays desired. The loaded carrier **C** is then moved into the refrigerated cart **Ca** with the plug **130** engaged with the receptacle **132** and the doors **D1** and **D2** closed.

The refrigeration means is then actuated which refrigerates the air in the cart, and the refrigerated air circulates through the louvers **129** and openings **129a** of the carrier **C** and into the carrier thereby refrigerating all the items including food to be heated on the tray. All the food is thus held in a refrigerated condition until needed.

When it is desired to prepare and serve the food on the tray nothing has to be added to the tray, nothing removed, nor does the tray have to be moved. The heaters **96** and **104** are actuated sufficiently prior to a need for heated food in dishes thereon. It has been found that a thirty-five minute heating period is satisfactory.

The doors **D1** and **D2** or **146** can be opened momentarily to actuate the switches. All the while the remaining food is being refrigerated and kept fresh. The heat generated in the dishes **70** and **82** is substantially maintained within the dishes.

When the food in the dishes is thoroughly heated the tray may be served. In the event service of the heated food is not desired immediately, the heaters may be automatically put on a lower or hold temperature by conventional means with the remaining foods continuing to be refrigerated for a later serving of the meal on the tray.

The tray, when supported on a table or the like, rests on the rib **47** and as a result the hot dishes **70** and **82** do not contact the surface of the supporting element.

While in general each heating means will be required for a complete meal it may be desired to heat only a selected dish on a tray and this may be accomplished by operating only the appropriate switch for a given heater.

As to the versatility of the apparatus, eggs may be precooked to the degree desired outside the apparatus, refrigerated and then be reheated in the apparatus for the same heating period as other foods by placing the precooked eggs on a piece of toast in the dish **70** which, due to the insulative quality of the toast, retards the heating of the eggs and results in highly palatable, tender eggs.

Relative to steaks, the same can be cooked to various degrees of doneness by controlling the amount of moisture. To achieve a "well-done" steak, the same is seared outside the apparatus for twenty seconds to carmelize the surface. The steak is then placed in dish **70**, and when desired, heated for the aforementioned thirty-five minute period with the result that the steak is well done.

To achieve a "medium" done steak it is also seared for twenty seconds and placed on a piece of toast in dish **70** with two teaspoons of water in the dish, and to achieve a "rare" steak it is also seared as above and placed on a piece of toast in the dish but with no moisture added. In normal practice, food is refrigerated before heating in the apparatus.

The apparatus produces, because of its gentle heating characteristics, food which is nutritionally as good as freshly cooked product, although it may be refrigerated for extensive periods prior to reheating.

Because the food is held in a refrigerated controlled environment, bacteriologically it is safe, as all food items are maintained at normal food refrigerated storage temperatures of about 40°F and only reheated from storage to serving temperature which is in excess of about 160°F.

L. Spanoudis; U.S. Patent 3,952,794; April 27, 1976; assigned to Owens-Illinois, Inc. describes a food service tray which allows selective heating and/or cooling of meal components carried thereby.

A base member of a thermally insulating material has a plurality of compartments divided into serially connected flow chambers having inlet and outlet openings. Food receptacles made of a thermally conductive material carry the meal components and are inserted into the compartments. Thermally insulating lids then cover the food receptacles.

A refrigerated fluid under pressure may be passed through all or part of the flow chambers to thereby maintain the contents of the food receptacles therein in a chilled condition. A heated fluid under pressure can then be passed through selected flow chambers to thereby rethermalize or heat the contents of the food receptacles in the heated flow chamber to a proper serving temperature.

Simultaneously with the heating of some food receptacles, the flow of refrigerated fluid can be continued to other flow chambers to maintain other meal components in a chilled condition for serving.

L.G. Williams; U.S. Patent 3,908,749; September 30, 1975; assigned to Standex International Corporation describes a food service system which comprises a cart that is movable from a central food preparation location to a remote food service location.

The cart is insulated and has a rack for a plurality of individual meal service trays. Each tray carries cold foods in suitable dishes and a heat-retaining food service unit for food which may be either hot when placed therein or may be heated by connecting each unit to a source of electrical energy at the service location.

The cart has conduit means which are connectable to a source of chilled air at the service location in order to keep the cold foods cold for a period of time while the food in the units is being heated to a desired serving temperature.

If desired, instead of heating the food in the units at the service location, the heat-retaining units may be heated prior to departure from the food preparation location and will maintain the food therein at elevated temperature for a period of time, even while the cart is connected to the chilled air source for keeping the cold foods at a refrigerated temperature.

Food Preparation

A process described by *S.F. Skala; U.S. Patent 3,888,303; June 10, 1975* provides a method for cooking or transferring heat in an improved manner by utilizing apparatus to obtain rapid heat transfer into thermal fluid, and delivery of the thermal fluid to rapidly attain controlled cooking temperatures, or chilling and freezing capabilities.

Controlled circulation of heat exchange fluid is maintained through components which include stored and travelling thermal fluids, thermal energy sources, temperature control devices, valves, pumps, and various houseware units. The travelling thermal exchange fluid is preferably a silicone oil which is fluid at subfreezing temperatures and resists chemical breakdown at high cooking temperatures. A source of thermal energy is preferably a thermostatically controlled resistance heating element which provides energy for a hot reservoir containing the thermal exchange fluid. A refrigeration unit provides energy for heat removal for a cold reservoir of the thermal exchange fluid.

The thermal exchange fluid at preset temperature levels, low or high, is delivered through a path to a valving assembly between the path and a double walled domestic appliance. The domestic appliance or houseware unit is coupled to the central appliance containing a hot and cold reservoir by the valving assembly which preferably includes selfsealing, quick connecting valve elements. Such valve elements allow two-way quick connect sealing to maintain the fluids in the path or line and in the houseware unit when uncoupled. When the houseware unit and central appliance are coupled, the thermal or heat exchange fluid flows freely between. To effectively reduce fluid loss at uncoupling, a sliding action valve is used.

A central appliance has insulated hot and cold reservoirs, respectively provided with heating and refrigerating units. The thermal exchange fluid is stored in these reservoirs and imparts desired temperature levels to thermal fluids which are delivered along closed paths directly to the houseware units or to precisely adjustable temperature control means prior to delivery to a houseware unit. The fluid enters double walled compartments or thermal chambers of houseware units through coupling and valving assemblies. The thermal fluid is moved in the system under urging of pumps. The thermal fluid moves through pipes and couplings, past heat exchangers, temperature regulators, coupling and valving assemblies between houseware units and lines of central appliance assembly.

Another process described by *S.E. Skala; U.S. Patent 4,024,904; May 24, 1977* relates to cooking by forced air heat exchange between a houseware and a thermal exchange fluid.

In the process, forced circulating air transfers heat between a heat exchanger and a houseware. The heat exchanger receives a regulated flow of thermal exchange fluid at one of several temperatures—hot, cold, or ambient. Thermal reservoirs which exchange heat with the thermal exchange fluid have large thermal capacity and rapidly circulating air effectively transfers heat to the houseware.

Capability of selecting the temperature of the thermal exchange fluid and of regulating the temperature of the circulating air provides functions and a degree of control which is not possible with conventional ranges. Some examples are the following. Moderate hot air temperatures avoid scorching of food. Air circulating at ambient temperature can cool food such as soup or coffee to serving temperature. Cold circulating air can freeze food which is useful for such applications as making ice cream. Another use of cold air is storage of food within an insulated pot until a preselected time at which a programmed cooking cycle can begin with hot circulating air.

ELECTRONIC APPLICATIONS

Semiconductor Cooling System

A process described by *B. Voboril, P. Reichel and P. Kafunek; U.S. Patent 3,989,095; November 2, 1976; assigned to CKD PRAHA, Czechoslovakia* relates to a cooling system for solid state power devices and in particular to heat sink apparatus for power semiconductor devices.

According to the process, there is provided a cooling system for a semiconductor device of the wafer type comprising an enclosed container forming an interior chamber. A wall portion of the container is formed of yieldably soft material adapted to contact against a surface of the semiconductor device. A porous member is disposed within the chamber adjacent the inner surface of the yieldable wall. A liquid medium partially fills the remainder of the chamber and immerses at least a portion of the porous member. The heat produced by the semiconductor device is transferred via the yieldable wall to the porous member wherein the latent heat causes the liquid to vaporize, the vapors thereafter condensing on the walls of the container and returning in drops to the liquid mass, which of course is absorbed by the porous member.

Preferably, at least the wall portion of the yieldable soft material is made of highly thermal conductive material. Also resilient means are provided biasing the porous member against the yieldable wall so that complete low resistance contact is made.

It is also preferred that the container be provided with external means for securing the yieldable wall in firm pressure contact with the semiconductor device so that extremely low resistance thermal contact is made. In this form the use of material which is also of high electrical conductivity enables the yieldable wall to be also used as the terminal for the semiconductor device.

Referring to Figure 8.7a, the apparatus comprises a hermetically sealed container 10 of any suitable shape forming an internal open chamber. The container 10 is abutted against a face of the semiconductor device 1 under a force F which ensures a firm contact therebetween. The wall portion 2 of the container which abuts against the semiconductor device 1 is made of soft, yieldable highly thermal conductive material, such as aluminum, copper or the like.

A porous member 3 made of ceramic, plastic or the like, is located within the container 10 abutting against the inner surface of the contacting wall 2. The porous member 3 is at least partially immersed in a liquid coolant 12 which does not completely fill the container 10. A spring 16 or other pressure means acts on the porous member 3 so as to ensure its firm flat contact with the inner surface of the contacting wall 2. The outer surface of the container 10 is provided with a plurality of radiating cooling fins 13 of suitably desirable convective design.

The semiconductor device **1** is of conventional type and is provided with a terminal abutting one face and requires only single sided cooling.

Figure 8.7: Semiconductor Cooling System

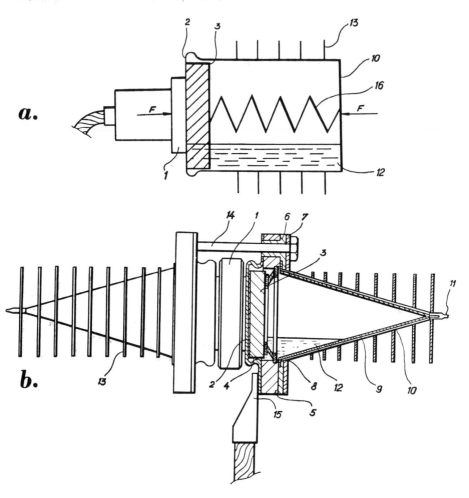

(a) Schematic longitudinal cross section of cooling apparatus for the single-sided cooling of a conventional semiconductor device.

(b) Longitudinal cross section of apparatus for the double sided cooling of a disc or wafer type semiconductor device.

Source: U.S. Patent 3,989,095

In Figure 8.7b a more specific form of cooling system is shown for a disc cell semiconductor device **1** requiring double sided cooling. Here, the container **10** may take the form of a conical hollow body, the exterior surface of which is provided with a plurality of fins of uniform diameter. The soft yieldable wall **2**, formed for instance of copper, is joined about its external circumference with a rigid hard ring **5** formed e.g., of steel. Preferably the wall **2** is brazed or welded about its entire circumference to the ring **5**. The rigid ring **5** is further welded or otherwise secured to the edge of the container **10** which is provided with a radially extending flange which thus seals the interior chamber formed by the container **10** and the soft yieldable wall **2**. The apex of the container **10** is provided with an inlet-outlet duct **11** enabling liquid coolant to be introduced into the chamber thus formed. After the liquid coolant is introduced, the enclosed chamber is evacuated to a predetermined vacuum and hermetically sealed by closing the duct **11** with a cap or weld in conventional manner.

The porous member **3** is inserted within the chamber adjacent the inner surface of the soft yieldable wall **2**. The disc porous member is biased and pressed against the wall **2** by a spring **8** such as a belleville spring fixed and secured in the rigid ring **5**. Preferably the porous member **3** is made of ceramic or similar porous material allowing absorption of the liquid **12** and having relatively high thermal conductivity but an extremely low or nonexistant electrical conductivity. If preferred, a layer **4** of material such as a metal foil, amalgam, or contact Vaseline may be interposed between the porous member **3** and the inner surface of the soft yieldable **2** in order to improve the heat transfer therethrough.

The amount of liquid medium **12** introduced into the container **10** is sufficient at least to partially immerse the porous member **3** but not sufficient to completely fill the container. Preferably the amount of liquid fills only a minor portion of the container. The inner surface of the container **10**, may also be provided with a coating **9** of smooth nonabsorbant material which enhances the condensation of the liquid coolant **12** into suitable drops ensuring the return of the condensation to the liquid mass.

The supporting ring **5**, to which the soft yieldable wall **2** and the edge of the container **10** is secured, forms a radially extending flange through which a plurality of circumferentially uniformly spaced apertures may be provided. In this manner the exterior of the cooling apparatus may be provided with means by which the soft yieldable surface may be secured in pressure contact against a face of the semiconductor device **1**. Tie rods **14** may be employed to connect a pair of coolant devices on either side of the semiconductor or one semiconductor with a retaining means provided on the opposite side of the semiconductor. The tie rod **14** extends through an insulating member **6**, e.g., made of ceramic or plastic, which extends through the apertures formed in the circumference of the ring **5**.

A metal washer **7** may be used below the head of the tie rod for strength and support. The member **6** extends through the ring **5** and through the hole in the radial flanged edge of the soft yieldable wall **2** as well as covering the flanged edge of the container **10**. In this way full electrical insulation of the tie rod from the yieldable wall **2** contacting the semiconductor device as well as from the ring **5** and the container **10** can be obtained.

Preferably, the soft yieldable wall is of such an area that it covers the entire surface of the semiconductor device **1** which it contacts. The porous member **3** is furthermore of such an area that it too is larger than the area of the contacting surface of the device **1** so that complete transference of heat from the entire surface of the semiconductor device can be made via the soft yieldable wall and the porous member **3**. It will be seen from Figure 8.7b that a terminal **15** can be joined as for example by brazing or welding with the exposed surface of the soft yieldable wall portion **2** radially extending about the ring **5**. This provides easy access and complete electrical conductivity through the highly conductive soft yieldable wall portion **2**. While one terminal is shown on only the coolant apparatus shown in the right portion of Figure 8.7b, a similar terminal can be placed on the coolant apparatus shown on the left section of the figure.

To enhance the stability of the porous member **3** and to prevent its movement within the container **10**, its surface adjacent the soft yieldable wall **2** may be scratched or grooved providing greater traction. The liquid coolant **12** may be, e.g., water, or similar easily vaporizable material. The semiconductor devices used in either of the examples shown above may be conventional wafer type devices formed of semiconductive material as, e.g., silicon, and have P-N junctions formed therein.

The wafer is preferably provided with associated supporting plates of a metal whose coefficient of thermal expansion approximates that of the semiconductive material of the wafer. These metal plates provide the contact surfaces against which the soft yieldable wall portion of the coolant apparatus described make contact. Such metal plates may be made, e.g., of tungsten or molybdenum.

The method of operating the apparatus for the cooling system of the process is as follows: The apparatus of either Figure 8.7a or Figure 8.7b is secured in proper position against the facial surface of the semiconductor device and is clamped by means of either the tie rods **14** or other clamping means so that the soft yieldable wall **2** is pressed against the surface with a sufficient force to provide a firm fixed and compressive abutment. The soft yieldable wall is so flexible that when compressed it can take on the contour of the surface of the semiconductor device and be sufficiently rolled so as to completely cover its surface.

The heat developed in the semiconductor device is transferred by conduction through the soft yielding wall portion **2** directly to the coolant fluid in the chamber of the container **10** and particularly to the coolant fluid absorbed by the porous member **3**. The coolant fluid is thus caused to vaporize, rises and expands within the chamber and then condenses on the inner walls of the chamber **10**, which walls are cooled by the stream of coolant air passing through and about the fins **13**. The coolant air may be ambient air or may be forced air produced by a fan or the like.

The condensed liquid coolant is then returned to the mass of liquid coolant **12** which is then transferred via capillary action and/or direct flow due to the shape of the chamber **10** to the contact surfaces between the soft yieldable wall portion **2** and the porous member **3** to fill the area of the porous member vacated by the initial vaporization of the liquid coolant **12**.

The major area at which vaporization occurs is of course the interface between the yieldable wall portion **2** and the porous member **3** since at this contact surface the degree of latent heat is at its highest. In this way heat is transferred from a heat source, namely the semiconductor device, to a heat output zone creating a vaporizable working medium which is then circulated through the chamber. The process remains in continuous cycle so long as the heat input zone is active.

Customized Nucleate Boiling Heat Transfer for Semiconductor Chips

R.C. Chu and K.P. Moran; U.S. Patent 4,050,507; September 27, 1977; assigned to International Business Machines Corporation describe a method for customizing the heat transfer from the walls of an electronic unit such as a semiconductor chip or wafer, wherein the required heat transfer characteristics of the unit are determined, and holes of a predetermined size and location are drilled in the walls of the unit with a high energy beam in accordance with the required heat transfer characteristic of the unit. The unit is then immersed in a suitable dielectric coolant so that nucleate boiling will start at approximately the desired wall temperature to obtain the required heat transfer from the unit.

The process also includes a heater means, located at or near the bottom of the heat transfer wall, to start bubble generation when the required temperature is reached.

Nucleate boiling is a phenomenon that takes place as the energy input in the form of heat from a solid surface to a surrounding liquid is increased, a point will be reached where vapor bubbles will form on the surface to be cooled. These bubbles form in preferred sites or nuclei. Initially, if the liquid temperature is below the saturation temperature of the liquid, the vapor bubbles will collapse. However, as the liquid temperature and energy input are increased, the bubbles will become more numerous. This process is referred to as nucleate boiling. The condition in which the liquid temperature is below the liquid saturation temperature is called subcooled boiling. As the bulk temperature approaches the saturation temperature, the process is called saturated boiling.

The boiling process is best illustrated by means of a boiling curve, Figure 8.8a, which is a log-log plot of heat flux versus the temperature difference between the heated surface and the saturation temperature of the liquid.

Figure 8.8: Nucleate Boiling Heat Transfer in Semiconductor Chip Manufacture

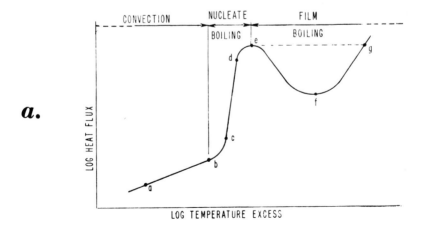

a.

(continued)

Figure 8.8: (continued)

b.

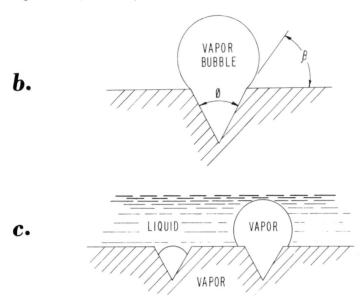

c.

d.

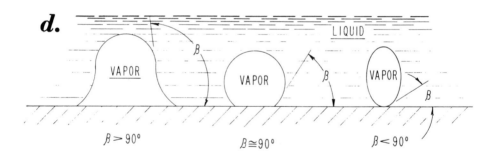

e.

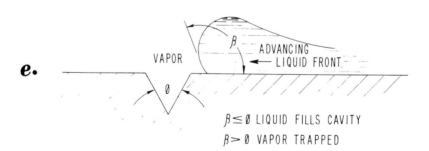

(continued)

Figure 8.8 (continued)

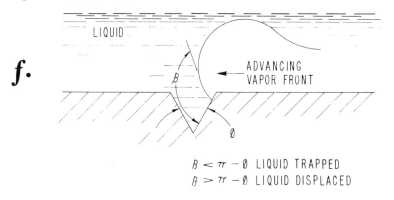

f.

LIQUID

ADVANCING
VAPOR FRONT

B

$\emptyset$

$B < \pi - \emptyset$ LIQUID TRAPPED
$B > \pi - \emptyset$ LIQUID DISPLACED

g.

HYSTERESIS IN BOILING

c

d

h

g

f

e

a

b

ΔT

h.

10
SILICON
CHIP

NUCLEATION
HEATER 20

SUBSTRATE
12

(continued)

Figure 8.8: (continued)

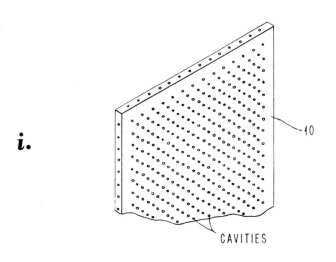

i.

CAVITIES

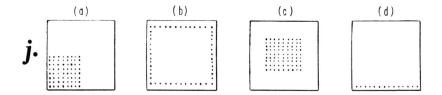

(a)   (b)   (c)   (d)

j.

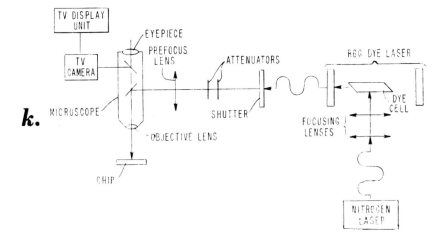

k.

(continued)

Figure 8.8 (continued)

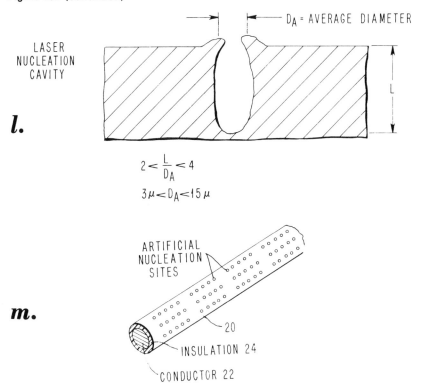

l.

LASER
NUCLEATION
CAVITY

D_A = AVERAGE DIAMETER

L

$$2 < \frac{L}{D_A} < 4$$

$$3\mu < D_A < 15\mu$$

m.

ARTIFICIAL
NUCLEATION
SITES

20

INSULATION 24

CONDUCTOR 22

(a) Graph showing a typical boiling curve
(b) Schematic representation of a vapor bubble formed from an artificial cavity
(c) Schematic representation of the growth of a vapor bubble from nucleation sites
(d) Further schematic representation showing characteristic bubble shapes with varying values of surface tension
(e) Advancing liquid front approaching a nucleation site with the various dimensions used in the theoretical explanation
(f) Schematic representation of an advancing vapor front with respect to a nucleation site
(g) Graphical representation of hysteresis in nucleate boiling
(h) Perspective diagram showing a multichip module having a nucleation heater in place thereon
(i) Blown up view of one of the chips of the multichip module shown in Figure 8.8h with artificial cavities formed on the surface walls
(j) Schematic representation of a number of chips having customized artificial cavities
(k) Schematic illustration of a laser drilling apparatus
(l) Schematic diagram of a typical laser nucleation cavity formed in the chips of Figure 8.8j
(m) View of a customized segment of the nucleation heater shown in Figure 8.8h

Source: U.S. Patent 4,050,507

For low values of heat flux the plot will be a straight line with a slope of approximately $\frac{5}{4}$, and the mode of heat transfer will be natural convection. The knee of the curve, b–c, represents the region where bubbles begin to form on the heated surface. From c to d the plot is again linear, but with a much steeper slope indicating a significant increase in heat transfer rate. At a heat flux designated at point d, the heater surface becomes crowded with vapor bubbles, and there is a decrease in the heat transfer rate. This point is known as the departure from nucleate boiling (DNB). If the heat flux is raised further to point e, the burnout point of heat flux will be reached. An insulating film will form on the heated surface at this point.

The region is an unstable one, and under certain conditions the temperature difference will change rapidly to point g. The surface temperature will be extremely high so that many materials will char or melt. For any practical cooling system to be used with electronic equipment, the operating point should be at or below point d.

There is general agreement in the heat transfer literature that nucleation is initiated by absorbed gas in the heated surface cavities, and that the liquid in the immediate vicinity of the heated surface must be superheated. A simplified model, represented by a conical cavity in the surface with a spherically shaped bubble emerging from it, is customarily used, as shown in Figure 8.8b.

The contact angle β is described as the angle measured from the solid surface through the liquid to a tangent from the bubble. The contact angle results from surface tension, and is a measure of how well the liquid wets the solid surface. The angle ϕ is the angle formed by the sides of the cavity and is an approximate indication of the roughness of the solid surface. Surface conditions have a pronounced affect on the shape of the boiling curve.

An expression has been developed for predicting the degree of superheat required to initiate the growth of a vapor bubble. This equation uses an equillibrium balance of hydrostatic and surface tension forces acting on a spherical bubble in combination with the Clausius-Clapeyron relationship between saturation temperature, pressure, enthalphy and specific volume. When heat is added, the vapor bubble begins to grow in the cavity. The growth is caused by evaporation at the liquid vapor interface in the vicinity of the heated wall. Initial bubble growth is dependent upon the wall superheat and surface tension. Once the bubble grows above the cavity surface (see Figure 8.8c), surface tension relationships on the surface and the bulk temperature of the liquid become important.

If the liquid is superheated, the bubble will grow until the buoyant force exceeds the surface tension force and the bubble will detach and rise through the liquid. If the liquid is highly subcooled, the bubble may collapse on the surface. The effects of surface tension on the bubble shapes are shown in Figure 8.8d. As the bubbles grow and detach, a quantity of vapor is generally trapped in the cavity serving as a nucleus for the formation of a succeeding bubble. If great care is taken to eliminate favorable nucleation centers, extremely high values of superheat can be reached before nucleate boiling occurs.

The conditions under which vapor may be removed from nucleation sites by an advancing liquid is shown in Figure 8.8e. After departure or collapse of a bubble, the liquid advances towards the cavity. The liquid will fill the cavity and no vapor will remain if $\beta \leqslant \phi$. If, however, $\beta > \phi$, some vapor will be entrapped below the liquid and remain within the cavity. This condition favors further nucleation. A liquid filled cavity can likewise be activated by an advancing vapor front, as shown in Figure 8.8f, if $\beta > \pi - \phi$.

High rates of heat transfer are achievable in nucleate boiling. These rates are a result of a combination of latent heat transfer and the turbulence produced by the growth, collapse or detachment of vapor bubbles. A bubble forms on the heated surface and grows into the liquid. The bubble will collapse or detach depending on the level of liquid subcooling. In any event, the bubble is removed from the surface and the resulting void is filled with high velocity, lower temperature, liquid. This violent liquid action in the vicinity of the heated surface is responsible for the higher rate of heat transfer associated with nucleate boiling.

Thermal hysteresis is a phenomenon that is characterized by a deviation from the boiling curve, that is, the temperature of the surface upon which the nucleate boiling is to take place goes beyond the temperature at which nucleate boiling should start. The vapor trapping phenomenon previously discussed has been advanced as a partial cause for the hysteresis effect. Figure 8.8g illustrates the action of thermal hysteresis with respect to the boiling curve. A surface boiling at point **c** for some time eventually becomes gas-free. Reduction in heat flux causes the boiling condition to follow path **cdba**, where; eventually boiling is completely eliminated.

A subsequent increase in heat flux may cause the path to go out to point **e** before boiling begins somewhat violently and causes a return to point **c**. Curves **g** and **f** represent alternate paths for differing amounts of initially active bubble patches. The extent of surface superheat attainable before nucleate boiling begins depends on the surface roughness; smoother surfaces show less superheat than the rougher surfaces. This suggests that smaller cavities are more effective in retaining vapor, while larger cavities tend to be penetrated by liquid and so are not likely to trap vapor.

The foregoing theory of nucleate boiling is applied, in this process, to semiconductor devices by drilling artificial nucleation site holes having a special shape by high energy beams such as lasers or electron beams, so that the hysteresis effect is minimized and the boiling curve is extended somewhat. These high energy beam drilled holes are applied to silicon chips **10**, which are mounted on substrate **12**, as shown in Figure 8.8h. The chips can be the well known silicon chips and, as shown, are mounted face down into the substrate **12**. Figure 8.8i shows a blown up view of one of the chips **10**, wherein the high energy beam holes have been drilled throughout the back surface of the chip and along the four adjacent sides thereof, only two adjacent sides of which are shown.

The desired cooling requirement is to maintain each of the chips **10** at a predetermined temperature. It should be appreciated that many of the chips **10** may require a higher power than others and, accordingly, a greater heat flux will be generated. The drilled holes having a special shape giving rise to controlled nucleate boiling are placed thereon so that the chip will have a higher cooling effect, and consequently maintain the desired temperature by removing the heat more effectively than from a chip having no artificial sites thereon. Also the chips **10** can have the artifical nucleation sites drilled in a certain location on the back of the chip, as shown in Figure 8.8j, so that a hot spot or section can have

a higher heat transfer or cooling rate than the rest of the chip, to thereby maintain a constant temperature gradient over the entire chip.

It is necessary to determine the heat characteristics of the chip so that the number and location of the high energy beam drilled holes can be decided. For example, (a) in Figure 8.8j was found to have a hot spot or section in the lower
left hand corner. Accordingly, that lower left hand section has high energy beam
drilled holes added to increase flux in that area. This arrangement of determining the heat characteristics of the chip and appropriately drilling the high energy
beam holes provides a chip having an even temperature thereover. Similarly, chip
(c) in Figure 8.8j is customized by including the high energy beam holes in the
middle of the chip where the high heat flux was required.

Chips (b) and (d) have the high energy beam nucleate boiling holes located along
the outer edges thereof. In chip (b) the line of holes will always be along the
bottom edge of the chip no matter which edge is at the bottom. Chip (d) requires that the chip be oriented vertically when submerged in a cooling liquid
with the row of holes at the bottom. These edge drilled holes perform a very
important function in the nucleate boiling phase of heat removal, in that they
tend to start the nucleate boiling at a lower temperature than the natural nucleation sites on the surface. The nucleate boiling starts at a lower temperature because of the improved nucleation holes obtained by drilling with a laser or electron beam.

If the chip, such as chip (d) in Figure 8.8j is oriented such that the surface is substantially vertical in the cooling fluid and the row of holes is at the bottom, the
vapor bubbles generated by the improved artificial nucleation sites rise and provide vapor or what is called reactivation of the natural nucleation sites to start
nucleate boiling at those sites.

The specially shaped artificial cavities are created by using a high energy beam
such as a laser beam or E beam equivalent. Actually, cavities of various diameters
and depths can be achieved by a combination of focus adjustments and pulsation
rates generated in the high energy beam generator. The laser system, which was
employed to drill the required holes, is shown schematically in Figure 8.8k. The
laser system consists of a nitrogen gas laser, a dye-laser and a modified microscope with a TV viewing camera and a display unit.

The nitrogen laser excites the dye-laser, the microscope focuses the dye-laser radiation onto the chip, and the television system provides means for visual observation. Two cylindrical focusing lenses are utilized between the nitrogen laser
and the dye cell. The output power, pulse width and spot quality of the dye-
laser can be controlled by adjusting the mirrors of its cavity, by altering the
above mentioned focusing lenses which focus the nitrogen laser radiation onto
the dye cell, or by changing the high voltage to the nitrogen laser.

A mechanical shutter and a prefocusing lens are located between the dye-laser and
the microscope. This arrangement is used for applying a preset number of pulses
and for adjusting to insure that the focal planes of the nitrogen and dye-laser ra-
diations coincide. The electronics for controlling the pulse repetition rate, the
peak power of the nitrogen laser and the number of pulses are not shown.

The shape and cross-sectional area of a laser drilled cavity is shown in the attached Figure 8.8l. The profile shown is unique and is quite different from that obtained by conventional drilling or machining operations. The particular shape is made possible because of the succession of multiple, repetitive high energy pulses. Each pulse enlarges the cavity diameter until a certain critical diameter and depth is reached. Further pulsation results in decreasing the cavity diameter until the desired cavity depth is achieved. This concave shape is possible only if the laser beam intensity is held constant. Thus, at constant beam intensity the effects of laser etching increase gradually to a maximum after which these effects decrease until either the desired or maximum obtainable depth is obtained.

In Figure 8.8l, the depth or length of the artificial cavity formed by the laser beam is designated as L. The average diameter is shown as D_A. It has been found by experimentation, that the unique laser drilled hole must have a length to diameter ratio $2 < L/D_A < 4$ to be effective. Also, the diameter should be in the range of $3\mu < D_A < 15\mu$. It has been found that a hole within these dimensional ranges is effective in dielectric liquid, such that complete wetting therein does not take place, and vapors are trapped so that site deactivation is minimized. Accordingly, nucleate boiling starts at these shaped sites at approximately the same temperature each time, and they are effective in extending the nucleate boiling beyond the usual DNB.

It will be appreciated that not only the hole dimensions are easily controlled, but the particular position of the hole can be programmed into the aiming of the beam device. Accordingly, the beam device can be stepped to particular positions so that a uniform hole location can be obtained for identical chips. Evidently, the high energy beam enters the surface and releases its energy below the surface to essentially burnout a clean opening thereunder. It can be seen that the opening of the hole tends to have a narrow entrance with a slightly built up section on either side.

Applying the foregoing theory explained in connection with a conical opening to the high energy beam drilled hole as shown in Figure 8.8l, it can be surmised that the liquid will tend to cross over the narrow opening, and thus not wet the entire inner hole surfaces to any great degree, thereby trapping some vapor in the hole to produce the nucleate boiling.

As has been mentioned in the foregoing theoretical discussion of nucleate boiling, temperature cycling of the nucleate boiling sites tends to exhaust the vapor in the site and cause temperature overshoot or hysteresis in the surface being cooled. Accordingly, to overcome this problem a nucleation heater **20** has been introduced along the bottom edge of the surface to be cooled or preferably along the bottom edge of the substrate containing the chips to be cooled as shown in Figure 8.8h. The individual heater element **20** on each chip or on the substrate just below the bottom row of chips performs the identical function in both situations, that is, the heater can be heated so that nucleate boiling takes place thereat causing the nucleate bubbles to break away from the heater and to rise along the surfaces of the chips.

These rising vapor bubbles, as mentioned previously, tend to wash the liquid from the artificial or natural nucleation sites, thus, starting nucleate boiling at a lower temperature than would normally be the case without the heater. It should be appreciated, that the providing of the vapor in the various nucleate boiling sites enables the boiling to start at approximately the desired temperature, thereby eliminating the hysteresis overshoot previously mentioned. The heater **20** is shown in detail in Figure 8.8m. The heater element **20** can be a resistance wire **22** containing the insulation **24** or no insulation for that matter. The heater element can be customized the same way as the chip can be customized, that is, the artificial nucleation sites or holes can be drilled therein with a laser or electron beam at the appropriate locations, and of the desired size.

The location of the holes is preferably in alignment with the above located holes on the chips so that the nucleate boiling vapor bubbles when rising along the surface of the chips **10** after leaving the heater **20** will be in line with the artificial nucleation sites on the chip. The heater **20** is energized during the start-up phase of the cooling process to start the nucleate boiling on the chips and is then turned off during the actual cooling phase to avoid the additional heat, and since the bubbles are no longer required.

Heater Bricks for Electrical Storage Heaters

A process described by *G.C. Eadie and C.F. Hinsley; U.S. Patent 4,006,734; February 8, 1977; assigned to British Steel Corporation, England* relates to heat storage media and is particularly though not exclusively concerned with heat storage media for use in electrical storage heaters. Such heaters include a heat storage medium, in suitable form such as in bricks, which is adapted to be heated by a resistive heating element using current supplied at specific low-cost off-peak periods and subsequently to release stored heat at a selected rate and over a different period.

The process for producing a heat storage medium comprises heating a mix of ferric oxide and an additive comprising a suitable compound of calcium at an elevated temperature effective to react the additive with the ferric oxide to improve volumetric heat capacity, reducing the reaction product to particulate form compacting the particles and sintering the compact.

A suitable calcium compound is an oxide of calcium or one capable of yielding the oxide at the elevated temperature.

The actual additive used is conveniently selected to improve the electrical resistivity of the medium, compared with that of ferric oxide alone in addition to improving the volumetric heat capacity.

Example 1: Finely divided iron oxide, typically Fe_2O_3 obtained as a by-product from a hydrochloric acid etch recovery plant and having an average analysis set forth in the table below was intimately mixed in a ratio characteristically of 5:1 by weight with finely divided strontium carbonate of standard commercial quality.

Component	Percent by Weight
FeO	5.93
Fe_2O_3	89.8
SiO_2	0.08
Al_2O_3	0.04
CaO	0.40

(continued)

Component	Percent by Weight
MgO	0.11
S	0.10
P	0.009
Mn	0.26

The mixture was fired in open trays for about forty-five minutes at some 1070°C to produce reaction between the ferric oxide and the strontium carbonate, before being allowed to cool to room temperature.

The resulting medium was wet ground for about eight hours to reduce it to finely divided form suspended as a slurry in water. After removal of the bulk of the water, by any conventional means, the medium now having the consistency of wet clay was pressed in a mold to the form required. In this example, the medium was pressed at about 450 bars to a compact, in the form of a brick, which was slowly dried to avoid cracking and subsequently sintered in air at 1300°C for 30 minutes. Both heating and cooling rates were graduated to avoid cracking due to thermally induced mechanical strain. The heat storage medium so produced displayed a volumetric thermal capacity of 2,800 kJ/l measures by a calorimetric method.

Example 2: Finely divided ferric oxide as in Example 1 is, in this case, intimately mixed with barium carbonate again in a mix ratio of 5:1 by weight. The mixture is fired at some 1000°C for about 20 minutes to produce a medium, which is subsequently processed to form heat storage bricks by a method similar to that of Example 1 also. The volumetric heat capacity of bricks made according to this embodiment was 1,650 kJ/l.

Example 3: Finely divided ferric oxide as in Example 1 was intimately mixed in a 1:1 weight ratio with calcium carbonate of laboratory quality in finely divided form. The resulting mixture is fired at 1000°C for 20 minutes to produce a medium which was briquetted as in previous examples and sintered at a temperature of about 1200°C. The volumetric heat capacity of the medium so produced was 2,700 kJ/l.

Example 4: Finely divided ferric oxide as in Example 3 was intimately mixed with calcium carbonate having a typical analysis set forth in the table below.

Component	Percent by Weight
$CaCO_3$	81.1
$Mg(OH)_2$	10.2
$MgCO_3$	2.2
$Fe(OH)_2$	0.6
$Ca(OH)_2$ + Al	5.9

The ferric oxide and calcium carbonate were intimately mixed in a weight ratio of 6:1 and fired at a temperature of about 1200°C for some 20 minutes. The medium produced was briquetted as in Example 1 and sintered at 1300°C for about 5 minutes. A volumetric thermal capacity of about 4,000 kJ/l was displayed by heat storage produced by this embodiment. It is believed that the improvement in thermal capacity over that of Example 3 results from incidental impurities in the calcium carbonate.

The end products of Examples 1 and 2 were ground and intimately mixed in a 1:1 weight ratio prior to pressing at 470 bars and final sintering at 1300°C for 30 minutes in air. The final medium displayed a volumetric thermal capacity of 3,050 kJ/l.

The heat insulation media produced possess certain advantages such as high electrical resistance and robustness compared with the use of ferric oxide alone and can be used to produce bricks of high volumetric heat capacity with values approaching or in some cases exceeding that of ferric oxide alone.

GLASS PROCESSING

Sterilizing Autoclave

G. Champel; U.S. Patent 3,897,818; August 5, 1975; assigned to Etablissements Joseph Lagarde SA, France describes a process and apparatus for rapidly cooling containers located in a chamber containing steam or steam and air, and particularly containers holding food products located in a sterilizing autoclave.

The apparatus shown in Figure 8.9 comprises an autoclave, generally indicated by reference numeral **1**, provided with a door **2** through which the containers may be introduced into the autoclave, and placed within an inner jacket **3**. The autoclave rests on feet **4a** and **4b** and is equipped with a fan **5**.

Figure 8.9: Rapid Cooling Technique for Containers in Steam Chamber

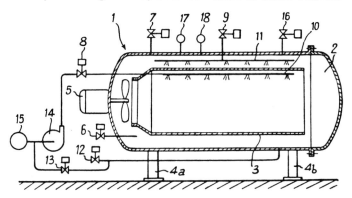

Source: U.S. Patent 3,897,818

The autoclave is provided with a steam inlet **6**, an inlet **7** for compressed gas, particularly air, valves **8** and **9** for admitting a fluid such as water, each associated with spray means in the form of tubes **10** and **11** having a plurality of perforations, with **10** so positioned that the liquid sprayed thereby comes into contact with the containers within the chamber **3** inside the autoclave, and **11** so positioned that the liquid atomized thereby cannot come directly into contact with the containers.

The device also comprises an outlet valve **12** and a recycling valve **13** supplying a pressure means **14**, such as a pump, which also receives liquid such as water from an external source **15** and supplies it to the valve **8** which supplies the cooling liquid to spray means **10**.

The device also comprises a valve **16** for connecting the autoclave to atmosphere. The autoclave is also provided with a thermostat **17** and pressure regulating means **18**, which may advantageously be program controlled, and the utility of which will be hereinafter explained.

One method of utilizing the device illustrated is described, specifically the case in which it is desired to cool containers such as glass bottles containing food products located in a sterilizing autoclave by running a cooling fluid thereover.

At the end of the sterilizing operation the atmosphere within the autoclave which may be stirred by the fan **5** contains only steam or a mixture of steam and a gas such as air, the steam having been introduced through the valve **6** and the gas through the valve **7** or, in the case of air, this gas may have been introduced into the autoclave during the closing of the door **2** when the containers were being introduced before sterilization. In the example described, air has been used as the gas and water as the liquid. In the bottom of the autoclave there is standing water of condensation coming from the steam which heated the bottles within the baskets within the jacket **3** inside the autoclave **1**, and the walls of the autoclave.

In the example described, at the end of the sterilization step, the prevailing temperature is 120°C and the gauge pressure within the autoclave is 1.8 bars.

Because at 120°C the gauge pressure due to the water vapor is 1 bar, 0.8 bar of the pressure is due to the presence of air. It is desired to decrease the temperature from 120°C to 70°C while maintaining the gauge pressure at 1.8 bars in a time of 3 minutes, for example.

The apparatus operates as follows: The thermostat **17**, which is program controlled, is programmed to actuate the valve **9** for introducing water to produce the desired decrease in temperature during the desired time.

Each time that the thermostat **17** opens the valve **9**, water is sprayed through the atomizing tube **11**, which water does not come into contact with the containers within the jacket **3**, but decreases the temperature in the autoclave **1**.

Since the pressure within the autoclave has a tendency to decrease, pressure control valve **18**, which is also program controlled, opens the valve **7** to admit compressed air to compensate for the decrease in pressure.

There is thus obtained a progressive decrease from 120° to 70°C in 3 minutes, the gauge pressure being kept at 1.8 bars, so that the absolute pressure is 2.8 bars.

However, at a temperature of 70°C the water vapor, considered alone, is under a pressure of minus 0.7 bar. If the absolute pressures are considered the pressure due to the vapor is thus 1 minus 0.7 or 0.3 bar, for a total pressure of 2.8 bars. Consequently, the absolute pressure due to the air is 2.8 minus 0.3, or 2.5 bars.

Thus, even if no additional compressed air were introduced through the valve **7**, the initial absolute pressure would fall only from 2.8 bars to 2.5 bars, or 0.3 bar. Without the apparatus described, the gauge pressure would fall from 1.8 bars to 1.8 - 1, or 0.8 bar, which would produce a decrease in pressure of 1 bar, that is to say the value of the pressure due to the vapor, since the water spray would condense only the vapor, thus eliminating its contribution to the total pressure.

It is of course understood that a programmed decrease from 120° to 70°C has been selected purely by way of example and could be, e.g., from 120° to 55°C, the part of the pressure due to the vapor being then even smaller, or about 0.1 bar.

During these 3 minutes at a constant pressure while the greater part of the vapor is replaced by compressed air, the temperature of the walls of the bottles has decreased from about 120° to 105°C. After this decrease in temperature at constant pressure the bottles are spray cooled by opening the valves **8** and **13** and starting the pumping means **14**. A part of the water of condensation passes through the valve **13** and rejoins the cold water arriving from the source **15** and the resulting mixture is pumped by the pressure means **14** into the circuit comprising the valve **8** and the spray means **10**.

If it is assumed that 70% of the total flow through the pressure means **14** consists of water at 10°C from the source **15** and 30% of water at 100°C from the autoclave through the valve **13**, the law of mixtures determines the temperature of the fluid reinjected into the autoclave at **10**, which temperature may be calculated to be about 37°C.

Thus the bottles at a wall temperature of 105°C receive first water sprayed at 37°C, i.e., at a difference in temperature of 105° - 37°, or 68°C. Without the second part of the apparatus which recycles the condensed water, the difference in temperature would be 105° - 10° or 95°C. Without the complete device according to the process the difference in temperature would have been 120° - 10° or 110°C. The thermal-shock, which is capable of breaking the bottles when there is a temperature difference of 110°C between the water and the bottles, is thus reduced to the extent that the temperature difference is brought to 68°C.

The outlet valve **12** may advantageously be a valve of the electrical level responsive type to evacuate the mixture of cold water and condensed water from the bottom of the autoclave and may advantageously maintain the level of the water within the autoclave constant.

Thus in a time which is shorter when the flow through the pressure means **14** is higher, the recycling water passed through the valve **13** cools and progressively attains a temperature near 10°C.

In accordance with the process, there is obtained, a greater homogeneity between the successive layers of bottles within chamber **3** because of the fact that the flow of water over these bottles is greater than the flow of cold water from the source **15**. The fan **5** may also be actuated for a certain time while cooling water is being supplied, thus reducing the thermal-shock by mixing the air around the containers when the atmosphere in the autoclave is still hot. In an advantageous manner the fan may stop between 30 seconds and 3 minutes after the beginning of the cooling, the time being modified in dependence upon the desires of the user.

Flat Glass Process

A process described by *Y.-W. Tsai; U.S. Patent 4,088,180; May 9, 1978; assigned to PPG Industries, Inc.,* relates to regenerative furnaces and their operation, and in particular to the type of regenerative furnace commonly employed in the manufacture of flat glass. The regenerators used in such furnaces are usually comprised of a gas-pervious bed of refractory material, such as a stacked arrangement of bricks, sometimes called checker packing, through which hot exhaust gases are passed during one cycle in order to heat the packing. In alternate cycles, the flow is reversed and the heat stored in the packing serves to preheat combustion air passing through the regenerator. The regenerators are generally employed in pairs, with one on either side of the combustion chamber. While one regenerator is absorbing heat from the exhaust gas, the other is heating incoming air.

In the process, there is provided one or more movable baffles beneath the packing near the flue entrance to each regenerator, which deflect a substantial portion of the incoming air flow during the firing cycle into the portion of the checker packing nearest the flue, thereby preferentially cooling that portion. When the cycle is reversed, the baffle is retracted so as to not interfere with the normal exhaust gas flow pattern. Since the flue end of the packing will thereby have been cooled more than other portions during the firing cycle, the subsequent uneven flow of exhaust gases will not cause an excessively unbalanced temperature rise at the flue end. Thus inordinate concentration of heat at the flue end is substantially reduced and thermal energy is more efficiently utilized.

Another set of baffles, which may be used either alone or in combination with the first-mentioned baffles, may be located in the plenum above the packing so as to discourage lateral flow along the plenum during the exhaust cycle, thereby alleviating the channeling of exhaust gases through the packing at the flue end. These latter baffles need not be movable.

OTHER APPLICATIONS

Electrostatographic Machine

A process described by *K. Mueller; U.S. Patent 4,085,794; April 25, 1978; assigned to Xerox Corporation* relates to a method and apparatus for removing heat from the pressure roller of the fuser assembly of an electrostatographic apparatus.

The process provides a heat transfer assembly including heat transfer means which contacts the pressure roller to remove heat by the passage of a heat transfer fluid about the assembly in response to a signal during operation of the machine. In this manner the temperature of the fuser roller may be maintained at a temperature sufficient (at a given pressure) to cause the toner to wet the surface of the paper image receiving members being processed, while cooling portions of the pressure roller in contact with the heat transfer means to minimize the effect of high temperatures on portions of the pressure roller not subjected to the passage of image receiving members.

Referring to Figure 8.10a document **D** to be copied is placed upon a transparent support platen **P** fixedly arranged in an illumination assembly, generally indicated by the reference numeral **10**, positioned at the left end of the machine.

Figure 8.10: Electrostatographic Machine

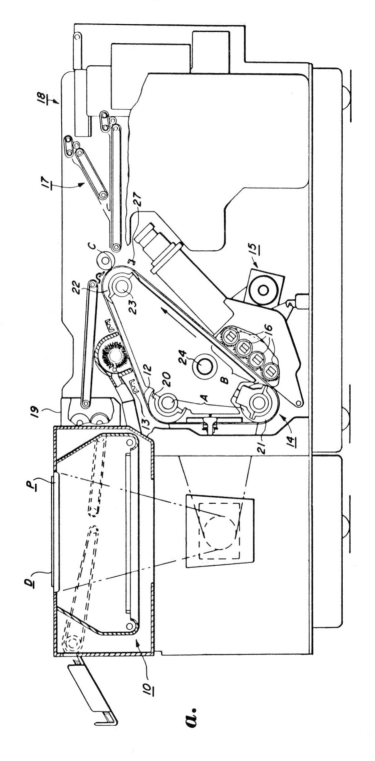

(continued)

Figure 8.10 (continued)

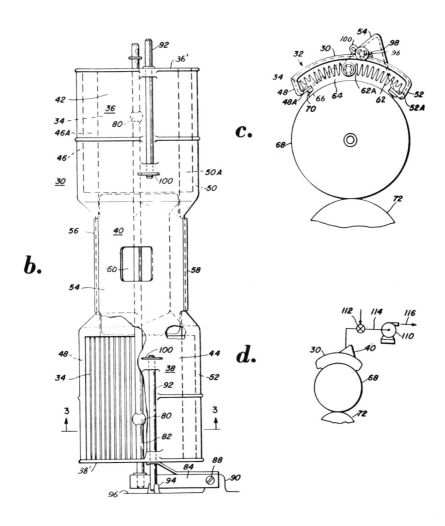

(a) Schematic sectional view of an electrostatic reproduction machine
(b) Plane view of the heat transfer assembly partially cut away to illustrate the configuration of the heat transfer means
(c) Cross-sectional elevational view of the heat transfer assembly taken along the lines 3–3 of Figure 8.10b
(d) Schematic flow diagram of the method of operation of the heat transfer assembly

Source: U.S. Patent 4,085,794

Light rays from an illumination system are flashed upon the document to produce image rays corresponding to the informational areas. The image rays are projected by means of an optical system onto the photosensitive surface of a xerographic plate in the form of a flexible photoconductive belt 12 arranged on a belt assembly, generally indicated by the reference numeral 14.

The belt 12 comprises a photoconductive layer of selenium which is the light receiving surface and imaging medium for the apparatus, on a conductive backing. The surface of the photoconductive belt is made photosensitive by a previous step of uniformly charging the same by means of a corona generating device or corotron 13.

The belt is journaled for continuous movement upon three rollers 20, 21 and 22 positioned with their axes in parallel. The photoconductive belt assembly 14 is slidably mounted upon two support shafts 23 and 24 with the roller 22 rotatably supported on the shaft 23 which is secured to the frame of the apparatus and is rotatably driven by a suitable motor and drive assembly in the direction of the arrow at a constant rate. During exposure of the belt 12, the portion exposed is that portion of the belt running between rollers 20 and 21. During such movement of the belt 12, the reflected light image of such original document positioned on the platen is flashed on the surface of the belt to produce an electrostatic latent image thereon at exposure station A.

As the belt surface continues its movement, the electrostatic image passes through a developing station B in which there is positioned a developer assembly generally indicated by the reference numeral 15, and which provides development of the electrostatic image by means of multiple brushes 16 as the same moves through the development zone.

The developed electrostatic image is transported by the belt to a transfer station C whereat a sheet of copy paper is moved between a transfer roller and the belt at a speed in synchronism with the moving belt in order to accomplish transfer of the developed image solely by an electrical bias on the transfer roller. There is provided at this station a sheet transport mechanism generally indicated at 17 adapted to transport sheets of paper from a paper handling mechanism generally indicated by the reference numeral 18 to the developed image on the belt at the station C.

After the sheet is stripped from the belt 12, it is conveyed into a fuser assembly, generally indicated by the reference numeral 19, wherein the developed and transferred xerographic powder image on the sheet material is permanently affixed thereto. After fusing, the finished copy is discharged from the apparatus at a suitable point for collection externally of the apparatus.

Further details regarding the structure of the belt assembly 14 and its relationship with the machine and support therefor may be found in U.S. Patent 3,730,623.

Referring to Figure 8.10b there is provided a heat transfer assembly generally indicated as 30 (which is positioned at 19 of the machine in Figure 8.10a), comprised of a cover, generally indicated as 32, and heat transfer elements, generally indicated as 34. The cover 32 may be formed of one piece construction and, as illustrated, is formed of side sections, generally indicated as 36 and 38, and an intermediate section, generally indicated as 40. The side sections 36 and 38 are formed of top walls **42** and **44**, front walls **46** and **48**, and back walls **50** and **52**, respectively. The intermediate section **40** is formed of a top wall **54**, a front wall **56** and a back wall **58**. The top wall **54** of the intermediate section **40** is provided with an orifice **60** to provide a conduit means for a heat transfer fluid, as more fully hereinbelow discussed. The end portions **36'**, **38'** of the side sections **36** and **38** opposite the intermediate section **40** are essentially unobstructive to provide an inlet means for the heat transfer fluid. The front walls **46** and **48**, and back walls **50** and **52** of the side sections **36** and **38**, respectively, are provided with inwardly extending tab sections **46A** and **48A**, and **50A** and **52A** to provide receiving means for each heat transfer element **34**.

As illustrated in Figure 8.10c, a heat transfer element **34** is formed with a plurality of fins **62** extending upwardly from a base **64**, and having an inner cylindrical surface **66** in contact with the cylindrical surface of a pressure roller **68** of the fuser assembly **19** of the machine. The end portions of the base **64** of a heat transfer element **34** are provided with slotted sections **70** which slidably engage the tab sections **46A** and **48A**, and **50A** and **52A** of the side sections **36** and **38** of the cover **32**, respectively. The pressure roller **68** contacts a fuser roller **72** as described above, during the image transfer operation of the machine.

Each heat transfer element **34** is held in place on the pressure roller **68** by two spherical section **80** provided on a shaft **82** above each element **34** and engaging the elements **34** between an intermediate finned section **62A** formed in the elements **34**. That portion of the element **34** engaging the spherical section may be dimpled for positive action. The ends of the shaft **82** are positioned in a shaft support means **84** rigidly affixed to an intermediate frame assembly of the electrostatographic machine such as by a screw **88** threaded into a tab section **90** thereof, it being understood that the opposite end of the shaft **72** is similarly supported.

The end sections **36** and **38** of the cover **32** are rigidly held in position by shafts **92** disposed in a shaft supporting section **94** formed in the main body portion **96** of the machine and extending through shaft receiving orifices **98** formed in the top walls of the side sections and held in place by securing means, such as by cotter pin **100**.

In operation (referring to Figure 8.10d), a fan **110** is activated on start-up of the machine, but is not placed in fluid communication with the assembly **30** until sensing means such as described in U.S. Patent 3,820,591, for example, to a signal that image receiving members of a dimension less than the maximum designed dimension therefor are being transported through the machine.

Such a signal activates a switching device which alters the configuration of valve **112** to cause air at ambient temperature to be introduced from the end portion of the side sections **36** and **38** about the fins **62** of the heat transfer elements **34** to remove heat connectively therefrom thereby lower the temperature of the surface of the pressure roller **68** in contact therewith. The thus heated air is withdrawn through the orifice **60** of the intermediate section **40** and is passed through conduit **114** and vented through conduit **116** to the atmosphere.

It is readily appreciated that the transfer assembly of this process may be easily assembled and disassembled for cleaning. It is to be understood that the positioning of a heat transfer element **34** by the spherical section **80** of the shaft **82** essentially effects a loading of the element **34** at one point thereby permitting the element to work like a universal. It is noted that the slotted sections **70** of an element **34** are dimensioned with respect to the tabs **48A** and **52A** of side section **38**, e.g., to provide a clearance therebetween to permit the elements to essentially float during rotation of the pressure roller.

Temperature Stabilization for Wrist Watch

J. Geiss and P. Eberhardt; U.S. Patent 3,863,707; February 4, 1975; assigned to OMEGA Louis Brandt & Frere SA, Switzerland describe a method for stabilizing the temperature of an instrument. In the process the instrument whose temperature has to be stabilized, is brought into contact with a container containing a temperature stabilizing element which presents, in the considered temperature range, at least one transformation point, so that the latent heat of transformation, which produces an elevation of the caloric capacity at the time of the passage through the transformation point, stabilizes the temperature of the instrument.

In Figure 8.11a there is schematically represented a wrist watch **1** placed in a box **2** which contains a temperature stabilizing element **3**. The container **2** can be worn in a manner similar to a wrist watch, because of the fact that it is mounted on a flexible wristlet **4**.

Figure 8.11: Temperature Stabilization for Wrist Watch

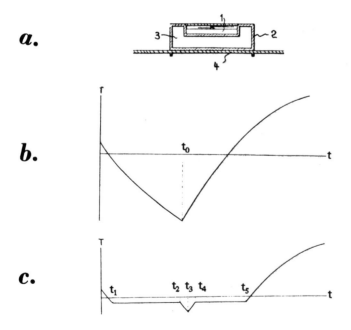

(continued)

Figure 8.11: (continued)

d.

(a) Schematic view of a watch associated to a container containing the stabilizing element
(b) Variation of the temperature of a measuring instrument not in contact with the temperature stabilizing means
(c) Variation of the temperature of the same measuring instrument in contact with a temperature stabilizing means with one temperature stabilizing element
(d) Variation of the temperature of the same measurement instrument in contact with a temperature stabilizing means containing two different temperature stabilizing elements

Source: U.S. Patent 3,863,707

The container as well as the dial of the watch, has predetermined optical properties, particularly a predetermined optical absorptance and emittance and optical reflectance. It is most important that the instrument, or in this case the watch movement is in tight contact with the container **2**, because this contact ensures good thermal conductivity between instrument and container. The fact that the box **2** contains a temperature-stabilizing element **3**, which has a transformation point (e.g., the melting point) in the considered temperature range, does not influence the stability temperature of the watch. This means that the temperature of the watch after a long journey in space, is not defined by the temperature-stabilizing element **3**. If the unit **1, 2** is submitted to a temperature change, the temperature of the temperature-stabilized measuring instrument will be constant at the transformation point of element **3**, until it has reached this temperature and executed the transformation. The temperature of the watch will vary along the graph in Figure 8.11c.

If the watch, which is not in contact with the temperature-stabilizing means, is suddenly exposed to cold, e.g., to the dark in space, its temperature will rapidly decrease until, at time t_0, it is suddenly exposed to heat, e.g., to the sun. Its temperature starts to rise rapidly, as is shown in Figure 8.11b. Due to these variations which may be several hundred degrees, the measuring instrument will certainly be damaged.

The watch movement **1** is placed in the box **2** which contains one temperature stabilizing element **3**. The temperature variations described above will effect quite different variations of the temperature of the watch, due to the fact that the container **2** with its filling **3** acts as a temperature stabilizing means. The temperature of the watch **1**, or any other instrument, will first decrease until it reaches, at the time t_1, the temperature of the transformation point of the element **3**. The transformation point of the element **3** may be the point or temperature of solidification or freezing of the element **3**. The temperature of the watch **1** remains at this temperature and does not change, until the whole element **3** is solidified, which is the case at the time t_2.

Then the temperature starts to decrease again, until the container **2** is submitted to sun radiation at the time t_3. The temperature of the unit **1, 2** then starts rising, until the abovementioned transformation point is reached at t_4. From then until the time t_s, the temperature of the container 2 and the measuring instrument 3 does not change because of the latent heat of melting of the temperature stabilizing element 3. At t_s, when the whole element 3 is melted, the temperature starts to rise again.

Accordingly, the temperature stabilizing element 3 has the effect of stabilizing the watch through very large fluctuations in temperature. The table below mentions some elements which may be used as temperature stabilizing elements.

Element	Melting Point (°C)	Density (g/cm^3)	Latent Heat of Fusion (cal/g)
Cs	28.6	1.9	3.9
Ga	29.8	5.9	19.2
Hg	-38.9	13.6	2.7
H$_2$O	0	1	79.7

The most convenient elements are the gallium and the water, because of the high specific heat of the water and the high density of the gallium. Because of the fact that the volume of both of these elements rises by freezing, it is necessary to provide a container 2 of an appropriate shape and manufacture to avoid its deterioration. Because of the high vapor pressure of the water, it may be indicated to provide the container 2 with a safety valve, if the temperature of the watch may reach about 100°C.

Other temperature stabilizing elements than the abovementioned can be used, for instance metallic alloys such as the ternary eutectic of Ga, In and Sn in the following composition: 62.5% Ga, 21.5% In, 16.0% Sn. The melting point of this alloy is 10.7°C.

The durations of temperature stabilization depend in a large manner on the absorptance and emittance of the casing 2 and the dial of the watch 1. Low powers give long durations of stabilization, because of the fact that the heat flux is low. If only one temperature stabilizing element is used, the temperature of the instrument to be temperature stabilized can only be stabilized in one direction. Depending on the initial temperature of the instrument, the temperature stabilizing element acts only for a rise or for a decrease of the temperature. So for instance a watch would be temperature stabilized during a heating period if its initial temperature is about 20°C and the container 2 is filled with gallium. The same watch would not be temperature stabilized, if the temperature would decrease. This could be achieved, if the container 2 would be filled with, e.g., water.

It is obvious that the watch can be temperature stabilized in both directions, if the container 2 contains both, gallium and water. In this case, the variation of the temperature of the watch 1 can be held between 0° and 29.8°C, which temperatures correspond to the melting point of the gallium and the water respectively.

It is most preferrable to choose the weight relation Ga/H$_2$O in such a way, that the relation of the total heat of transformation of both elements, corresponds approximately to the relation of the respective heat flux at the transformation points. So it is possible to obtain superior and inferior temperature stabilization durations which are about equal. To achieve this, e.g., 59 g of gallium and 10 g of water can be put into the container 2. The volumes of both elements are 10 cm^3.

The inconveniences arising from the use of only one temperature stabilizing element can be diminished by using a container **2** and a dial of the watch **1**, each having a predetermined and appropriate absorptance and emittance, so that the equilibration temperature of the measuring instrument **1** falls in an admissible temperature range. Thus, it is possible to limit the maximum temperature of the watch **1** at a value which is below 80° to 100°C for a perpendicular sun radiation of the dial. It has to be noted, that in practice, such a radiation will not last very long, so that the maximum temperature of the watch will not be reached.

Tankless Domestic Water Heater

R.D. Cooksley; U.S. Patent 3,968,346; July 6, 1976 describes a compact and fast acting tankless heater which has an elongated tubular casing having water inlet and outlet conduits at opposite ends. An elongated generally helically twisted inner tube is disposed coaxially within the casing to define helical channel means between the tube and casing for conducting water longitudinally from the casing inlet to the casing outlet.

The inner tube has a water inlet at one end communicating with the water inlet of the casing and the other end of the tube is closed. An elongated electric heating means is disposed within the inner tube for heating the water. The inner tube has a plurality of longitudinally spaced openings for conducting hot water and steam from the inner tube into the helical channel means to combine with the water flowing through the helical channel means from the casing inlet to the casing outlet. An additional electric heater sleeve can be provided to surround the exterior of the casing.

The heat exchanger may be compactly made to provide hot water on a continuous or intermittent basis as necessary to meet the domestic hot water requirements.

COMPANY INDEX

The company names listed below are given exactly as they appear in the patents, despite name changes, mergers and acquisitions which have, at times, resulted in the revision of a company name.

INVENTOR INDEX

U.S. PATENT NUMBER INDEX

NOTICE

HEAT PUMP TECHNOLOGY
FOR SAVING ENERGY 1979

Edited by M.J. Collie

Energy Technology Review No. 39

With energy costs escalating and fossil fuel supplies diminishing, the heat pump is becoming attractive for residential space heating and cooling because of its high efficiency. This book compares its overall efficiency to that of electrical resistance and fossil fuel heating, covering air-source and water-source pumps, and single-package and split-system units.

It reports on computer-simulated studies of residential heating in 9 cities in which a heat pump replaced a gas or oil furnace, and on tests of heat pumps in buildings in Washington, D.C., Hanover, New Hampshire, and Albuquerque, New Mexico. Data on the Annual Cycle Energy House, in which a heat pump, thermal storage and solar assistance provide space heating, cooling and water heating, attest to energy saved, as do the solar-assisted heat pumps in other areas.

Because early heat pumps had design deficiencies, improvements were mandated. The latter chapters examine modifications such as capacity control to alter refrigerant flow.

Following is a condensed table of contents with **examples of some** important subtitles.

ISBN 0-8155-0744-5

348 pages

PASSIVE SOLAR ENERGY DESIGN AND MATERIALS 1979

Edited by J.K. Paul

Energy Technology Review No. 41

Passive solar energy is a dynamic system whereby a building collects and stores energy to heat and cool itself. Unlike active solar systems employing complicated technology, passive systems collect, store and distribute thermal energy by natural radiation, conduction and convection. Sophisticated design and wise selection of building materials eliminate the need for fans, pumps or heat exchangers.

This volume describes over 100 buildings using specially designed components for the 3 passive systems, direct gain, indirect gain and isolated gain. Movable insulation for nocturnal shielding of glass areas and polymer ceiling tiles containing material to store and release heat are innovations discussed. Hybrid arrangements, utilizing simple mechanical aids to operate an otherwise passive system, are included.

A select group of case studies gives the reader an in-depth view of passive solar design. This book will interest producers of architectural and building materials, chemicals, and components, as well as architects and designers.

A partial and condensed table of contents with chapter headings and **examples of some** subtitles follows. Over 200 illustrations.

1. STATE OF THE ART
Background—Origins and Developments
Passive Solar Approaches
Direct Gain
Thermal Storage Wall
Solar Greenhouse
Roof Pond
Convective Loop

2. COMPONENTS—WINDOW TREATMENTS
Self-Inflating Curtain
Window Quilt Insulating Shade
Insulated Window Shutter
  Summer Research
Comprehensive Window Design
  Site, Appendages, Accessories
Energy Transport Control in Windows
Transparent Heat Mirrors
Single and Multiple Glass Glazing Media
Solar Membrane and Cloud Gel
Transparent Insulation
Optical Shutter
Clearview Solar Collector

3. COMPONENTS—THERMAL STORAGE
Wall Panel—Testing and Energy Balance
Solar Collector Storage Panel Testing
Poroplastic/Wax in Collector Wall

Waterwall Design
Solar Window—Calculating Heat Loss

4. MISCELLANEOUS COMPONENTS
Freon-Activated Controls—Skylid
Hinged Shutters and Nightwall Clips
Beadwalls—Economics and Evaluation
Thermic Diode Solar Panels
Heat Pipe Flat Plate Collectors
Weather-Responsive Building Skins
Thermocrete
Skytherm Roof Pond
Hinged Sky Lites
Peakshaver Solar Collector

5. CASE STUDIES
Kelbaugh House, Princeton, NJ
Benedictine Warehouse, Pecos, NM
First Village, Santa Fe, NM
Hunn House, Los Alamos, NM
Atascadero House, Atascadero, CA
  House and Solar System Design
  Cost and Performance
  Owner Observations

6. APPLICATIONS—DIRECT GAIN
Systems in Arizona, Arkansas, California, Colorado, Wyoming, Delaware, Minnesota, Ohio, New Enland, Wyoming:
  Project at Guadalupe, AZ
  Airport Terminal, Aspen, CO
  Peabody House, Dover, MA
  Kalwall Corp. Building, Manchester, NH

7. APPLICATIONS—INDIRECT GAIN
Mass Trombe Wall
  Delap House, Fayetteville, AR
  Caivano House, Bar Harbor, ME
  Beatrice Mongeau Houses, Pittsboro, NC
Water Trombe Wall
  Jantzen House, Carlyle, IL
  Whitcomb House, Los Alamos, NM
Roof Pond
  Hammond House, Winters, CA

8. APPLICATIONS—ISOLATED GAIN
Sunspace
  Hull Residence, Prescott, AZ
  Naumann House, Boulder, CO
  Spence-Urban House, W. Des Moines, IA
  Eccli House, New Paltz, NY
Thermosiphon
  Davis House, Albuquerque, NM

9. RULES OF THUMB

ISBN 0-8155-0746-1

386 pages

COGENERATION
OF
STEAM AND ELECTRIC POWER
1978

Edited by Robert Noyes

Energy Technology Review No. 29

The cogeneration concept, as presented in this book, refers to the generation of both process steam and electricity on site by heating water with a primary fuel, such as coal, oil or gas.

When fuel was cheap, industries were not interested in capital outlays necessary for cogeneration equipment. Now, however, the picture has changed. Refineries, chemical plants, paper mills and other big energy users can get maximum value from their fuel by a "topping cycle," viz. producing steam and running it through a turbine generator before using the steam for processing operations. Also, cogeneration can be accomplished through a "bottoming cycle," utilizing process steam prior to electrical power generation. Both methods are discussed in detail.

In the past few years, the U.S. Government has been placing much emphasis on the need for conserving energy. One of the ways of saving considerable amounts of energy is the development of cogeneration plants by those industries which use significant amounts of electricity and thermal energy for process heat and/or space heating. Since electric power generation produces heat as a by-product, any thermal energy which can be used produces a fuel savings of from 10 to 30%, depending on the applications and methods involved. The studies indicate that for the greatest potential development of cogeneration, government support may be required.

Various technological studies, such as material prepared for the National Science Foundation and the Federal Energy Administration, were the basis for this review. A partial, condensed table of contents follows here.

ISBN 0-8155-0706-2

WIND POWER 1979
RECENT DEVELOPMENTS

Edited by D.J. De Renzo

Energy Technology Review No. 46

Wind power could become a sizeable supplier of clean, inexhaustible energy. Heretofore, high cost has hindered this growth, but wind energy conversion systems are becoming more cost-competitive with other systems. Solutions to the attendant technological problems are also evolving.

This new book presents developments occurring since 1975, when Noyes Data Corporation published its first book on wind power. While the first chapter reviews the status of wind power conversion in 1975, subsequent chapters give an up-to-date assessment of geographical distribution of wind power in the United States, of technological advances, and of large- and small-scale use.

An appendix provides names and addresses of many manufacturers, researchers, and distributors dealing with wind energy conversion systems.

The partial table of contents below lists chapter headings and **examples of some** subtitles.

ISBN 0-8155-0759-3

347 pages

GEOTHERMAL ENERGY

Recent Developments

1978

Edited by M.J. Collie

Energy Technology Review No. 32

The heat underneath the Earth's crust seems a highly desirable energy source alternative to oil and gas. The Earth does not give up her energy readily, still geothermal energy activities are progressing on a broad front worldwide wherever possible.

Operational fields are increasing in numbers, and new and improved technologies, especially new turbine designs and deep-well pumps, coupled with heat exchangers at the surface, are leading to much greater operating efficiency.

Yet, as a rule, geothermal power stations, with a few exceptions, are inefficient in converting thermal energy to electric energy, because of the overall low temperatures. Also the highly mineral-laden waters in the geothermal reservoirs increase operating expenses through scale formation, thus complicating the process of extracting the heat.

Geothermal energy is not expected to play a large overall role in the U.S. But it can have regional significance, especially in the West. For developing countries, however, geothermal energy, especially in volcanic regions, may be the only economic form. Electricity systems there are too small to justify nuclear power stations, but the comparatively small size of geothermal power stations fits the scale of electricity supply systems in these countries.

This energy Technology Review is based on studies conducted under the auspices of various government agencies and the last chapter constitutes a survey of the recent patent literature.

ISBN 0-8155-0727-5

445 pages

INSULATION GUIDE FOR BUILDINGS AND INDUSTRIAL PROCESSES 1979

Edited by L.Y. Hess

Energy Technology Review No. 43

The question today is not whether to insulate but "How?" and "How much?" This definitive guide provides the answers. It assesses thermal insulation materials and systems for residential and industrial building applications, appraising building insulation as used for different purposes in Part A.

Recent rapid rises in fuel prices have confirmed the economic value of insulating as a fuel conserving practice. At the same time that maximum insulation of buildings effects considerable savings in costs, it confers the added benefit of increasing the availability of fuel, thus providing industry with more fuel to be used for productive purposes.

Part B of this book explores the use by industrial plants and utilities of thermal insulation to conserve energy, protect personnel and maintain process temperatures. It considers the effective temperature range and optimum thickness of insulating materials, their applicability in industrial processes, and the resultant economic rewards, noting the high rate of return on investment from insulating. It also evaluates sources of information on heat transmission.

The following condensed table of contents with chapter headings lists examples of **some important subtitles.**

ISBN 0-8155-0752-6   200 pages

COAL LIQUEFACTION PROCESSES 1979

by Perry Nowacki

Chemical Technology Review No. 131
Energy Technology Review No. 45

With the precarious position of petroleum supplies in the world market and the unfavorable balance of payments caused by overdependence on foreign sources, it may be expedient to exploit coal further as a source of liquid fuels. In the United States coal is the most abundant natural energy resource. Through its conversion to liquid fuels, it could supply a portion of the energy required for transportation, heating, etc.

Coal can furnish a complete range of liquid fuels, as well as chemical feedstocks. To achieve maximum yields, however, judicious adjustment of temperature, pressure, time and catalyst, as elaborated in this book, is essential. The text reviews the histories of important domestic and foreign coal liquefaction processes—pyrolysis, solvent extraction, and catalytic and indirect liquefaction. It details the chemistry of the processes and their inherent problems, also discussing economics, developmental status, and environmental impacts.

Chapter headings and **examples of some** important subtitles follow below:

1. **INTRODUCTION**

2. **COAL RESOURCES**

3. **COAL LIQUEFACTION CHEMISTRY**

4. **OVERVIEW OF PROCESSES**

5. **PYROLYTIC PROCESSES**
   Char Oil Energy Development (COED)
   Clean Coke
   TOSCOAL
   Occidental Flash Pyrolysis
   Coalcon Hydrocarbonization
   Oak Ridge Hydrocarbonization
   Flash Hydropyrolysis of Lignite
   Flash Hydrogenation
   Lurgi-Ruhrgas Flash Carbonization

6. **SOLVENT EXTRACTION**
   Solvent-Refined Coal (SRC) Process
     Pilot Plants—Ft. Lewis & Wilsonville
   Consol Synthetic Fuel Process
     Cresap Pilot Plant
   Exxon Donor Solvent Process
   CO-Steam Process
   Supercritical Gas Solvent Extraction
   UOP Solvent Extraction

7. **CATALYTIC LIQUEFACTION**
   H-Coal Process
     Trenton, N.J. Pilot Plant
   Synthoil Process
   Clean Fuel from Coal Process
   Zinc Halide Coal Liquefaction
   Dow Coal Liquefaction

8. **INDIRECT LIQUEFACTION**
   Fischer-Tropsch Process
     SASOL Commercial Plant, South Africa
   Mobil M-Gasoline Process

9. **COAL LIQUIDS FROM H-COAL**
   Analysis and Properties of Products
   Distillate Separation & Characterization

10. **SOLVENT-REFINED EXTRACT—COMMERCIAL, EXPANDED-BED HYDROPROCESS**

11. **ECONOMICS OF SRC PROCESS**
   Reclaiming, Preparing, Liquefying Coal
   Hydrogenation of Fuel Oil and Naphtha
   Stack Gas Desulfurization
   Hydrogen Manufacture
   Capital Investment and Costs

12. **ENGINEERING ASSESSMENT OF SYNTHOIL PROCESS**
   Feed Slurry Pumping and Heating
   Reactor Design
   Removal of Spent Catalyst
   Low-Temperature Carbonization

13. **FISCHER-TROPSCH DESIGN & COST**
   Recovery of Heat, Liquids, Sulfur
   Gasification and Methanation
   Water Systems

14. **EQUIPMENT AND MATERIALS**
   Pumps, Heat Exchangers, Fired Heaters
   Vessels and Pipe Systems
   Instrumentation and Metallurgy

15. **ENVIRONMENTAL, HEALTH AND SAFETY ISSUES**
   Solid Waste Management
   Hazards in Product Transfer and Storage
   Federal Legislation

16. **COMPARISON OF PROCESSES**
   Technical and Economic Aspects
   Probability of Commercial Success

ISBN 0-8155-0756-9

339 pages

FEDERAL ENERGY
INFORMATION SOURCES
AND DATA BASES 1979

by Carolyn C. Bloch

Determining the proper agency as a source of energy information is often more perplexing than perusing that information once it is obtained. Federal agencies amass myriad amounts of data and systematically store it away for reference. This information can serve as an easy, economical tool for scientists, managers, and engineers concerned with all phases of energy studies and projects, provided they have the know-how to obtain it. This time-saving directory of sources offers the assistance they need.

The directory comprises all agencies that deal in some capacity with energy—civilian, military, and legislative. It supplies the address and a capsule description of each source, delineating its field of emphasis, services offered and, many times, availability of publications.

The following cites main source categories and **examples of some** specific sources.

I. CABINET DEPARTMENTS

1. DEPARTMENT OF ENERGY
Libraries
    Government-Owned/Contractor-
    Operated Laboratory and Facility
    Libraries
DOE Centers and Systems
    Bonneville Power Administration
    Lawrence Livermore Laboratory
Energy Information Administration

2. DEPARTMENT OF AGRICULTURE

3. DEPARTMENT OF COMMERCE
National Technical Information Service
    Foreign Translations
Patent Office
    Off. of Technol. Assessment & Forecast
District Offices
National Bureau of Standards
    National Standard Data Reference System
    Nat. Oceanic Atmospheric Admin.

4. DEPARTMENT OF DEFENSE
Library and Analysis Centers
Navy
Army
    Army Mobility R&D Command, Ft. Belvoir
Air Force
    Weapons Lab., Kirtland AFB Tech. Lib.

Tri-Service Industry Information Center
How DDC Serves the General Public
Service to Legal Profession

5. DEPARTMENT OF HOUSING AND URBAN DEVELOPMENT
Nat. Solar Heating & Cooling Inf. Center

6. DEPARTMENT OF INTERIOR
U.S. Geological Survey
    Energy Resources Data Systems

7. DEPARTMENT OF TRANSPORTATION
Information Systems, TRISNET

II. ADMINISTRATIVE AGENCIES

1. ENVIRONMENTAL PROTECTION AGENCY
Information Sources and Systems
    Air Pollution Technical Center
    Solid Waste Inf. Retrieval System

2. GENERAL SERVICES ADMIN.
Data Bank on U.S. Resources

3. NASA
Library Network (NALNET)
Industrial Application Centers
National Space Science Data Center

4. NATIONAL SCIENCE FOUNDATION
Directorate for Scientific, Technological
    and International Affairs

5. NUCLEAR REGULATORY COM. LIB.

III. QUASI-GOVERNMENT AGENCIES

1. SMITHSONIAN SCIENCE INFORMATION EXCHANGE

IV. CONGRESSIONAL OFFICES

1. CONGRESSIONAL BUDGET OFFICE

2. CONGRESSIONAL RESEARCH SERVICE

3. GENERAL ACCOUNTING OFFICE

4. GOVERNMENT PRINTING OFFICE

5. LIBRARY OF CONGRESS
Science & Technology Div., Ref. Section

6. OFFICE OF TECHNOLOGY ASSESSMENT

V. INDEX

ISBN 0-8155-0764-X

115 pages